D1082924

An Introduction to
Abstract Algebra

An Introduction to
Abstract Algebra

Department of Mathematics

California State College at Fullerton

Fullerton, California

INTERNATIONAL TEXTBOOK COMPANY

Scranton, Pennsylvania

Preface

Groups, rings, modules, fields and Galois theory, together with some intrusions, such as homology theory, make up what is usually called abstract algebra. This book is intended to serve two main purposes: an introductory text in abstract algebra at either the sophomore or junior level and a text in some of its more advanced topics suitable for seniors. It can be used for either or both of these purposes.

Chapters 1, 2, 3, the first half of Chapter 4, Chapter 5, and the first half of Chapter 7 could very well be used for an introductory course, and the remainder of the book for the more advanced portions. In this way the reader can continue with the same text from an introductory to a possible follow-up course, with the not inconsiderable advantages of retaining the same notation and maintaining a convenience of reference. Many variations in the choice of material for the two courses are certainly possible within the framework of the book.

A chapter (Chapter 3) on linear algebra is included so that the book may, to some extent, be self-contained. While this chapter does not present as exhaustive a treatment of linear algebra as a good independent text on this subject, it is believed that there will be found sufficient material here for many purposes. Chapters 2 and 4 treat elementary group theory with some degree of thoroughness. Chapter 5 deals with rings and their homomorphisms, while Chapter 6 is concerned with factorization and elementary ideal theory. Chapter 7 renders to the module a thoroughness of discussion which this very important algebraic system deserves. Chapters 8 and 10 introduce the reader to some of the more modern concepts of abstract algebra, such as graded algebras, tensor algebras, categories, functors, homology groups, and the attempt is made here to exhibit the relationship between homological algebra and algebraic topology. Chapter 9 contains an introduction to Galois theory. The purpose of Chapter 11 is to provide an introduction to some of the important work that has been done in the study of the structure of general rings and, in this way, to emphasize the nature of structure problems in algebra.

It is believed that the book provides a sufficient variety of topics to make it possible for both teacher and student alike to select their own preferences for the two courses of study.

A deliberate attempt has been made to keep proofs concise, in the hope that readers may not be discouraged by excessive length, but rather encouraged to fill in their own details. A proof is so much a personal matter that a reader must actively participate in its achievement. One man's proof may be another man's bewilderment. It is important not to deprive the reader of doing some thinking of his own and thus to increase both his understanding and satisfaction.

The author must bear full responsibility for any errors in the book. He hopes there are not any serious ones.

DENNIS B. AMES

Fullerton, California
February, 1969

Contents

8. Algebras . . . 256

9. Field Theory and Galois Theory . . . 284

10. Homological Algebra . . . 313

An Introduction to
Abstract Algebra

chapter 1

Introduction

It is an oversimplification to say that algebra is made up of sets of elements, mappings, and binary operations. However, this modest but incomplete assertion does tell quite a lot about the nature of algebra. We let it stand for a beginning and consider the terms involved in it. The sets of elements with their binary operations are known as **algebraic systems**. The set of integers, for example, with the binary operations of addition and multiplication forms an algebraic system. Each algebraic system is defined precisely by a set of axioms. This is known as the **axiomatic method.** What is an axiom of an algebraic system? It is a property assumed to be true of the system being defined. The axioms of a system constitute the set of rules under which the "game" for this particular system has to be played, and the advantage of such precision is obviously enormous.

After an algebraic system has been defined, a very natural type of mapping arises, known as a homomorphism, and this mapping turns out to be of fundamental importance. A homomorphism is a mapping of an algebraic system into a like algebraic system, and it has the property of preserving their binary operations. Such a mapping has many other interesting properties, as we shall shortly see. We also make mention of the important part of algebra, called *structure theory,* which is occupied with the decomposition of algebraic systems into simpler subsystems. This brief summary is not seriously intended to encompass the whole story of algebra's fulfillment. Perhaps, however, it is enough for the time being.

When we speak of abstract algebra we are thinking of a theory (a body of definitions, axioms, and theorems) that deals with a variety of algebraic systems in which neither the nature of the elements is specified nor the binary operations identified. It is this generality that gives abstract algebra its power, since to specify is to limit.

1-1 SETS

We start with a naïve and intuitive idea of a set of objects as being merely a collection or family of objects that is defined in such a way that we must be able to state of an arbitrary object that it is either in the set or does not belong to it. Thus we may speak of the **set of real numbers**, meaning the entire collection of all real numbers. We may not at the moment be able to decide whether log 2, say, is a real number, but we must be able to assert that log 2 is either a real number or isn't one, and hence that if it is then it is in the set, whereas if it is not then it is not in the set. (As a matter of fact, log 2 can be proved to be a real number.) This, for out present purpose, is the meaning of the term **set**. We might add that there is in existence an axiomatic theory of sets which gives some precision to the theory, but it will not concern us here.

It is customary in algebra to speak of the objects of a set as its *elements*. If x is an element of a set A we write

$$x \in A,$$

and this means then that the element x belongs to or is contained in the set A. We write $x \notin A$ to mean x is not in A. Examples of sets are at hand everywhere: the set of all books in some library, the set of all integers greater than 2, the set of all stars that are less than a billion light years away from the earth.

A set C is called a **subset** of a set A if every element of C is an element of A. This means that $x \in C$ implies $x \in A$. If C is a subset of A we denote this by $C \subset A$, which is read as "C is contained in A." Another way of expressing this is to write $A \supset C$, which is read as "A contains C." Note that $C \subset C$! Thus a set is a subset of itself.

If $C \subset A$ and $A \supset C$, then we write this as $A = C$, meaning every element of C is an element of A, and vice versa. Thus if we want to prove that $A = C$, we can do so by showing that $A \subset C$ and $C \subset A$.

If $C \subset A$ and $C \neq A$, then C is called a **proper** subset of A.

Of course a set would be defined if we could list all its elements. Since this is not always possible, nor even always convenient, we introduce the following notation for a set. If X is a set whose elements either have a certain property P or do not have P then we write the subset A of elements of X that do have P, as follows:

$$A = \{x \in X \mid x \text{ has } P\}.$$

The subset B of X defined by $B = \{x \in X \mid x \text{ does not have } P\}$ is called the **complement of A with respect to X.** The vertical bar is a symbol for "such that" or "for which." For instance, if X is the set of real numbers, and T is the subset of X consisting of all real numbers that have the property of being solutions of the equation $1 = 2 \sin x$, we write

$$T = \{x \in X \mid 1 = 2 \sin x\}.$$

We shall make frequent use of two operations on sets, to wit:

1. The **intersection** of two sets A and B, written $A \cap B$, is the set consisting of all elements which are common to both A and B, that is, all elements which are in both A and B:

$$A \cap B = \{a \in A \mid a \in B\} = \{b \in B \mid b \in A\}.$$

2. The **union** of two sets A and B, written $A \cup B$, is the set of all elements that are either in A or B or possibly in both:

$$A \cup B = \{x \mid x \in A \text{ or } x \in B\}.$$

Thus if A is the set of all people over 28 years old and B is the set of all people who have green eyes and are more than 28 years old, then $A \cup B$ is the set of all people who either have green eyes or are more than 28 years old. One alone of these physical attributes is enough to put that person in $A \cup B$, and the possession of two of them certainly puts that person in $A \cup B$.

If $C \subset A$ and B is any set, then it is easy to see that: $B \cap C \subset B \cap A$ and $B \cup C \subset B \cup A$.

If A and B are two sets then it is easy to show that the set operations intersection $\cap$ and union $\cup$ are commutative; that is,

$$A \cap B = B \cap A \quad \text{and} \quad A \cup B = B \cup A.$$

These follow at once from the definition above of eqality of two sets.

Moreover, if A, B, C are three sets, then it follows easily that these set operations are associative; that is,

$$(A \cap B) \cap C = A \cap (B \cap C) \quad \text{and} \quad (A \cup B) \cup C = A \cup (B \cup C).$$

We prove next that

$$A \cap (B \cup C) = (A \cap B) \cup (A \cap C).$$

Let $x \in A \cap (B \cup C)$ then either $x \in A \cap B$ or $x \in A \cap C$ and hence $x \in (A \cap B) \cup (A \cap C)$. Hence $A \cap (B \cup C) \subset (A \cap B) \cup (A \cap C)$. Next let $y \in (A \cap B) \cup (A \cap C)$; then $y \in A \cap B$ or $y \in A \cap C$. Now $B \subset B \cup C$ and hence $A \cap B \subset A \cap (B \cup C)$. Similarly, $A \cap C \subset A \cap (B \cup C)$. Hence in either case, $y \in A \cap (B \cup C)$. Therefore $(A \cap B) \cup (A \cap C) \subset A \cap (B \cup C)$. Thus from the definition of equality we have

$$A \cap (B \cup C) = (A \cap B) \cup (A \cap C).$$

In the same way we can prove that

$$A \cup (B \cap C) = (A \cup B) \cap (A \cup C).$$

These last two identities are called the **distributive laws.** They state that the set operations union and intersection are distributive with respect to each other.

For the time being these set operations will suffice for our purposes.

At this point we adopt for its logical convenience the convention that there is such a thing as the *empty set* or *null set*, which is denoted by $\emptyset$. It is the set that contains no objects. This is a convenience since we may not know in advance whether some set we are defining has any objects in it or not. For instance when searching for the intersection $A \cap B$ of two sets, we choose to regard it as a set, yet it may turn out to be empty. Of course when we define a set, about the first thing we would want to examine would be whether it is empty or not.

If A and B are two nonempty sets and if $A \cap B = \emptyset$ then the sets A and B are said to be **disjoint.**

If A and B are any two nonempty sets, we next define the set $A \times B$, called the **cartesian product** of A and B. $A \times B$ is the set of all ordered pairs or couples (a, b) where $a \in A$ and $b \in B$. The term ordered means the first member of the pair belongs to A and the second member to B. This notation is already familiar to the reader, since it is used in analytic geometry to signify the points (x,y) of the plane in a cartesian coordinate system. These would be points of the set $R \times R$ where R is the set of real numbers.

EXERCISES

1. If A and B are sets, define the set $A \setminus B$ by

$$A \setminus B = \{x \in A \mid x \notin B\}.$$

If B is a subset of A then $A \setminus B$ is called the **complement** of B in A and is denoted by B'. If B and C are subsets of the set A prove the following:

(a) $(B \cup C)' = B' \cap C'$
(b) $(B \cap C)' = B' \cup C'$
(c) $(B \setminus C)' = C \cup B'$
(d) $B = C'$ if and only if $B \cup C = A$ and $B \cap C$ is the empty set.

2. If A and B are subsets of a set S, the symmetric difference of A and B is defined by

$$A * B = A \cup B \setminus A \cap B.$$

(a) Prove the symmetric difference is associative.
(b) Find $(A * B)'$ in terms of A, A', B, B'.
(c) Prove $A \cap ((A \cap B) * B)$ is the empty set.

3. If A and B are abritrary sets, prove
(a) $(A')' = A$.
(b) $A \subset B$ implies $B' \subset A'$.
(c) $A \setminus (A \setminus B) = A \cap B$.

4. For sets A, B, C, prove

$$A \cup (B \cap C) = (A \cup B) \cap (A \cup C).$$

5. If A, B, C are arbitrary sets, prove that

$$C \setminus (A \cup B) = (C \setminus A) \cap (C \setminus B).$$
$$C \setminus (A \cap B) = (C \setminus A) \cup (C \setminus B).$$

(These are known as *De Morgan's formulas*).

1-2 MAPPINGS

A **mapping** f of a nonempty set A into a nonempty set B, symbolized by $f: A \to B$, is a function on A to B, that is, a rule that assigns to each element of A a unique element of B. This is known as a *single-valued mapping*, in that to each $a \in A$ corresponds a unique element $b \in B$. It is the only kind of mapping we shall use in this book. A is called the **domain** of the function f and B is called its **range** or **codomain**. We denote by af the unique element of B into which the element $a \in A$ is mapped by the mapping f. af is called the **image** of a under f.

A more sophisticated definition of a mapping f of $A \to B$ is that it is a subset of $A \times B$ (that is a set of elements (a, b), $a \in A$, $b \in B$) such that each $a \in A$ occurs in one and only one ordered pair (a, b) of the subset. On the other hand, any $b \in B$ may possibly occur in many different pairs. In this notation, note that the element b of (a, b) is of course af.

In all mappings the sets involved are assumed to be nonempty.

Illustrative Example. Let A be the set $A = \{\alpha, \beta, \gamma, \delta\}$ and B the set $B = \{a, b, c\}$. A mapping f of $A \to B$ is defined by $\alpha f = a$, $\beta f = c$, $\gamma f = c$, $\delta f = a$. Thus f is the subset (α, a), (β, c), (γ, c), (δ, a) of $A \times B$.

Even this very simple example brings out several noteworthy facts about a mapping: (1) different elements of **A** may map into the same element of B, and (2) not every element of B necessarily has to be the image of some element of A.

Two salient properties are required of a mapping f of $A \to B$: (1) f must map every element of A into an element of B, and (2) the image in B of an element of A must be unique.

We can also have mappings of a set into itself. For n an integer the mapping f defined by $nf = 2n$ is a mapping of the set of integers into itself.

Composite Mappings

If f maps $A \to B$ and g maps $B \to C$, then we can form the composite mapping $f \circ g$ which maps $A \to C$ by way of B; that is, $A \xrightarrow{f} B \xrightarrow{g} C$. In detail this means that if $a \in A$ then $af \in B$ and $(af)g \in C$. This leads at once to the precise definition of the composite mapping $f \circ g$. We define $f \circ g$ as that mapping from $A \to C$ for which

$$a(f \circ g) = (af)g, \quad \text{for all} \quad a \in A.$$

If $A \neq C$, observe that the mapping $g \circ f$ does not exist. It has no meaning. For g maps $B \to C$, while f maps $A \to B$. Thus two mappings f and g are composable, in the sense $f \circ g$, if and only if the range of f is contained in the domain of g.

Frequently we shall simply write fg for the composite $f \circ g$.

If f maps $A \to B$ and $b \in B$, we use the symbol $f^{-1}b$ to designate the set of all elements of A that are mapped under f into the element b of B. Thus

$$f^{-1}b = \{a \in A \mid af = b\}.$$

This is to be read as the set of all $a \in A$ for which (or such that) $af = b$.

If C is a subset of B, this notation readily extends to the subset $f^{-1}C$ of A defined by

$$f^{-1}C = \{a \in A \mid af \in C\}.$$

$f^{-1}C$ is called the **inverse image** of C under f.

If f maps $A \to B$ and C is a subset of A, then $Cf = D$ is a subset of B. In general, $C \subset f^{-1}D$ and $C \neq f^{-1}D$. For it is possible there exists an element x of A for which $x \notin C$ and yet $xf \in D$, in which case $x \in f^{-1}D$ and $C \subset f^{-1}D$.

Note that, for the present, f^{-1} by itself has no meaning. However, we might expect it to mean the inverse mapping of f, if such a mapping exists. We will soon look into this.

We emphasize that if f and g are mappings of a set A into a set B, then $f = g$ if and only if $af = ag$ for *every* $a \in A$.

Two very natural questions arise in connection with a mapping f of $A \to B$: (i) If $b = af$, is a the only element of A that is mapped into the element b of B by f? (ii) Is every element of B the image of some element of A?

These questions lead to the following two properties of a mapping that will be of particular importance to us in our later work.

Definition. A mapping f of $A \to B$ is called **injective** (one-to-one) if distinct elements of A are always mapped into distinct elements of B. This means that if a and a' are in A and $a \neq a'$, then $af \neq a'f$.

We isolate the following simple criterion for an injective mapping, since it is a very common way of proving injectivity.

Lemma 1. A mapping f of $A \to B$ is injective if and only if $af = a'f$ implies $a = a'$.

Proof: Assume f is injective and let $af = a'f$. Then $a = a'$. For if $a \neq a'$, then, since f is injective, $af \neq a'f$, which is a contradiction. Conversely, assume that whenever $af = a'f$ we have $a = a'$. To show f is injective, suppose $a \neq a'$. Then if $af = a'f$, we would have $a = a'$. Hence if $a \neq a'$ we have $af \neq a'f$, which proves f is injective.

Definition. A mapping f of $A \to B$ is called **surjective** if every element $b \in B$ is the image under f of at least one element $a \in A$—that is, $b = af$ for at least one $a \in A$.

The symbols Af or im f are used to denote the image of A under f.

Since $Af \subset B$, another way of saying f is *surjective* is to write $Af = B$. This means f maps A onto all of B; no element of B is left out.

Definition. A mapping f of $A \to B$ that is both injective and surjective is called **bijective**. Another name for a bijection is *one-to-one correspondence* of the sets A and B.

We shall use the notation 1_A to denote the identity mapping on a set A. It is the mapping for which $a 1_A = a$, for all $a \in A$.

If f is a bijective mapping of $A \to B$, then to each $b \in B$ corresponds a unique element $a \in A$ determined by $b = af$. Thus the mapping f induces a mapping f^{-1} of $B \to A$ defined by $bf^{-1} = a$, where $af = b$. Since $a(ff^{-1}) = (af)f^{-1} = bf^{-1} = a$ we see that $ff^{-1} = 1_A$. Similarly it follows that $f^{-1}f = 1_B$. Thus a bijective mapping f of $A \to B$ determines a mapping f^{-1} of $B \to A$ such that $ff^{-1} = 1_A$ and $f^{-1}f = 1_B$. The mapping f^{-1} is called the **inverse** of the mapping f (likewise, of course, f is the inverse of f^{-1}).

We now prove that f^{-1} is a bijective mapping of $B \to A$. If a is any element of A, then $a = a 1_A = a(ff^{-1}) = (af)f^{-1}$, and therefore f^{-1} is surjective. Next if $bf^{-1} = b'f^{-1}$, then $(bf^{-1})f = (b'f^{-1})f$, that is $b(f^{-1}f) = b'(f^{-1}f)$ and hence $b = b'$, so that f^{-1} is injective.

THEOREM. If f is a mapping of $A \to B$, then the inverse mapping f^{-1} of $B \to A$ exists if and only if the mapping f is bijective.

Proof: All that is left for us to prove of this theorem is that if f^{-1} exists, then f is bijective. This follows at once, as above, from the two equations $ff^{-1} = 1_A$ and $f^{-1}f = 1_B$.

In some illustrative examples we assume a knowledge of addition and multiplication of integers and fractions (rational numbers), although their formal treatment does not come until later. Since this is done only for the purposes of illustration it should not be objectionable.

Example. If A and B are two sets, then the mapping of $A \times B \to A$ defined by $(a, b) \to a$, $a \in A$, $b \in B$, is surjective but is not injective unless $B = \{b\}$. For if $b \neq b'$, (a, b) and (a, b') both map into a. This map is called the **projection** of $A \times B$ on A.

Example. Let Z be the set of integers $0, \pm 1, \pm 2, \ldots$.

1. The mapping f of $Z \to Z$ defined by $n \to 2n - 1$ that is $nf = 2n - 1$, $n \in Z$, is injective but not surjective.

2. The mapping g of $Z \rightarrow Z$ defined by $ng = n + 1$ is bijective.

3. The mapping h of $Z \rightarrow Z$ defined by $(2n)h = n$, $(2n - 1)h = n$ is surjective but not injective. For while h maps Z onto all of Z, we do have, for instance, $4h = 2 = 3h$, and so h is not injective.

Example. Let Q be the set of all rational numbers—that is, fractions of the form m/n where m and n are integers and $n \neq 0$.

The mapping f of $Q \rightarrow Q$ defined by $xf = 2x - 1$, $x \in Q$, is a bijection. For $2x - 1 = 2y - 1$ implies $x = y$ so that f is injective, while if y is any element of Q then $\left(\dfrac{y + 1}{2}\right)f = y$ so that f is surjective. Its inverse f^{-1} is given by $xf^{-1} = \dfrac{x + 1}{2}$, $x \in Q$. For $x(f \circ f^{-1}) = (xf)f^{-1} =$

$(2x - 1)f^{-1} = \dfrac{2x - 1 + 1}{2} = x$ and

$$x(f^{-1} \circ f) = (xf^{-1})f = \left(\frac{x + 1}{2}\right)f = 2\left(\frac{x + 1}{2}\right) - 1 = x.$$

To compute f^{-1}, start with the obvious fact that since $xf = 2x - 1$, then $(2x - 1)f^{-1} = x$. Now putting $y = 2x - 1$, we have $yf^{-1} = \dfrac{y + 1}{2}$ for all $y \in Q$. (We can use x or y or any symbol as a representative element of Q.)

Example. Let Z be the set of integers and define a mapping f of $Z \rightarrow Z$ by

$$xf = \frac{x}{2} \text{ for } x = 0, \pm2, \pm4, \pm6, \dots$$

$$xf = x + 1 \text{ for } x = \pm1, \pm3, \pm5, \pm7, \dots.$$

Then $f^{-1}\{0\} = \{0, -1\}$, and

$$f^{-1}\{1, 2, 3\} = \{2, 4, 1, 6\}.$$

Thus f is not injective, but it is surjective.

We showed that if a mapping f has an inverse f^{-1} then f is bijective. We can now generalize this to prove the following lemma, in which the method of proof is the same.

Lemma 2. Let f be a mapping from $A \rightarrow B$.

1. If there exists a mapping g from $B \rightarrow A$ such that $f \circ g = I_A$, then f is injective.

2. If there exists a mapping h from $B \rightarrow A$ such that $h \circ f = I_B$, then

f is surjective. (g is called a *right inverse* of f and h is called a *left inverse* of f. If $g = h$, then g is called an *inverse* of f and f is now bijective.)

Proof: (1) Let $af = a'f$ for two elements $a, a' \in A$. Then $a = aI_A = a(f \circ g) = (af)g = (a'f)g = a'(f \circ g) = a'I_A = a'$. Hence f is injective. (2) Let b be any element of B. Then $b = bI_B = b(h \circ f) = (bh)f$ and $bh \in A$. Hence f is surjective.

This last lemma is often very useful.

Let f be a mapping of $A \to B$ and let C be a subset of A. We can define a mapping g of $C \to B$ by $xg = xf$, for all $x \in C$. This mapping g is called the **restriction** of the mapping to C. That is, g is a restriction of f to C if $g = f$ on C, where $C \subset A$.

Likewise if f is a mapping of $A \to B$ and if A is a subset of a set D, then a mapping h of $D \to B$ for which $yh = yf$ whenever $y \in A$, is called an **extension** of f to D. That is h is an extension of f to D if $h = f$ on A, where $A \subset D$.

If C is a subset of the set A, we can define a mapping ι of $C \to A$ by $x\iota = x$, for all $x \in C$. This mapping is called the **inclusion** mapping of $C \to A$.

Finite and Infinite Sets

Let n be a positive integer. A set S is said to have cardinal number n if there is a bijective mapping of S into the set $1, 2, \ldots, n$ of positive integers. (We can also express this by saying the two sets are in one-to-one correspondence.)

A nonempty set whose cardinal number is a positive integer is called a **finite set**, and a set that is neither empty nor finite is called an **infinite set**. It is not hard to show that a set is infinite if and only if there is a bijective mapping of the set into one of its proper subsets. For example the set E of even integers is a proper subset of the set Z of integers, and, for $n \in Z$, $n \to 2n$ is a bijective mapping of $Z \to E$.

Two nonempty sets (finite or infinite) are said to have the same cardinal number if there exists a bijective mapping of one set into the other set. If the sets are infinite the cardinal number is known as a **transfinite cardinal number**. Thus the set Z of all integers has the same cardinal number as the subset of even integers. There is an arithmetic of transfinite cardinals and the reader is urged to consult a treatise on the theory of sets. We shall not need this arithmetic in this book.

A set that has the same cardinal number as the set of positive integers is called a **countable** or **denumerable** set. It can be proved that the set of rational numbers is countable and that the set of real numbers has a greater cardinal number than the set of positive integers.

EXERCISES

1. Z is the set of integers. Which of the following mappings of $Z \to Z$ are injective or surjective or bijective?

(a) $n \to |n|$.
(b) $n \to n + 2$.
(c) $n \to n^2$.
(d) $n \to n^3$.
(e) $n \to 2n + 1$.
(f) $n \to n^3 + 1$.
(g) $n \to n$, for n even.
$\quad n \to -n$, for n odd.

(h) $n \to n!$ for $n \geq 0$.
$\quad n \to n$ for $n < 0$.
(i) $2n \to n$.
$\quad 2n - 1 \to 0$.
(j) $n \to n$, $n < -2$.
$\quad n \to -n$, $-2 \leq n \leq 3$.
$\quad n \to n$, $n > 3$.

2. Find both composite mappings in Exercise 1 for the pairs

(a) $1(a)$, $1(e)$;   (c) $1(i)$, $1(f)$;
(b) $1(g)$, $1(h)$;   (d) $1(c)$, $1(j)$.

3. Let f be a mapping (function) of the sets $A \to B$. If C and D are subsets of A, prove

$$(C \cup D)f = Cf \cup Df.$$
$$(C \cap D)f \subset Cf \cap Df.$$

If H and K are subsets of B, prove

$$f^{-1}(H \cup K) = f^{-1}H \cup f^{-1}K.$$
$$f^{-1}(H \cap K) = f^{-1}H \cap f^{-1}K.$$

4. In the accompanying illustration, the diagram of sets and mappings is commutative, which means $f = gh$. If f is injective, prove that g is injective and if f is surjective prove that h is surjective.

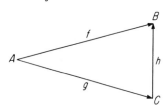

5. For $A \xrightarrow{f} B \xrightarrow{g} A$, prove that if fg is injective and gf is surjective, then f is bijective.

1-3 BINARY OPERATIONS

Let A be a nonempty set. A mapping f of $A \times A \to A$ defines what is called a **binary operation** in the set A. This means that if x and y are any two elements of A, then $(x, y)f$ is an element of A. Thus starting with any two elements of A the mapping f sends them into a third element of A. Thus we can write

$$(x, y)f = x * y,$$

and think of $x * y$ as being the unique element of A manufactured out of the two elements x and y of A. The asterisk $*$ is used to symbolize the "combining" of these two elements (that is the binary operation) to form an element of A.

For example, if A is a set of real numbers and x and y are real numbers that are in A, then $x * y$ could possibly mean the sum $x + y$ or the product xy or the difference $x - y$, or if A is a set of positive integers $x * y$ might be x^y. We prefer to attach no specific meaning to $x * y$, since it is precisely this generality that we want to preserve. In this way we derive results that are valid for many different interpretations of the binary operation symbolized by the asterisk $*$. Frequently, for simplicity, we write xy for $x * y$ and still intend that it is to be regarded as a general binary operation, and this is true sometimes even when we call it a product of x and y. It is the author's responsibility and obligation to stipulate unequivocally when a specific concrete binary operation is being used.

It is assumed in a set A with a binary operation that equality "$=$" has the following three properties:

1. It is reflexive, that is $x = x$ for all $x \in A$.
2. It is symmetric, that is if $x = y$ then $y = x$.
3. It is transitive, that is if $x = y$ and $y = z$ then $x = z$.

Example 1. Let Z be the set of all integers 0, ± 1, ± 2, ... We give examples of binary operations in Z:
   (a) The mapping of $Z \times Z \to Z$ defined by $(m, n) \to m + n$.
   (b) The mapping of $Z \times Z \to Z$ defined by $(m, n) \to mn$.
   (c) The mapping of $Z \times Z \to Z$ defined by $(m, n) \to m - n$.
   (d) The mapping of $Z \times Z \to Z$ defined by $(m, n) \to m + mn + n$.

Example 2. Let P be the set of positive integers 1, 2, 3, ... Examples of binary operations in P are:
   (a) The three binary operations 1(a), 1(b), 1(d) (but not 1(c), since it is not a mapping of $P \times P \to P$; for example $(2, 3) \to 2 - 3 = -1$ does not belong to P).
   (b) The mapping of $P \times P \to P$ defined by $(m, n) \to m^n$.
   (c) The mapping of $P \times P \to P$ defined by $(m, n) \to m$.

Definition. Let A be a set. A binary operation in the set A is called **commutative** if

$$x * y = y * x, \text{ for all } x, y \in A.$$

Examples 1(a), 1(d) are commutative binary operations and so is 2(a), however 1(c), 2(b), 2(c) are not.

Definition. Let A be a set. A binary operation on the set A is called **associative** if

$$(x * y) * z = x * (y * z)$$

for all $x, y, z \in A$.

Examples 1(a), 1(b), 2(c) are examples of associative binary operations, while the binary operations in 1(c), 1(d), and 2(b) are not associative.

If a binary operation is not known to be associative then we must distinguish between $(x * y) * z$ and $x * (y * z)$, since they may differ if the binary operation is not associative. On the other hand, if a binary operation is known to be associative then we can write the triple product as $x * y * z$ without any danger of ambiguity.

Observe that in Examples 1(a) and 1(b) the binary operations are both commutative and associative; while in 1(c) and 2(b) they are neither commutative nor associative. In Example 1(d) the binary operation is commutative but not associative; while in 2(c) it is associative but not commutative.

A very important example of a binary operation that is associative but not commutative is **map composition.** Let E be the set of all mappings of a set A into itself. Then if $f, g \in E$, clearly the composite mappings $f \circ g$ and $g \circ f$ exist and belong to E. Moreover in general we cannot expect these two mappings to be equal—that is, $f \circ g \neq g \circ f$. However, map composition is associative. This is easy to prove. Let $f, g, h \in E$. Form $(f \circ g) \circ h$. It also belongs to E. Let $a \in A$, then

$$a[(f \circ g) \circ h] = [a(f \circ g)]h = [(af)g]h$$
$$= (af)g \circ h = a[f \circ (g \circ h)].$$

This is true of every $a \in A$. Hence

$$(f \circ g) \circ h = f \circ (g \circ h).$$

We use the notation $(A, *)$ to designate a set A and a binary operation $*$ defined on A. An element $e \in A$ is called a **neutral element** of the binary operation $*$ of $(A, *)$ if $e * a = a * e = a$, for all $a \in A$. If $(A, *)$ has a neutral element, then $y \in A$ is called a **right inverse** of the element $x \in A$ if $x * y = e$, and $z \in A$ is called a **left inverse** of $x \in A$ if $z * x = e$.

If $(A, *)$ has a natural element e, then it is unique. For if e' is a second neutral element of $(A, *)$ then $e = e * e' = e'$.

If the binary operation is associative and if $x \in A$ has both a left inverse z and a right inverse y, then $y = z$. For $zx = e$ implies that $(zx)y = ey$ and hence $z(xy) = y$. But $xy = e$ and $ze = z$, hence $z = y$. We call an element $u \in A$, which is both a left and a right inverse of $x \in A$, the **inverse of x.** That is $ux = e = xu$. Furthermore, if the binary operation is associative then the inverse of an element is unique.

For if v is also an inverse of x, that is $vx = e = xv$, then $(ux)v = ev = v$ and also $u(xv) = ue = u$. Hence since $(ux)v = u(xv)$ we have $v = u$.

We introduce here, in passing, the names of three important algebraic systems, deferring until the next chapter their formal definitions. A set $(A, *)$ with an associative binary operation is called a **semigroup**. If, in addition, $(A, *)$ has a neutral element then $(A, *)$ is called a **monoid**. If each element of a monoid $(A, *)$ has an inverse, then $(A, *)$ is called a **group**.

Example. The positive integers 1, 2, 3, ... form a semigroup under addition (but not a monoid) while under multiplication they form a monoid. (We are assuming that ordinary addition and multiplication of integers is associative.) The nonnegative integers 0, 1, 2, 3, ... form a monoid under addition, while all the integers 0, ± 1, ± 2, ... form a group under addition, but not under multiplication.

Example. If A is any set, then the set B of all bijections of $A \to A$ forms a group under map composition whose neutral element is the identity mapping I_A. For map composition is associative and every bijection has an inverse.

EXERCISES

1. A is a set and a binary operation is defined in A by a mapping f of $A \times A \to A$. Let B be a set that is bijective to A and let g be the bijection of $A \to B$. Denote by $g \times g$ the mapping of $A \times A \to B \times B$ defined by $(a, a') \to (ag, a'g)$ where $a, a' \in A$. Let h be the mapping of $B \times B \to B$ defined by $(g \times g)h = fg$. The accompanying diagram is thus commutative.

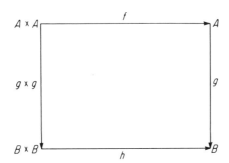

(a) Prove that h defines a binary operation in B that is commutative and/or associative if the binary operation in A is commutative and/or associative.

(b) If the binary operation in A has a neutral element e, prove that eg is the neutral element for the binary operation in B.

(c) If $a \in A$ has an inverse $a^{-1} \in A$, show that the element $b = ag$ of B has the inverse $b^{-1} = a^{-1}g$.

(d) Show that if A has the structure of a group, then B is a group.

2. Let Z be the set of integers with ordinary multiplication as the binary operation. Find the binary operations defined by the following bijections g of Z.

(a) $Z \rightarrow Z$ where $ng = n - n_0, n \in Z$ and n_0 is a fixed integer.

(b) $Z \rightarrow Z$ where $ng = n - 1, n \in Z$.

(c) $Z \rightarrow E$ where E is the set of even integers and $ng = 2n, n \in Z$.

(d) $Z \rightarrow E$ where $ng = 2n - 2, n \in Z$.

1-4 CARTESIAN PRODUCT

We now generalize the notion of a cartesian product to an arbitrary family of sets.

If $X_1, X_2, \ldots, X_n$ are n nonempty sets (distinct or not) we define their *cartesian product* $\prod_{i=1}^{n} X_i$ as the set of all elements of the form $(x_1, x_2, \ldots, x_n)$, where $x_i \in X_i$, $i = 1, 2, \ldots, n$. If $X_1, X_2, X_3, \ldots$ is an infinite sequence of sets, their *cartesian product*, written $\prod_{i=1}^{\infty} X_i$ is defined as the set of all sequences $(x_1, x_2, \ldots, x_n, \ldots)$ where $x_i \in X_i$, $i = 1, 2, \ldots$.

The concepts of union and intersection of sets can be readily generalized to an arbitrary family of sets. Let $\{X_i\}$, $i \in I$, designate a family of sets X_i indexed by the set I. The set I may be finite or infinite. We define

$$\bigcup_{i \in I} X_i = \{x_i \mid x_i \in X_i \quad \text{for some } i \in I\},$$

$$\bigcap_{i \in I} X_i = \{x_i \mid x_i \in X_i \quad \text{for each } i \in I\}.$$

If I is an arbitrary nonempty set and $\{X_i\}$, $i \in I$, is a family of nonempty sets (distinct or not), the cartesian product of this family (indexed by the set I) is defined to be the set of all mappings (functions) of the index set I into the union of the family such that each $i \in I$ is mapped into a unique element of the set X_i. Thus an element f of the cartesian product is a mapping of $I \rightarrow \bigcup X_i$ such that if $x_i \in X_i$, $i \in I$. The element f of the cartesian product is often written as $(x_i)_{i \in I}$, $x_i \in X_i$, for each f determines a unique $(x_i)_{i \in I}$ and conversely. Thus if I is the finite set of natural numbers $1, 2, \ldots, n$ then

$$(x_i)_{i \in I} = (x_1, x_2, \ldots, x_n).$$

We denote the cartesian product by

$$\prod_{i \in I} X_i.$$

The mapping f can be called a *choice function* in that it chooses or determines an element x_i from each X_i. That such an unlimited or "infinite" choice is to be assumed as possible is essentially one form of the so-called **axiom of choice**.

Axiom of Choice

The cartesian product of a nonempty family of nonempty sets is nonempty.

The axiom states that we can make a choice of an element from each set of a nonempty family of nonempty sets. The axiom is often used in proofs of theorems without the fact being explicitly acknowledged. It has been proved that the axiom of choice is equivalent to the well-ordering axiom and to Zorn's axiom. (See Sec. 1-5.)

The mapping of the cartesian product onto one of its member sets

$$\prod_{i \in I} X_i \to X_j, \quad j \in I$$

defined by

$$(x_i)_{i \in I} \to x_j$$

is called the jth projection of the cartesian product on X_j. Here x_j is that element in $(x_i)_{i \in I}$ that belongs to the set X_j. It follows from the axiom of choice that a projection of the cartesian product onto one of its member sets is surjective.

1-5 EQUIVALENCE RELATIONS

If A is a set, we define a relation R on A to be a *subset* of $A \times A$. Thus R is a set of elements that are ordered pairs of the form (a_1, a_2), where $a_1, a_2 \in A$. We write $a_1 R a_2$ if and only if $(a_1, a_2) \in R$.

Example. Let Z be the set of integers $0, \pm 1, \pm 2, \ldots$, then the symbol $<$ (less than) is a relation on Z. Here R is the set of all ordered pairs (x, y) of integers such that $x < y$. R is a subset of $Z \times Z$.

A relation R on a set A is called *reflexive* if aRa for all $a \in A$; that is, the subset R of $A \times A$ contains all elements of the form (a, a), $a \in A$. R is called *symmetric* if xRy implies yRx; that is, if $(x, y) \in R$ implies $(y, x) \in R$. R is called *transitive* if xRy and yRz together imply xRz; that is, if (x, y) and (y, z) belong to R then $(x, z) \in R$.

A relation R on a set A that is reflexive, symmetric, and transitive is called an **equivalence relation** on the set A.

Example. Define the relation $x \equiv y$ (modulo 2) on the set Z of integers to mean $x - y$ is divisible by 2. It is called a *congruence relation* and we can form such a relation for any integer $n \geq 1$. Such a congruence relation is clearly an *equivalence relation*.

Example. The relation $<$ (less than) on Z is seen to be transitive but is neither reflexive nor symmetric.

Example. The relation $\leq$ (less than or equal to) on the set Z of integers is reflexive and transitive but is not symmetric.

Let R be an equivalence relation on a set A. Let $a \in A$. Form the set α of all elements of A that are equivalent to the element a. Note $a \in \alpha$, since R is reflexive. Let $b \in A$ and form the set β of all elements of A that are equivalent to the element b. If α and β have a common element x, then aRx and xRb and hence, since R is transitive, aRb and therefore $\alpha = \beta$. Thus α and β are either identical or else $\alpha \cap \beta = \phi$ and they are disjoint. The sets $\alpha, \beta, \gamma, \ldots$ are called the **equivalence classes** of R on A, and any two of them are either disjoint or identical. Moreover, every element of A is in one and only one equivalence class. Two elements $x, y \in A$ are in the same equivalence class if and only if xRy. All this means that an equivalence relation R on a set A partitions the set A into a set of disjoint equivalence classes whose union is A. The set of equivalence classes is called the *quotient set* of A modulo R and is usually designated by A/R. Of course two equivalence classes do not necessarily contain the same number of elements.

EXERCISE

1. If a set A is the union of a family of nonempty disjoint subsets, prove that these subsets are the equivalence classes of an equivalence relation on A and define the relation.

Posets and Zorn's Axiom

A (binary) relation, written $\leq$, on a set S is said to be *antisymmetric* if for all $x, y \in S$,

$$x \leq y \text{ and } y \leq x \text{ imply } x = y.$$

A (binary) relation $\leq$ on a set S is called a *partial order* of S if it is reflexive, antisymmetric, and transitive. A set with a partial order is called a **poset**.

Example 1. The set S of all subsets of a nonempty set E with the inclusion relation $\subset$ is a poset. If $\alpha, \beta \in E$ then $\alpha \subset \beta$ means, as usual, α is a subset of β. It is easy to verify that this yields a partial order of S. The significance of the term partial is that there exist subsets α and β of E such that neither is a subset of the other.

A subset T of a poset S is said to be *totally (linearly) ordered* if for every pair of elements $x, y \in T$, either $x \leq y$ or $y \leq x$. Thus in Example 1, S is partially but not totally ordered.

A subset U of a poset S is said to have an *upper bound* b if $b \in S$ and $x \leq b$ for all $x \in U$.

A poset S is said to be *inductively ordered* if every totally ordered subset of S has an upper bound.

A *maximal element* of a poset S is an element $m \in S$ such that if $x \in S$ and $m \leq x$, then $x = m$.

Example 2. Let S be the set of positive integers from 1 to 12. For $m, n \in S$ define a partial order $m \leq n$ of S by $m \mid n$; that is, m is a divisor of n. Clearly it is reflexive, for $m \mid m$ for every $m \in S$; it is antisymmetric for $m \mid n$ and $n \mid m$ imply $m = n$; it is transitive, for $m \mid n$ and $n \mid r$ imply $m \mid r$. Hence S is a poset. The subset 1, 5, 10 of S is totally ordered and has the upper bound 10. In fact it is clear that every totally ordered subset has an upper bound, and so S is inductively ordered. It is readily proved that 7, 8, 9, 10, 11 and 12 are all maximal elements of S.

Example 3. In Example 1, if T is a totally ordered subset of S, then the union of all the subsets of E that belong to T is an upper bound of T. Hence S is inductively ordered. Since E is a subset of itself, E is an element of S and clearly it is the only maximal element of S.

We shall make use of the following axiom in our later work.

Zorn's Axiom

If S is a nonempty poset that is inductively ordered, then S contains at least one maximal element.

EXERCISES

1. Let E be a finite set. Let S be the set of all proper subsets of E and order S by inclusion $\subset$. Prove S is an inductively ordered poset and find all the maximal elements of S. If E contains n elements, how many maximal elements does S contain?

2. A subset of a group G that forms a group under the same binary operation as that of G is called a *subgroup* of G. Assume G is an infinite group and let $x \in G$. Let S be the set of all subgroups of G that do not contain x.

(a) Prove that with the relation inclusion $\subseteq$, S is a poset.

(b) Prove that the union of a totally ordered subset T of S is a subgroup of G and an upper bound of T.

(c) Prove there is a largest subgroup of G that does not contain x.

1-6 THE INTEGERS

The rudimentary starting point in the development of the number system is with the five axioms for the natural numbers (the positive integers) first set forth by the mathematician G. Peano. They are:

1. 1 is a natural number.

2. For each natural number x there exists a unique natural number x', called the successor of x.

3. 1 is not the successor of any natural number.

4. If $x' = y'$ then $x = y$.

5. If S is a set of natural numbers with the two properties (*a*) $1 \in S$ (*b*) if $x \in S$ then $x' \in S$, then S contains all the natural numbers.

We are not going to follow through from these axioms to arrive logically at all the properties of the natural numbers, and from there on to all the properties of the integers 0, ± 1, ± 2, ..., for it is too lengthy a process. Yet it is remarkable and interesting that we can start with just these five axioms and develop mathematically the properties of the set of all integers. This means that all that is assumed known about the integers are these five axioms about the positive integers; everything else, without exception, has to be (and actually can be) derived from these axioms. It may mildly startle the reader to realize the implication of the foregoing: definitions have to be given of addition and multiplication, and their properties of being commutative and associative binary operations have to be derived. In the course of this we even ascertain that the successor of 2 is 3, although this might be suspected, and that $x' \neq x$, as again might be suspected. The first axiom ensures that the set of natural numbers is nonempty, and in fact axioms (2) and (3) will ensure there are actually infinitely many of them. Axiom (5) is called the **finite induction axiom,** and using it and axiom (4), it is easy to prove $x' \neq x$. The reader should do this by assuming T is the set of all natural numbers for which $x' \neq x$, and then using axiom (5) to show T is the set of all natural numbers. Another interesting and instructive exercise for the reader is to prove that $x = y$ implies $x' = y'$.

The short cut taken here is to define the set of all the integers as an algebraic system, which will then be called a **system of integers**.

First let Z be a set of elements satisfying the following axioms.

1. Z has a binary operation addition, which is both commutative and associative and which has a neutral element 0. Moreover every element $z \in Z$ has an additive inverse, denoted by $-z$, in Z. Thus $z + (-z) = 0 = (-z) + z$. (For $x, y \in Z$ we put $x - y$ for $x + (-y)$ and thus $x - y \in Z$.)

2. Z has a binary operation multiplication which is both commutative and associative and which has a neutral element $1 \neq 0$.

3. The two binary operations of the set Z are related by the *distributive law*

$$x(y + z) = xy + xz, \text{ for all } x, y, z \in Z.$$

Up to this point Z is an example of what we shall later call a *commutative ring* with identity 1. We shall prove later the following two theorems about such a ring:

For all $x, y \in Z$,

$$(-x)y = -xy = x(-y)$$

and

$$(-x)(-y) = xy.$$

Actually these easily follow from the definition of $-z$ as the additive inverse of z and from the distributive law.

Now suppose Z contains a subset P that is stable (closed) under addition and multiplication and such that every $z \in Z$ satisfies one and only one of the following three mutually exclusive conditions: either $z \in P$ or $-z \in P$ or $z = 0$. The effect of this is to separate the nonzero elements of Z into two disjoint subsets P, called the **set of positive elements** of Z, and P', called the **set of negative elements** of Z. P' consists of all nonzero elements z of Z for which $-z \in P$. Clearly $Z = P \cup P' \cup \{0\}$. The set Z is then said to be *ordered* with P as its subset of positive elements.

We now add to the preceding three axioms for the set Z the two following axioms:

4. Z is ordered with P designating its subset of positive elements.

5. (*Finite Induction Axiom.*) If S is a subset of positive elements of Z such that (*a*) $1 \in S$ (*b*) if $z \in S$ then $z + 1 \in S$, then $S = P$.

A set Z which satisfies all five axioms is called a **system of integers**. We assume the existence of such a system and defer until later the proof that a system of integers is unique up to what is called an **order-isomorphism**. This means that any other set Z' satisfying the same set of axioms as the set Z can be placed in one-to-one correspondence with Z in such a way that this correspondence will preserve addition, multiplication, and the ordering. In the light of this latter remark we shall consider that we are entitled to speak of the system of integers, for two such systems are algebraically indistinguishable and differ only in notation. We shall, however, return to this in a later chapter.

Ordering in the usual, intuitive sense means that an element of Z is greater than, equal to, or less than a second element of Z. For $x, y \in Z$, define $x > y$ (x greater than y) if and only if $x - y \in P$— that is, $x - y$ is positive. Hence $x > 0$ signifies $x \in P$. Clearly, if $x > y$ and if $z > 0$, then $xz > yz$, while if $z < 0$ then $yz > xz$. These follow from the two properties of the set P. Moreover, if x and y are two distinct elements of Z, then either $x > y$ or $y > x$. This latter fact is described by saying that Z is **totally ordered.**

We show next that the elements of Z satisfy the *cancellation law of multiplication* which states that if $x \neq 0$, then $xy = yz$ implies $y = z$. If $y \neq z$ suppose $y - z > 0$. If $x > 0$, then $x(y - z) > 0$ and hence

$xy > xz$, which contradicts $xy = xz$. If $x < 0$, then $xz > xy$, again a contradiction. Thus $y \neq z$ implies a contradiction, and hence $y = z$.

Since the cancellation law of multiplication is a property of Z, this makes Z an example of what we shall later (Chapter 5) define as an *integral domain*.

We shall now refer to the elements of Z as integers.

If $x \neq 0$ is an integer then x^2 is a positive integer. For if $x > 0$ then $x^2 > 0$, and if $x < 0$ then $-x > 0$ and hence $(-x)(-x) > 0$. Since $(-x)(-x) = x^2$, then $x^2 > 0$. In particular $1 = 1^2 > 0$.

Absolute Value

If x is an integer, the absolute value $|x|$ is defined by (a) if $x > 0$, $|x| = x$; (b) if $x < 0$, then $|x| = -x$; (c) if $x = 0$, then $|x| = 0$.

Thus under all circumstances,

$$-|x| \leq x \leq |x|.$$

Treating each of the three possibilities (1), (2), and (3) of the definition, separately, it is easy to show that for all integers x and y,

$$|xy| = |x|\,|y|$$
$$|x + y| \leq |x| + |y|.$$

1-7 FINITE INDUCTION

We do not enter here into a detailed study of the integers and their arithmetic, but adopt the compromise of allowing the ensuing outline to suffice for our immediate needs. Least of all do we consider the philosophical meaning of an integer—thus avoiding intricate and wearisome metaphysical investigations. This is certainly one advantage of the axiomatic method. Many of the properties of the integers are those of any abstract-ordered integral domain, and their discussion is deferred until this appropriate time. Later the reader will be able without difficulty to identify the system of integers as an ordered principal ideal domain (plus the induction axiom) and hence enjoying all the properties of such an algebraic system. However, the property of finite induction and its consequences are peculiar to the system of integers. Since this axiom and the well-ordering theorem are extensively employed in the proofs here and in subsequent chapters, we will discuss them now.

THEOREM 1. 1 is the least positive integer.

> **Proof:** Let S be the set of all positive integers ≥ 1. Then $1 \in S$. Let $x \in S$. Then $x \geq 1$. Since $x + 1 > x$, $x + 1 \geq 1$ and therefore $x + 1 \in S$. Hence S contains all the positive integers.

THEOREM 2. If $x < y$, then $x + 1 \leq y$ (that is there is no integer between x and $x + 1$).

Proof: If $x < y$ then $y - x > 0$. Hence, by the previous theorem, $y - x \geq 1$ and therefore $y \geq x + 1$.

THEOREM 3 (Well-Ordering). Any subset of positive integers contains a least (smallest) positive integer. (This property of the positive integers is described by saying the set of positive integers is *well-ordered*.)

Proof: Let S be a nonempty subset of positive integers. If $1 \in S$ then, by Theorem 1, 1 is the least positive integer in S. Now suppose $1 \notin S$. Let T be the set of all positive integers that are less than every member of S. $1 \in T$. There must exist an $x \notin T$ such that $x + 1 \notin T$, for if not, T would contain all positive integers and this would contradict S being nonempty. We show $x + 1$ is the least positive integer in S. Since $x + 1 \notin T$, there exists $y \in S$ such that $x + 1 \geq y$. But $x < y$. Hence by Theorem 2, $y = x + 1$ and so $x + 1 \in S$. If z is any element of S, then $z > x$ and hence $z \geq x + 1$, by Theorem 2. Hence $x + 1$ is the least positive integer in S.

It is an interesting fact that the well-ordering theorem implies the finite induction principle. For let S be a set of positive integers such that $1 \in S$ and such that if $x \in S$, then $x + 1 \in S$. Let T be the set of all positive integers not in S. If T is nonempty, then it has a least member y and $y \neq 1$. Hence $y > 1$, and so $y - 1 > 0$. Now $y - 1 < y$ and therefore $y - 1 \in S$. But this implies $y \in S$, which contradicts $y \in T$. Thus T must be empty, and hence S contains all the positive integers.

Second Form of Finite Induction Principle

Let S be a set of positive integers with the two properties: (a) $1 \in S$. (b) For x a positive integer, if all positive integers $< x$ are in S, then $x \in S$. Then S contains all the positive integers.

Proof: Let S be a set of positive integers satisfying the conditions (*a*) and (*b*). Let T be the set of positive integers that are not in S. If T is nonempty, then it contains a least positive integer y. Since $1 \notin T, y > 1$. Since all positive integers $< y$ are in S, this makes y belong to S which contradicts $y \in T$. Hence T must be empty and therefore S contains all the positive integers.

Applications of the induction axiom occur most frequently in the following form. Suppose $S(n)$ stands for a statement made involving the positive integer n. If it can be verified directly that the statement is true for some fixed positive integer n_0, that is $S(n_0)$ is true, and if it can be proved that the truth of the statement $S(n)$, $n > n_0$, implies the truth of the statement $S(n + 1)$, then the statement is true for all $n \geq n_0$. For this simply means that if S is a subset of positive integers for which the statement is true and n_0 is the least positive integer in S, then if $n \in S$ implies $n + 1 \in S$, we know that S contains all positive integers $\geq n_0$. Incidentally while in the induction axiom itself $n_0 = 1$, it is seen, however, that the induction principle can operate starting with any positive integer. The assumption that $S(n)$ is true is referred to as the *induction hypothesis*.

Applications of the second form of the finite induction principle occur in a similar manner, only now the induction hypothesis is that the statement $S(k)$ is true for all $k < n$, where of course $k > n_0$.

As illustrations of the use of finite induction, the following two important algebra theorems are now proved.

General Associative Law

Let A be an algebraic system with an associative binary operation which we shall denote for simplicity of notation as multiplication. Let $a_1 a_2 \ldots a_n$ be a product of n factors from A. For $n = 3$ we know $(a_1 a_2)a_3 = a_1(a_2 a_3)$ and we want to show that in computing the product for any $n \geq 3$ it is immaterial in what order the binary operations are successively performed. This is what is called the **general associative law**.

Let $(a_1 a_2 \ldots a_i)(a_{i+1} \ldots a_n) = b$ and $(a_1 \ldots a_j)(a_{j+1} \ldots a_n) = c$, $i < j$, be the final multiplications in two distinct ways of computing the general product. Now use the second form of the induction principle. The theorem is true for $n = 3$ and if it is assumed true for products with $< n$ factors, we have by the induction hypothesis that

$$b = [a_1 \ldots a_i][(a_{i+1} \ldots a_j)(a_{j+1} \ldots a_n)]$$
$$c = [(a_1 \ldots a_i)(a_{i+1} \ldots a_j)][a_{j+1} \ldots a_n].$$

From the fact that the law holds for $n = 3$, it follows that $b = c$. Hence the general associative law is valid for any finite number of factors.

General Commutativity Law

Let A be an algebraic system with an associative and commutative binary operation which is again denoted by multiplication.

We want to show

$$a_1 a_2 \ldots a_n = a_{i_1} a_{i_2} \ldots a_{i_n} \text{ for all } n \geq 2,$$

where $a_{i_1}, \ldots, a_{i_n}$ is any rearrangement of the a_i, $i = 1, 2, \ldots, n$. Since the commutative law is assumed true for $n = 2$, we can commute a_1 in succession with each of its predecessors in the product $a_{i_1} a_{i_2} \ldots a_{i_n}$ so as to bring it to the front position. Hence

$$a_{i_1} \ldots a_{i_n} = a_1 a_{j_2} \ldots a_{j_n},$$

where $a_{j_2}, \ldots, a_{j_n}$ is some rearrangement of $a_2, \ldots, a_n$. We now use induction as before, the hypothesis here being that the law is true for all integers k such that $2 \leq k < n$. Hence $a_{j_2} \ldots a_{j_n} = a_2 a_3 \ldots a_n$. But clearly this implies

$$a_{i_1} \ldots a_{i_n} = a_1 a_2 \ldots a_n.$$

Hence the general commutative law holds for all integers ≥ 2.

1-8 PROPERTIES OF THE INTEGERS

Definition. An integer x is said to *divide* an integer y, written $x \mid y$, if there is an integer z such that $y = xz$. x is called a **divisor** of y, and y is called a **multiple** of x.

THEOREM 4 (Euclidean Algorithm). If a and $b, b > 0$, are integers, then there exist unique integers q and r such that

$$a = bq + r, \text{ where } 0 \leq r < b.$$

Proof: Let S be the set of all integers of the form $a - bx$, where x runs through all integers. Clearly, S contains positive and negative integers and possibly 0. Let r be the least nonnegative integer in S. For $x = q$, let $a - bq = r$. We must show $0 \leq r < b$. Suppose $r \geq b$. Then $a - bq \geq b$ and hence $a - b(q + 1) \geq 0$. Also, $a - b(q + 1) < a - bq = r$. Hence $0 \leq a - b(q + 1) < r$. Since $a - b(q + 1) \in S$, this contradicts the definition of r. Hence $0 \leq r < b$. To prove the uniqueness of q and r, assume there exist integers q' and r' such that $a = bq' + r'$ with $0 \leq r' < b$. Then $b(q - q') = r' - r$. If $r' \neq r$, suppose $r' > r$, then since $b > 0$ we have $q > q'$. But $r' - r < b$ and this denies $r' - r$ being a multiple of b. Hence $r' = r$, and therefore $q = q'$ by the cancellation law of multiplication.

Definition. The *greatest common divisor* (G.C.D.) of two integers a and b is the positive integer d such that (1) $d \mid a$ and $d \mid b$ (2) if $c \mid a$ and $c \mid b$ then $d \mid c$.

That d is unique follows from the fact that if $d' > 0$ satisfies the two conditions then $d \mid d'$ and $d' \mid d$ and therefore $d' = d$. That the G.C.D. exists follows from the next theorem.

THEOREM 5. If a and b are two nonzero integers then there exists integers s and t such that $as + bt$ is the G.C.D. of a and b.

Proof: Let S be the set of all integers of the form $ax + by$, where x and y run through all integers. S contains positive integers. Let d be the least positive integer in S. For $x = s$ and $y = t$ let $as + bt = d$. Next we show $d \mid a$. By the Euclidean algorithm, $a = qd + r, 0 \leq r < d$. Then $qas + qbt = qd = a - r$. Hence $r =$

$a(1 - qs) - bqt$, and so $r \in S$. Unless $r = 0$, this contradicts the definition of d. Thus $a = qd$. Similarly, $d \mid b$. Furthermore, it is seen from $as + bt = d$ that every common divisor of a and b must divide d. Hence d is the G.C.D. of a and b.

Definition. A positive integer $x > 1$ is called a **prime** if its only positive divisors are 1 and x.

Definition. Two nonzero integers are called **coprime** or **relatively prime** if their G.C.D. is 1.

THEOREM 6. Let the integer p be a prime. If $p \mid ab$ where $a \neq 0$, $b \neq 0$, then $p \mid a$ or $p \mid b$. (This is the *fundamental property of being prime*.)

> **Proof:** Let $p \mid ab$ and assume $p \nmid a$. Since p is a prime this means a and p are coprime. Hence for integers s and t, $as + pt = 1$. Therefore $abs + bpt = b$. Since p divides each term on the left, we have $p \mid b$.

> **Corollary.** If p is a prime and $p \mid a_1 a_2 \ldots a_n$, then p divides some a_i.

> **Proof:** We use finite induction. By the theorem we know this is true for $n = 2$. Assume the theorem true for a product of $n - 1$ factors. Now $p \mid (a_1 a_2 \ldots a_{n-1}) a_n$ and hence $p \mid a_n$ or $p \mid a_1 a_2 \ldots a_{n-1}$. By the induction hypothesis, the theorem is true in either case. Hence it is true for a product of n factors. Hence the theorem is true for all $n \geq 2$.

THEOREM 7 (Fundamental Theorem of Arithmetic). Every integer > 1 is a product of primes and this product is unique up to the order of its factors.

> **Proof:** We first show $a > 1$ is a product of primes. If a is a prime, that is certainly true. If a is not a prime, let $a = bc$ where $b > 1$, $c > 1$. Now use the second form of the finite induction principle and assume that any integer > 1 and $< a$ is a product of primes. Then by the induction hypothesis the integers b and c are products of primes. But since $a = bc$, this means a is the product of primes. Hence every integer > 1 is a product of primes.

To prove uniqueness of this factorization again use induction, but this time the first form of it. Assume the uniqueness for all integers > 1 that can be expressed as a product of $< k$ prime factors, where $k > 1$. If $k = 1$ the integer itself is a prime to start with and this is certainly its unique prime factorization. Let $a > 1$ be a product of k primes, $a = p_1 p_2 \ldots p_k$, and suppose the integer a has a second prime factorization $a = q_1 q_2 \ldots q_r$. Then $p_1 p_2 \ldots p_k = q_1 \ldots q_r$. Now $p_1 \mid q_1 q_2 \ldots q_n$ and hence p_1 divides some q_i. By reindexing the q_i, if necessary, we can suppose $p_1 \mid q_1$. Since both are primes, $p_1 = q_1$. Now cancel them and we have

$$p_2 \ldots p_k = q_2 \ldots q_r = b.$$

By the induction hypothesis, the uniqueness is true for this latter integer b, since it has a factorization as a product of $< k$ primes. But since $a = p_1 b$, this means the factorization for a is unique and $r = k$, and each q_i is equal to some p_j. Hence by the induction principle the theorem is now true for all integers > 1 and the only way in which two factorizations of the same integer > 1 can differ is in the order of the factors.

There is nothing to prevent some or all of the p_i being equal in the prime factorization,

$$a = p_1 p_2 \ldots p_k.$$

If we group the equal ones together then a can be written as the product of powers of distinct primes:

$$a = p_1^{\alpha_1} \ldots p_m^{\alpha_m}.$$

The *least common multiple* (L.C.M.) of two nonzero integers a and b is defined to be the positive integer k such that (1) $a \mid k$ and $b \mid k$ (2) if $a \mid w$ and $b \mid w$ for some integer w, then $k \mid w$.

THEOREM 8. The L.C.M. k of two integers a and b exists and is unique.

Proof: Let S denote the set of all common multiples of a and b. Now S contains ab and $-ab$ and hence contains a positive integer. Let k be the least positive integer in S. We claim k is the L.C.M. of a and b.

First it is evident that $a \mid k$ and $b \mid k$. Now suppose $a \mid w$ and $b \mid w$ for some integer w. By the Euclidean algorithm

$$w = qk + r, \qquad 0 \leqq r < k.$$

However, if $r > 0$ we have a contradiction, since $w - qk$ obviously belongs to S and $r < k$. Hence $r = 0$ and therefore $k \mid w$. This proves k is the L.C.M. of a and b. The uniqueness of k follows quickly. For if k' is a L.C.M. of a and b, then since k and k' are both positive, and since $k \mid k'$ and $k' \mid k$, it follows that $k' = k$.

EXERCISE

1. If d is the G.C.D. and k is the L.C.M. of two integers a and b, where $ab > 0$, then $dk = ab$.

All these facts about the G.C.D. and the L.C.M. of two integers can also be easily deduced from the fundamental theorem of arithmetic.

Write each integer as the product of powers of distinct primes and compare the factors in each product.

The notions of G.C.D. and L.C.M. readily extend to any finite number of integers.

1-9 MATRICES

For purposes of later illustrative examples of algebraic systems and in order that these illustrations be independent of formal work in linear algebra, we include here a brief discussion of matrices that will suffice for these purposes.

A rational number is a number of the form a/b, where a and b are integers and $b \neq 0$. We shall assume a knowledge of the ordinary addition and multiplication of these fractions. If m and n are positive integers then an $m \times n$ matrix A over the rational numbers is an arrangement of mn rational numbers a_{ij}, $i = 1, \ldots, m$, $j = 1, \ldots, n$ into m rows and n columns and is denoted by

$$A = \begin{bmatrix} a_{11} & a_{12} & \ldots & a_{1n} \\ a_{21} & a_{22} & \ldots & a_{2n} \\ \ldots & \ldots & \ldots & \ldots \\ a_{m1} & a_{m2} & \ldots & a_{mn} \end{bmatrix} = (a_{ij})$$

where the a_{ij} are rational numbers and a_{ij} itself is the rational number in the ith row and jth column. The particular arrangement of the rational numbers and the numbers themselves are determined by some external means or scheme, but this will not concern us here.

A matrix is called *square* if $m = n$—that is, if the number of rows is the same as the number of columns.

We restrict ourselves to the definitions of addition and multiplication of square matrices. Let (a_{ij}) and (b_{ij}) be two $n \times n$ matrices over the rational numbers. We define

$$(a_{ij}) + (b_{ij}) = (a_{ij} + b_{ij})$$
$$(a_{ij})(b_{ij}) = (c_{ij})$$

where

$$c_{ij} = \Sigma a_{ik} b_{kj}$$
$$= a_{i1} b_{1j} + a_{i2} b_{2j} + \cdots + a_{in} b_{nj}.$$

Thus

$$\begin{bmatrix} 2/3 & 1/2 \\ 0 & -1 \end{bmatrix} \begin{bmatrix} 1/4 & 2 \\ 3 & 3/5 \end{bmatrix} = \begin{bmatrix} 5/3 & 49/30 \\ -3 & -3/5 \end{bmatrix}.$$

Straightforward computations show that matrix addition is both commutative and associative, while matrix multiplication is associative but not in general commutative. The $n \times n$ matrix whose entries are all 0 is called the *zero matrix*. The $n \times n$ matrix for which $a_{ii} = 1$, $i = 1, 2, \ldots, n$ and $a_{ij} = 0$, $i \neq j$ is called the **identity matrix**. All its entries on the principal diagonal are 1, and those off the principal diagonal are 0. Denote the identity matrix by I. It is easy to verify that if A is any $n \times n$ matrix then $IA = A = AI$.

An $n \times n$ matrix (a_{ij}) is said to be *nonsingular* if it has an *"inverse"* matrix (b_{ij}), that is if there exists an $n \times n$ matrix (b_{ij}) such that

$$(a_{ij})(b_{ij}) = I = (b_{ij})(a_{ij}).$$

Of course (a_{ij}) is the inverse of (b_{ij}) so that (b_{ij}) is nonsingular.

This brief discussion of matrices will suffice to render the illustrative examples, appearing later, independent of a formal knowledge of linear algebra. We merely point out that later we shall see that the rational numbers form what is called a **field** (that is, an integral domain in which every nonzero element has a multiplication inverse). The matrices above were taken over the rational field but the setting could have been an arbitrary field.

EXERCISES

1. In the Peano axioms in Sec. 1-6, x' stood for the successor of x. Prove (a) $x' \neq x$ and (b) $x = y$ implies $x' = y'$.

2. If x and y are integers prove $|xy| = |x| \, |y|$ and that $|x + y| \leq |x| + |y|$.

3. Which of the following are equivalence relations?

(a) Two straight line segments are equivalent if and only if their lengths differ by more than one inch.

(b) Two plane triangles are equivalent if and only if a side of one triangle is equal to a side of the other.

(c) Z is the set of integers and two integers are equivalent if their sum is even.

(d) For Z the set of integers and (x, y), $(x', y') \in Z \times Z$ define (x, y) and (x', y') as equivalent if and only if $xy' = x'y$.

4. If x and y are relatively prime positive integers such that xy is a perfect square, prove with the aid of the fundamental theorem of arithmetic that x and y are perfect squares.

5. Prove by induction:

(a) $8^n - 3^n$ is divisible by 5 for all positive integers n;

(b)
$$2^2 + 4^2 + \cdots + (2n)^2 = \frac{2n}{3}(n + 1)(2n + 1).$$

6. For nonzero integers m and n, prove (a) $-m \leq n \leq m, m > 0$, if and only if $|n| \leq m$; (b) $|mn| = dh$, where d is the G.C.D. of m and n and h is the L.C.M. of m and n.

7. If p is a prime, prove that all but two of the coefficients of the binomial expansion of $(1 + x)^p$ are divisible by p.

8. Use induction on n to prove that

$$n^p \equiv n \pmod{p}.$$

This is read as n^p is congruent to n modulo p and signifies that $n^p - n$ is divisible by p.

If $p \nmid n$, show $n^{p-1} \equiv 1 \pmod{p}$. This is called **Fermat's theorem.**

9. For elements x and y in a ring prove by induction that
(1) $x^n \cdot x = x^{n+1}$, where $x^1 = x$.
(2) if $xy = yx$ then $xy^n = y^n x$ for all positive integers n.
(3) if $xy = yx$ then $(xy)^n = x^n y^n$ for all positive integers n.

10. Two integers a and b are said to be congruent modulo the positive integer m, written

$$a \equiv b \pmod{m}$$

if and only if $m \mid a - b$. Prove this is an equivalence relation on the set of integers.

11. Prove the congruence $ax \equiv b \pmod{m}$ always has solutions for x, if a and m are relatively prime. Prove that any two solutions are congruent modulo m.

12. A set S is said to be *countable* if there is a bijection of S with the set of positive integers. Prove that the union of two countable sets is countable, and that any nonempty subset of a countable set is either finite or countable.

13. For m a positive integer > 1, let ϕ_m denote the number of positive integers $m_1, m_2, \ldots, m_{\phi_m}$ less than m and coprime to m. (E.g., $\phi_6 = 2$, $\phi_{10} = 4$.) Let n be a positive integer coprime to m. Then $nm_i = q_i m + r_i$, $0 \le r_i < m$, $i = 1, 2, \ldots, \phi_m$. Prove $r_i = m_i$, $i = 1, 2, \ldots, \phi_m$ and prove that $n^{\phi_m} \equiv 1 \pmod{m}$, where $(n, m) = 1$. This is *Euler's theorem.* If $m \equiv p$, a prime, it specializes to Fermat's theorem, $n^{p-1} \equiv 1 \pmod{p}$, $p \nmid n$.

14. A subgroup H of a group G is a subset of G that forms a group under the same binary operation as that of G. Let A and B be subgroups of a commutative group G whose binary operation is addition. Define a mapping f of the cartesian product $A \times B \to G$ by $(a, b)f = a + b$, $a \in A$, $b \in B$. Prove that f maps $A \times B$ into a subgroup of G.

chapter 2

Groups

In this chapter we begin the study of our first algebraic system, the group. The concept of group is fundamental in algebra, and throughout this chapter we shall derive the properties of a group as defined by a set of axioms. This will enable us to obtain results that are applicable to many concrete situations. Later in the chapter we study the properties of a particular kind of mapping (called a **homomorphism**) of a group into a group. Again, as we shall see, this type of mapping is of fundamental importance in the development of the theory.

The semigroup, monoid, and group are the simplest types of algebraic systems in that they are characterized by having a single binary operation, the binary operation being required to be associative.

2-1 DEFINITIONS AND ELEMENTARY PROPERTIES

Definition 1. Any nonempty set with an associative binary operation is called a **semigroup.**

Examples of a semigroup are (1) the positive integers under addition or under multiplication, and (2) the even integers under multiplication.

Definition 2. A **monoid** is a semigroup that has a neutral element.

Examples of a monoid are (1) the positive integers under multiplication (but not under addition), (2) the nonnegative integers under addition, and (3) the set of all mappings from any set A to itself under map composition (4) the set of rational numbers under multiplication.

Definition 3. A **group** is a monoid in which every element has an inverse element.

Examples of a group are (1) the set Z of all integers under addition (but not under multiplication); (2) the nonzero rational numbers under multiplication (but not under addition); (3) the even integers under addition; (4) the positive rational numbers under multiplication; (5) the integers 1 and -1 form a group of two elements under multiplication; (6) all the rational numbers under addition (but not under multiplication).

A very important example of a group is the set of all bijective mappings from any set A to itself under map composition. If f and g are such bijections, then fg is a bijection. For if $c \in A$ then, since g is surjective, $c = bg$, $b \in A$, and since f is surjective $b = af$, $a \in A$. Hence $c = (af)g = a(fg)$, which proves fg is surjective. Moreover, if a, $b \in A$ and $a(fg) = b(fg)$ then $(af)g = (bf)g$. Since g is injective $af = bf$, and since f is injective $a = b$. Thus fg is injective and hence is a bijection of $A \to A$. The identity mapping I_A of $A \to A$ is the neutral element and we know every bijection has an inverse bijection. Thus all the bijections from $A \to A$ form a group.

Definition 4. A semigroup or monoid or group is called **commutative** or **abelian** (after the mathematician Niels Henrik Abel) if its binary operation is commutative.

All the groups in the examples from (1) to (4) are abelian. However, the group of bijections, where the set A has three or more elements, is not an abelian group, for in general $fg \neq gf$.

If a group is abelian, there is actually no loss of generality in assuming the binary operation to be addition—that is, addition in the sense of a commutative and associative binary operation—in which case we write the neutral element as 0 and the inverse of an element x as $-x$. We also write $2x$ for $x + x$ and, in general, nx for the sum $x + x + x \cdots + x$ to n terms. Define $1 \cdot x = x$. In this sense $-nx$ means $n(-x)$. We call such a group an **additive abelian group.**

If a group is not necessarily an abelian group, we frequently assume for convenience the binary operation to be multiplication, which we take to be associative but not necessarily commutative. In this case it is customary to denote the neutral element by the usual letter e (some writers use 1) and the inverse of the element x by x^{-1}. We define $x^0 = e$ and write, for $n > 0$, x^n for $x \cdot x \cdots x$ to n factors. In this sense x^{-n} means $(x^{-1})^n$. Moreover $x \cdot x^n = x^{n+1}$ and by induction we can prove that $x^m \cdot x^n = x^{m+n}$ and $(x^n)^m = x^{nm}$, where n and m are integers. Such a group is called a **multiplicative group.**

These notations are very commonly employed in the literature on groups. This literature is a vast one for groups have been and still are the subject of study and research. There are many problems awaiting solutions. The reader is strongly advised to visit a good mathematics library and inspect the section devoted to groups. He may be amazed at its size. Being one of the most fundamental of algebraic systems, the group concept, as is to be expected, permeates all of algebra. Later we shall find that nearly all algebraic systems basically are groups satisfying additional conditions.

The semigroup and monoid will not figure in our later work and so we list now the axioms of a group, independent of these concepts.

Definition 5. A nonempty set G with an *associative* binary operation, designated (for simplicity) as multiplication, is a *group* if it satisfies the following axioms:

1. G contains a neutral element e; that is, $ex = x = xe$ for all $x \in G$.

2. Each $x \in G$ has an inverse $x^{-1} \in G$; that is, $x^{-1} \cdot x = e = x \cdot x^{-1}$.

We have seen in the first chapter that a neutral element e of a group is unique and that each inverse x^{-1} of an element $x \in G$ is unique. When we speak of an **abstract group** we mean any set of elements with any associative binary operation that satisfies the above set of axioms.

We next derive some elementary and highly useful properties of the elements of an abstract group, using multiplication as the binary operation.

A. If $x \in G$ and $x^2 = x$, then $x = e$.

An element x for which $x^2 = x$ is called an **idempotent.** This states then that the only idempotent in a group is the neutral element.

If $x^2 = x$ then $x^{-1} x^2 = x^{-1} x = e$. Also $x^{-1} x^2 = (x^{-1} x)x = ex = x$. Hence $x = e$.

B. If $xy = xz$ or $yx = zx$ then $y = z$.

This property is called **cancellation** and the statement is that cancellation is always valid in a group. Assume $xy = xz$. Now $x^{-1}(xy) = (x^{-1}x)y = ey = y$, and similarly $x^{-1}(xz) = z$. Since $x^{-1}(xy) = x^{-1}(xz)$, we have $y = z$. Similarly, we can show that $yx = zx$ implies $y = z$.

C. $(x^{-1})^{-1} = x$ for all $x \in G$.

For $(x^{-1})^{-1} \cdot x^{-1} = e$, hence $(x^{-1})^{-1}x^{-1}x = ex = x$. But

$$(x^{-1})^{-1} \cdot x^{-1}x = (x^{-1})^{-1}e = (x^{-1})^{-1}.$$

Therefore $(x^{-1})^{-1} = x$.

D. $(xy)^{-1} = y^{-1}x^{-1}$.

For $(y^{-1} x^{-1})(xy) = y^{-1}(x^{-1} xy) = y^{-1}(x^{-1} x)y = y^{-1}ey = y^{-1} \cdot y = e$. Similarly $(xy)(y^{-1}x^{-1}) = e$. Hence $y^{-1}x^{-1}$ is the inverse of xy. Hence $(xy)^{-1} = y^{-1} x^{-1}$.

In general we have

$$(x_1 x_2 \cdots x_n)^{-1} = x_n^{-1} x_{n-1}^{-1} \cdots x_1^{-1}$$

which is the rule for finding the inverse of a product of any finite number of elements.

Lemma 1. A set G with an associative binary operation is a group if and only if the equations $ax = b$ and $ya = b$ have solutions in G for all elements a and b of G.

Proof: By solutions it is meant that there exist elements x and y in G such that $ax = b$ and $ya = b$, where a and b are any given elements in G.

If G is a group, then $a(a^{-1}b) = b$ and $(ba^{-1})a = b$ and therefore $x = a^{-1}b$ and $y = ba^{-1}$. Conversely, assume $ax = b$ and $ya = b$ have solutions for all a and b in G. Then the equation $ax = a$ has a solution. Call it e, so that $ae = a$. Let b be any element of G, and let $y \in G$ be the solution of $ya = b$. Then $be = yae = ya = b$. This proves e to be what is called a *right neutral element* of G. In the same way we can show that G has a left neutral element e', that is $e'x = x$ for all $x \in G$. Since $e' = e'e = e$, we see that e is the neutral element of G. Let a be any element of G. The equations $ax = e$ and $ya = e$ have solutions in G. Hence $y = ye = yax = ex = x$. Thus $ax = e$ and $xa = e$ and hence x is the inverse a^{-1} of a. Thus G is a group.

2-2 SUBGROUPS

Definition. A nonempty subset H of a group G is called a **subgroup** of G if H is a group under the *same* binary operation as that of G.

The unit group $1 = \{e\}$ and G itself are subgroups of G and referred to as improper subgroups.

Examples of subgroups are (1) the even integers form a subgroup of the additive group of integers and (2) the positive rational numbers form a subgroup of the multiplicative group of nonzero rational numbers.

The set Q of rational numbers is a group under addition. The subset Q' of Q consisting of all positive rational numbers is a group under multiplication, but Q' is not a subgroup of Q.

THEOREM 1. A nonempty subset H of a group G is a subgroup of G if, and only if, (1) $x, y \in H$ imply $xy \in H$ and (2) $x \in H$ implies $x^{-1} \in H$.

Proof: If $x \in H$, then $x^{-1}x = e \in H$. Thus H satisfies the axioms for a group. The converse is obvious.

Corollary. A nonempty subset H of a group G is a subgroup of G if and only if $x, y \in H$ implies $xy^{-1} \in H$.

Proof: $xx^{-1} = e \in H$ and if $x \in H$, then $ex^{-1} = x^{-1} \in H$. Hence if $x, y \in H$ then $x, y^{-1} \in H$ and therefore $x(y^{-1})^{-1} = xy \in H$. Again the converse is obvious.

The following simpler test for a subgroup is valid if the group is finite.

THEOREM 2. A nonempty subset H of a finite group G is a subgroup of G if and only if $x, y \in H$ implies $xy \in H$ (that is, if and only if H is closed or stable under the binary operation of G).

Proof: If $H = \{e\}$, then H is the unit subgroup 1. Let $x \in H$, $x \neq e$. Then $x, x^2, x^3, \ldots$, are all in H, and since H is finite, there must be repetitions among them—that is, $x^m = x^n$, $m > n$. Hence $x^{m-n} = e$ and so $e \in H$. Also, $x \cdot x^{m-n-1} = x^{m-n} = e$. Hence $x^{m-n-1} = x^{-1}$ and therefore $x^{-1} \in H$. Thus H is a subgroup of G.

THEOREM 3. The intersection $\bigcap_{i \in I} H_i$ of any family, H_i, $i \in I$ (I is the index set of the family) of subgroups of a group G is a subgroup of G.

Proof: The intersection is never empty, for it always contains the neutral element e. If $x, y \in \bigcap_{i \in I} H_i$, then xy belongs to each subgroup H_i of the family, and hence $xy \in \bigcap_{i \in I} H_i$. If $x \bigcap_{i \in I} H_i$, then x^{-1} belongs to each subgroup H_i of the family and hence $x^{-1} \in \bigcap_{i \in I} H_i$. These results prove $\bigcap_{i \in I} H_i$ is a subgroup of G.

Generators of a Group

Let S be a nonempty subset of a group G. The intersection of all subgroups of G that contain S form a subgroup $[S]$ of G, called the **subgroup generated by the set S**. The set S is called a **system of generators** of the subgroup $[S]$. There are subgroups that contain S, for G itself is one.

If G is written as a multiplicative group, the elements of $[S]$ are the finite products of the form $x_1 x_2 \cdots x_k$ where each x_i, $i = 1, 2, \ldots, k$, is either an element of S or the inverse of some element of S. For clearly the set of all elements of this type form a group, and this group contains S. Moreover, any subgroup containing S would have to contain all elements of this type. Thus the subgroup $[S]$ generated by the set S is the smallest subgroup of G that contains S. Every group has a set of generators, for example, all its elements or all its elements without the neutral element.

Illustrative Example 1. If x is an element of a multiplicative group G, then the subgroup $[x]$ generated by x would consist of all elements of the form $x^0 = e, x^n$ and x^{-n}, where n is any positive integer. A group that is generated by a single element is called a **cyclic group**. Thus $[x]$ is a cyclic subgroup of G. Obviously a cyclic group is always abelian. It may be a finite or infinite group. If $x^n \neq e$ for any positive integer n, then $[x]$ is an infinite cyclic group, If, however, $x^n = e$ and n is the smallest positive integer for which this is true, then the cyclic group $[x]$ is finite and contains n elements.

The union of two subgroups A and B of a group G is not in general a subgroup of G. The subgroup $[A \cup B]$ generated by $A \cup B$ is the smallest subgroup containing both A and B and is called the *join $A \vee B$*

of A and B. If G is an additive abelian group then the set $A + B$ of all elements $a + b$, $a \in A$, $b \in B$ is seen to be a subgroup of G and $A + B = A \vee B$.

EXERCISES

1. Prove that the set (m) of all integral multiplies of a positive integer m form a group under addition.

2. G is the set of all pairs (x, y) of real numbers for which $x \neq 0$. A binary operation is defined in G by $(x, y) \circ (u, v) = (xu, yu + v)$. Verify that G is a group by testing all the axioms.

3. Determine which of the following are groups, monoids or semigroups and whether they are commutative or not.
   (a) The set of all positive rational numbers under multiplication.
   (b) The set of all negative integers under addition.
   (c) The set of all nonnegative integers under addition.
   (d) The set S of all mappings of a set A into itself under map composition.
   (e) The set of all bijections of A to A.

4. A permutation of a set S is a bijection of S to itself. Prove all the permutations of the set $S = \{a, b, c, d\}$ form a group under map composition. Find its order.

5. If R is the set of real numbers, prove the following are groups.
   (a) The set T of all transformations f from $R \to R$ of the form $xf = ax + b$, where $a \neq 0$, $b \in R$ are fixed for this f, and $x \in R$ is arbitrary, under map composition.
   (b) The same set T under map addition, where a is arbitrary.
   (c) The set S of all periodic functions f of $R \to R$ of period $k \in R$, that is $(x + k)f = xf$, $x \in R$, under addition. If $f,g \in S$, is $f \circ g$ a periodic function?
   (d) The set B of all transformations f from $R \to R$ of the form $xf = \dfrac{ax + b}{cx + d}$, $ad \neq bc$, where a, b, c, d are fixed real numbers for this particular f and x is an arbitrary real number, under multiplication.
   (e) The same set B under map addition.
   (f) The set of all mappings f of the real plane into itself of the form $(x, y)f = (ax + by, cx + dy)$, where a, b, c, d are fixed real numbers for this f and $\begin{vmatrix} a & b \\ c & d \end{vmatrix} = ad - bc = 1, a^2 + b^2 = c^2 + d^2 = 1$.

6. Prove that under map composition the set S_n of all permutations on n symbols forms a group. Is it commutative? (A permutation is a bijective mapping of the set of n symbols into itself.)

7. Find all the subgroups of S_3 and S_4.

8. Prove the set of all complex numbers $z = x + yi$, $i = \sqrt{-1}$ for which $|z| = \sqrt{x^2 + y^2} = 1$ form a group under multiplication.

9. If a, b, c, d are rational numbers in the sets T and B of Exercises 5(a) and 5(b), are these subsets of T and B groups? Prove your answer.

10. If x is an element of order n (that is, n is the least positive integer for which $x^n = e$) of a group G prove that the order of an element yxy^{-1}, $y \in G$, is a divisor of n.

2-3 COSETS

We start with a group G, written as a multiplicative group, and take H to be a subgroup of G. Let $x \in G$. The set Hx of all elements of the form hx, $h \in H$, is called a **right coset** of H in G. We can form such a right coset for each element of G. The set xH, $x \in G$, is called a **left coset** of H in G. If G is written as an additive group, then we write the right cosets as $H + x$. Since in this case G is usually assumed a commutative group, it follows that $x + H = H + x$.

In general, $x \notin H$ and it is easy to see in this event that neither the right coset Hx nor the left coset xH can be a subgroup of G. For Hx could not contain the neutral element e, since if $x \notin H$ then $x^{-1} \notin H$. On the other hand if $x \in H$ then $Hx = xH = H$. For if $x \in H$, then certainly $Hx \subset H$, and if h is any element of H then $h = (hx^{-1})x$ and hence $h \in Hx$, that is $H \subset Hx$. Therefore $Hx = H$ and similarly $xH = H$.

Another important question about cosets is this. If x and y are distinct elements of G, is it possible for Hx and Hy to have any elements in common? We answer this question by proving the next theorem.

THEOREM 4. Two cosets Hx and Hy are either identical or disjoint and $Hx = Hy$ if and only if $xy^{-1} \in H$ (or equivalently if and only if $yx^{-1} \in H$).

Proof: If $z \in (Hx) \cap (Hy)$, then $z = h_1 x$ and $z = h_2 y$ for some $h_1, h_2 \in H$. Hence $h_1 x = h_2 y$ and therefore $x = h_1^{-1} h_2 y \in Hy$ and $y = h_2^{-1} h_1 x \in Hx$. Hence $Hx \subseteq Hy$ and $Hy \subseteq Hx$, and therefore $Hx = Hy$. Thus if Hx and Hy have one element in common then they are identical. The equation $h_1 x = h_2 y$ implies that $xy^{-1} = h_1^{-1} h_2 \in H$, a condition independent of z. Conversely, if Hx and Hy are two cosets such that $xy^{-1} \in H$, then $(xy^{-1})^{-1} = yx^{-1} \in H$, and therefore $x \in Hy$ and $y \in Hx$. Hence as before $Hx = Hy$. A similar theorem of course is true for left cosets, $xH = yH$ if and only if $y^{-1}x \in H$ (or equivalently if and only if $x^{-1}y \in H$).

If H is a subgroup of a group G, define a relation xRy, $x, y \in G$, on G by xRy if and only if $xy^{-1} \in H$. This is an equivalence relation on G, for

    1. xRx means $xx^{-1} = e \in H$.

    2. xRy means $xy^{-1} \in H$ and therefore $(xy^{-1})^{-1} = yx^{-1} \in H$, hence yRx.

    3. xRy and yRz means $xy^{-1} \in H$ and $yz^{-1} \in H$ and therefore $(xy^{-1})(yz^{-1}) = xz^{-1} \in H$, that is xRz.

This equivalence relation partitions G into a set of equivalence classes which here are the right cosets of H in G. Moreover, G is the disjoint union of the distinct right cosets (that is, the union of disjoint sets). In the light of this we could have concluded more briefly that, as equivalence classes, the right cosets of H in G are either identical or disjoint.

We designate by G/H the set of distinct right cosets of H in G.

Two right cosets (or left cosets) Hx and Hy have the same number of elements. For $hx \to hy$ is clearly a bijection between Hx and Hy. In particular then the number of elements in any coset is the same as the number of elements in the subgroup H. Moreover a right coset and a left coset have the same number of elements, for $hx \to xh$ is a bijection of Hx onto xH.

    Definition 6. The **order** $\#G$ of a group G is the number of elements in G.

By this we shall mean that if G is a finite group of n elements, then $\#G = n$, whereas if G is an infinite group we mean its order is infinite.

    Definition 7. If H is a subgroup of G, the number of distinct right cosets of H in G (that is, the number of elements in the set G/H) is called the **index** of H in G.

We write $|G:H|$ for the index of H in G. Recalling that 1 is used to designate the subgroup of G consisting of the neutral element e only, we see that $\#G = |G:1|$.

    Illustrative Example 2. Let S be a set of four elements x, y, z, w, and let S_4 denote the set of all bijective mappings of $S \to S$. Under map composition we know that S_4 forms a group. It is known as the symmetric group of degree 4. The order of S_4 is the number of such mappings on these four elements. We can determine the order of S_4 as follows. There is a choice of four elements into which the element x can be mapped, a choice of three remaining elements into which y can be mapped, two into which z can be mapped, and one into which w can be mapped. Hence the order of S_4 must be $4 \times 3 \times 2 \times 1 = 4! = 24$. Let H be that subset of S_4 comprising all mappings that transform x into itself—that is, that leave x fixed. If $f, g \in H$ then $x(fg) = (xf)g = xg = x$, so that $fg \in H$. Also if $f \in H$, then $x = xf$, and therefore $xf^{-1} = xff^{-1} = x$. Thus f^{-1} belongs to H. Hence H is a subgroup of S_4.

Consider the right cosets of H in S_4. They have the form Hf, where $f \in S_4$. Suppose now that f is a mapping that maps x into y. Then $f \notin H$, and we can see that all mappings of the coset Hf will map x into y. For if $h \in H$, then $x(hf) = (xh)f = xf = y$. Next we prove that all mappings in S_4, that map x into y, are in the coset Hf. Let g be such a mapping, then $x(gf^{-1}) = (xg)f^{-1} = yf^{-1} = x$. Thus $gf \in H$, and so the cosets Hf and Hg are identical. This means $g \in Hf$. Therefore the coset Hf can be described as the set of all mappings in S_4 that map x into y. It is now clear that there are in all four distinct right cosets of H in S. If f_1, f_2, f_3 are mappings that map x into y, z, and w respectively, then these four cosets can be written as H, Hf_1, Hf_2, Hf_3.

Thus the index of H in G is 4. We point out that the next theorem proves $\#H = 6$, since $\#G = 24$ and $|G:H| = 4$.

THEOREM 5 (Lagrange). If H is a subgroup of a group G, then $|G:H| = \#G/\#H$.

This states that the order of a subgroup is a divisor of the order of the group.

Proof: Since each coset of H in G contains $\#H$ elements and since G is the disjoint union of the distinct cosets of H in G, then it follows at once that $\#G = |G:H| \cdot \#H$.

In particular, if G is a finite group of order n then the order of any subgroup of G is a divisor of n. The converse is not true. We shall point out later the fact that there need not exist a subgroup of order m for every divisor m of n, that is, the converse of Lagrange's theorem is false.

Corollary. If $\#G = p$, where p is a prime, then G is a cyclic group.

Proof: Let $a \in G$, $a \neq e$. Then the cyclic group $[a]$ is generated by the element a is a subgroup of G. Now $[a] \neq 1$ and since G (by Lagrange's theorem) can have no proper subgroups, then $[a] = G$.

Definition 8. The *order of an element* $x \in G$ is defined to be $\#[x]$, that is the order of an element x is equal to the order of the cyclic subgroup generated by x.

If the element x is of finite order, then the least positive integer m (there is one, since the positive integers are well-ordered), such that $x^m = e$, is the order of the element x. An element that is not of finite order is said to be of *infinite order*. Clearly all elements of a finite group have finite orders.

Corollary. The order of an element of a finite group is a divisor of the order of the group.

Proof: If $x \in G$, where G is a finite group, then the order of x is $\#[x]$, and, by Lagrange's theorem, we know $\#[x] \mid \#G$.

A group is called a **torsion group** or **periodic group** if all its elements are of finite order. Of course every finite group is a torsion group. If the only element of finite order in a group is the neutral element, the group is said to be **torsion-free**. A group is called **mixed** if it has both elements of finite order and elements of infinite order.

Example. The additive group of integers is a torsion-free group.

Example. The multiplicative group of nonzero rational numbers is a mixed group. The element -1 has the order 2, for $(-1)^2 = 1$. However, it and the neutral element 1 are the only elements that are of finite order.

Illustrative Example 3. In the illustrative example 2 let $f \in G$ and suppose f maps $x \to y$, $y \to w$, $z \to z$, $w \to x$. We find f^2 maps $x \to w$, $y \to x$, $z \to z$, $w \to y$, and f^3 maps $x \to x$, $y \to y$, $z \to z$, $w \to w$. Thus f^3 is the identity map on S and is therefore the neutral element of G. Hence the element f of G has the order 3, and we also see that its order is a divisor of the order 24, of G.

2-4 NORMAL SUBGROUPS

The most important type of subgroup is the normal subgroup. A subgroup H of a group G is called a **normal (or invariant) subgroup** of G if and only if $xH = Hx$ for all $x \in G$; that is, if and only if $x^{-1}Hx = H$ for all $x \in G$. We write $H \triangle G$ to mean that H is a normal subgroup of the group G. Clearly the improper subgroups 1 and G are normal subgroups of G. A group may or may not have proper normal subgroups. A group is called **simple** if it has no proper normal subgroups.

Example. A cyclic group of prime order has no proper subgroups and hence, *a fortiori*, is simple.

If $H \triangle D$ then $Hx = xH$ for all $x \in G$. This means that for any $h \in H$, there exists $h' \in H$ such that $hx = xh'$, it does not imply that every $h \in H$ commutes with the element x.

Probably the most useful criterion for a normal subgroup is provided by the next theorem.

THEOREM 6. $H \triangle G$ if and only if $x^{-1}hx \in H$ for all $x \in G$ and for all $h \in H$.

    Proof: If $H \triangle G$, then $x^{-1}Hx = H$ for all $x \in G$. Hence $x^{-1}hx \in H$ for all $h \in H$. Conversely, assume $x^{-1}hx \in H$ for all $x \in G$ and all $h \in H$. Then $x^{-1}Hx \subseteq H$, and hence $xHx^{-1} \subseteq H$ for all $x \in G$. Therefore $H = x^{-1}(xHx^{-1})x \subseteq x^{-1}Hx$. Hence $H = x^{-1}Hx$ for all $x \in G$ and $H \triangle G$.

If G is an abelian group and H is any subgroup of G, we see that $xH = Hx$ elementwise (that is, $hx = xh$ for every $h \in H$). Hence every subgroup H of an abelian group is a normal subgroup. As mentioned before, an abelian group is often written as an additive group, in which the cosets are written as $x + H (= H + x)$ instead of $xH (= Hx)$.

Illustrative Example 4. Let S_3 be the group of bijective mappings of the set $S = \{x, y, z\}$ into itself. (S_3 is the symmetric group of degree 3.) There are exactly two mappings in S_3 that leave no element of S fixed. They are $f\colon x \to y, y \to z, z \to x$ and $g\colon x \to z, y \to x, z \to y$. If I is the identity mapping on S, then we find that I, f, and g form a commutative subgroup K of S_3. For we find that $f^2 = ff = g$, $g^2 = f$, $fg = I$, and $gf = I$. The other mappings in S_3 are $r\colon x \to x, y \to z, z \to y$, $s\colon x \to z, y \to y, z \to x$ and $t\colon x \to y, y \to x, z \to z$. Each of these is a mapping that leaves exactly one element of S fixed. The six mappings I, f, g, r, s, t constitute the group S_3. This can also be methodically checked out, for it can be readily proved that

$$fr = s, \quad fs = t, \quad ft = r, \quad rf = t, \quad sf = r, \quad tf = s,$$

$$gr = t, \quad gs = r, \quad gt = s, \quad rg = s, \quad sg = t, \quad tg = r,$$

$$tr = g, \quad rt = f, \quad ts = f, \quad st = g, \quad rs = g, \quad sr = f,$$

$$r^2 = s \quad = t^2 = I, \quad f^3 = g^3 = I.$$

We can easily show, by Theorem 6, that K is a normal subgroup of S_3. Consider for example rfr^{-1}. Since $r^2 = I$, $r^{-1} = r$, and therefore $rfr^{-1} = rfr = (rf)r = tr = g \in K$. Similarly, we find that sfs^{-1}, tft^{-1}, rgr^{-1}, sgs^{-1}, and tgt^{-1} all belong to K. Of course an element such as gfg^{-1} belongs to K, since it is a product of elements of K. Hence K is a normal subgroup of S_3.

Another way of showing that K is a normal subgroup of S_3 is the following. We note that K is the subgroup consisting of the identity mapping I and the two mappings that leave no element of S fixed. If r is any mapping that leaves an element fixed and if $f \neq I$ is an element of K, then consider the mapping rfr^{-1}. If this composite mapping leaves an element fixed, say x, then $x(rfr^{-1}) = x$, and hence $x(rf) = xf$, and therefore $(xr)f = xr$. But this implies that f leaves the element xr fixed, which would contradict f being in K. Hence rfr^{-1} cannot leave any element of S fixed and so must belong to K. Of course if $f = I$ then $rIr^{-1} = I \in K$. Therefore, by Theorem 6, K is a normal subgroup of S_3.

Illustrative Example 5. The subgroup H in the illustrative example 2 of the previous section is not a normal subgroup. To see this, let h be the mapping in H which maps $x \to x, y \to w, z \to y, w \to z$ and let f be the mapping which maps $x \to y, y \to z, z \to w, w \to x$. Then $x(fhf^{-1}) = z$, and hence $fhf^{-1} \notin H$. Thus by Theorem 6, H is not a normal subgroup.

Illustrative Example 6. Let S be an infinite set. The bijections of $S \to S$ form a nonabelian group with map composition as the binary operation. A bijection that leaves fixed all but a finite number of elements of S is called a **finite bijection.** We often express this by saying a finite bijection is one that leaves **almost all** elements of S fixed. Thus to each finite bijection α there is associated a finite subset, denoted here by S_α, of elements of S that are moved by α. Let H be the set of all finite bijections in G. Then if $\alpha \in H$, we have $x\alpha = x$ for all $x \in S$ which are not in S_α. We shall show that H is a subgroup of G. First observe that $\alpha \in H$ implies $\alpha^{-1} \in H$. This follows at once from the definition of the inverse of a bijection. In fact we have $S_{\alpha^{-1}} = S_\alpha$. Next we see that, if $\alpha, \beta \in H$, then the product (map composition) $\alpha\beta$ moves at most those elements that are in the union $S_\alpha \cup S_\beta$. This union is a finite set and so $\alpha\beta \in H$. Thus H is a subgroup of G.

Next let us prove that H is a normal subgroup of G. Let $\alpha \in H$, $\beta \in G$, and consider $\beta\alpha\beta^{-1}$. If $x \in S$ and $x \notin S_\alpha$, then $x(\beta\alpha\beta^{-1}) = x$. Hence $\beta\alpha\beta^{-1}$ can move only a finite number of elements. It is therefore a finite bijection and so belongs to H. This is enough to prove that H is a normal subgroup of G.

If α is not a finite bijection, the right coset $H\alpha$ can be described as the coset of all bijections β in G for which

$$x\beta = x\alpha \quad \text{for almost all } x \in S.$$

For if $\beta \in H\alpha$, then $\beta = h\alpha$, where $h \in H$. Hence $x\beta = xh\alpha = (xh)\alpha = x\alpha$, for almost all $x \in S$.

2-5 FACTOR GROUPS

Our first use of the concept of a normal subgroup is to construct a new group out of a group G and a normal subgroup H of G. This is one of the most important uses of a normal subgroup.

Let $H \triangle G$ and consider the set G/H of the distinct right cosets Hx of H in G. Since H is normal, we can think of these cosets as being either right or left cosets. From the (group) product $(Hx) \cdot (Hy)$ of two cosets Hx and Hy. It is the set of all elements of the form $h_1 x h_2 y$ where h_1, $h_2 \in H$. Now

$$h_1 x h_2 y = h_1 x h_2 x^{-1} x y = h_1 (x h_2 x^{-1}) x y = h_1 h_3 x y, \text{ where}$$

$$h_3 = x h_2 x^{-1} \in H.$$

Hence $(Hx)(Hy) \subset Hxy$. Also $hxy = (hx)(ey) \in (Hx)(Hy)$, since $e \in H$. Hence $Hxy \subset (Hx)(Hy)$. Thus $(Hx)(Hy) = Hxy$. We now use this fact to define a binary operation in the set G/H. For Hx, $Hy \in G/H$, define $Hx \cdot Hy = Hxy$. This binary operation is easily seen to be associative. The neutral element is the element H of G/H, for $H \cdot Hx = Hx = Hx \cdot H$. The inverse of Hx is Hx^{-1}, and hence the set G/H of cosets of H in G, with this binary operation, is a group. It is called the **quotient** or **factor group** of G with respect to H. We emphasize that H must be a normal subgroup of G in order for the set G/H of distinct cosets to form a group.

We have written the factor group G/H in terms of right cosets. However, the same notation is used for the factor group when written in terms of left cosets. There are as many left cosets as right cosets. This is seen from the bijection $Hx \rightarrow x^{-1}H$ of right cosets to left cosets. This mapping is surjective, since any left coset zH is the image of Hz^{-1}. Furthermore, $x^{-1}H = y^{-1}H$ implies $xy^{-1} \in H$, hence $(xy^{-1})^{-1} = yx^{-1} \in H$ and therefore $Hx = Hy$, and thus the mapping is also injective.

It is worth noting that if H is a subgroup of index 2 in G; that is, $|G:H| = 2$, then $H \; \Delta \; G$. For if $G/H = \{H, xH\}$ and $x \in G$, then $G/H = \{H, Hx\}$ and $xH = Hx$ for all $x \in G$.

Illustrative Example 7. In Illustrative Example 4 it was proved that K is a normal subgroup of S_3. Hence the factor group S_3/K exists. Its elements, written as right cosets, are K, Kt. (Note $Kr = Ks = Kt$.) We can check that these actually form a group under the definition of the multiplication of right cosets. We see that K is the neutral element of S_3/K, and we have $(Kt)^2 = Kt^2 = K$, $(Kg)(Kf) = Kgf = KI = K$ so that S_3/K is a group.

As a further and important illustration of the concept of factor group, we next consider the commutator subgroup of a group.

Commutator Subgroup

Let G be a group. If $a, b \in G$, then the element $aba^{-1}b^{-1}$ is called the **commutator of a and b.** Of course if G is abelian, then $aba^{-1}b^{-1} = e$ for all $a, b \in G$. The commutator of a and b is therefore some measure of the deviation of a and b from being commutative. Since $(aba^{-1}b^{-1})^{-1} = bab^{-1}a^{-1}$, we see that the inverse of the commutator of a and b is the commutator of b and a. The subgroup of G generated by all elements of G of the form $aba^{-1}b^{-1}$ (it consists therefore of all finite products of commutators of elements of G) is called the **commutator subgroup** $G^{(1)}$ of G.

THEOREM 7. $G^{(1)} \; \Delta \; G$.

Proof: Let $c_1 c_2 \ldots c_r$ be a product of commutators and let x be any element of G. Then $x(c_1 c_2 \cdots c_r)x^{-1} = (xc_1x^{-1})(xc_2x^{-1}) \cdots (xc_rx^{-1})$, and so, in order to prove $G^{(1)} \; \Delta \; G$, it suffices to prove that $t = xcx^{-1} \in G^{(1)}$, where c is a commutator and $x \in G$. Now $c = x^{-1}tx \in G^{(1)}$ and $t^{-1}c = t^{-1}x^{-1}tx \in G^{(1)}$. Hence $t^{-1} \in G^{(1)}$ and therefore $t \in G^{(1)}$.

THEOREM 8. A factor group G/H of a group G is abelian if and only if $H \supseteq G^{(1)}$.

Proof: $(xH)(yH) = xyH = yxH = (yH)(xH)$ if and only if $x^{-1}y^{-1}xy \in H$ for all x and y of G.

EXERCISES

1. Show that the set (m) of multiples of a positive integer m forms a subgroup of the additive group Z of integers. Describe the cosets in $Z/(m)$ and check that they form a group under addition.

2. If $m = p$, a prime, prove that, under a very natural definition of multiplication of the cosets in Z/Z_p, the nonzero cosets of Z/Z_p form a multiplication group. Why must p be a prime?

3. The additive group of integers Z is a subgroup of the group C of complex numbers $m + ni$, where m and n are integers. Find C/Z and write down an expression for its distinct cosets.

4. Let C be the same group as in exercise (3) and let D be the subgroup of C for which m and n are both even integers. Find C/D and show it has just the four distinct cosets $D, 1 + D, i + D, 1 + i + D$. Verify that these do form a group, and that every element of C is in one and only one of these cosets.

5. K is the additive group of all complex numbers $a + bi$, a and b real, and R is the subgroup of all real numbers. Find K/R.

6. C_1 is the multiplication group of all nonzero complex numbers, and R_1 is the subgroup of C_1 of all nonzero real numbers. Find C_1/R_1 and determine the condition that two cosets of C_1/R_1 be equal. Show the elements $\sqrt{3} - 1 + 2i$ and $1 + (\sqrt{3} + 1)i$ of C_1 are in the same coset.

7. f is a mapping $G \rightarrow G'$ of two groups. For any $x' \in Gf, f^{-1}(x') = \{x \in G \mid xf = x'\}$ is clearly a nonempty subset of G.

(a) If $y' \in Gf$ and $y' \neq x'$, show that $f^{-1}(x')$ and $f^{-1}(y')$ are disjoint (no common elements).

(b) Show that every $x \in G$ is in one of these subsets and hence that they form a quotient set Q of G.

(c) Prove that the mapping g of $Q \rightarrow Gf$ defined by $qg = x'$, where $q = f^{-1}(x')$, is a bijection.

(d) Hence show that g defines a binary operation in Q which makes Q a group. (See Exercise 1, Sec. 1-3.)

(e) If π is the mapping of $G \rightarrow Q$ defined by $x\pi = q$, where $x \in q$, show π is surjective and that $f = \pi g$.

(f) Identify the quotient set Q in relation to G and f.

2-6 EXAMPLES OF FINITE GROUPS

Example 1. A group is completely determined if the product of any two of its elements is known. If this is the case we can tabulate the elements of a finite group by means of a *group table* which serves to exhibit these products. A group table is simply a multiplication table, headed by a row of all the elements of the group, starting with the neutral element. The elements are listed in the same order in a column at the left side of the table. The entry in the ith row and jth column of the table is the product of the ith element in the left column and the jth element in the heading. An example will offer the best explanation.

Consider the group G of order 6 that is generated by two of its elements x and y which satisfy the three relations $x^3 = e$, $y^2 = e$, $(xy)^2 = xyxy = e$, where e is the neutral element of G. From the relations it follows readily that $xyx = y$ and hence that $yxyx = e$, $x^2y = yx$, $yx^2 = xy$. We can now construct the group table for this group.

	e	x	x^2	y	xy	yx
e	e	x	x^2	y	xy	yx
x	x	x^2	e	xy	yx	y
x^2	x^2	e	x	yx	y	xy
y	y	yx	xy	e	x^2	x
xy	xy	y	yx	x	e	x^2
yx	yx	xy	y	x^2	x	e

For example, the entry in the 4th row and 3rd column of the table is the product $(xy)x^2 = (xyx)x = yx$. Note that the proper subgroups of this group are the cyclic groups $\{e, x, x^2\}$, $\{e, y\}$, $\{e, xy\}$ and $\{e, yx\}$. Since the index of $\{e, x, x^2\}$ in the group is 2, it is a normal subgroup. It is easy to verify that it is the only one of the four subgroups that is normal.

Each row or column of a group table contains each element of the group once and only once. This is no accident. For if G is a finite group and $a \in G$, then the elements aw, for all $w \in G$, run through all the elements of the group without repetition. For if z is any element of G, then $z = a(a^{-1}z)$ and $w = a^{-1}z$, and if $ay = az$ then $y = z$. Observe that symmetry about a principal diagonal of a group table indicates the group is abelian.

Example 2. The **symmetries** of a geometrical figure are bijections of the set of points of the figure onto itself, which preserve distance. This means the distance between any two points of the figure is unchanged by the symmetry. Such symmetries are called **isometries** in geometry. Two symmetries performed in succession lead naturally to their product, which is again a symmetry. Moreover, this product is simply a composition of mappings and hence is associative. The identity bijection is a symmetry and, since every bijective mapping has an inverse bijective mapping, it follows that the set of symmetries of a geometric figure forms a group under map composition. One type of symmetry is the reflection. A **reflection** is a symmetry in which the points of the figure that stay fixed under the symmetry lie on a line or a plane, often called the **mirror**.

Consider the symmetries of the letter H. There are 4 of them, the identity bijection I, the reflections K and L about the lines K and L, and the product $KL = LK$. It is obvious that $K^2 = L^2 = I$. Hence the symmetries of the letter H form an abelian group of four elements. (It is known as the *Klein 4-group.*)

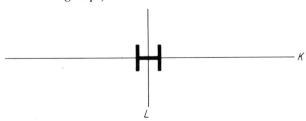

A symmetry of a regular polygon is a bijection of the set of points of the polygon onto itself, which leaves unchanged the distance between any two points of the polygon. It seems intuitively evident that a symmetry maps vertices of the polygon into vertices, and hence that a symmetry is determined by its effect on the vertices of the regular polygon. These symmetries will consist of rotations about a perpendicular axis through the center of the polygon and of reflections about certain lines. (The reflections can be regarded as rotations about their mirrors.) Clearly one symmetry is the identity bijection, which maps every point of the polygon into itself. As one might conjecture, the set of symmetries of a regular polygon with map composition as the binary operation forms a group.

Example 3. Let us find the symmetries of an equilateral triangle.

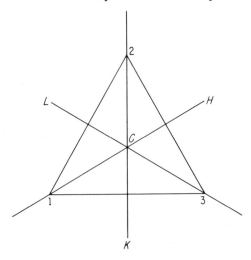

Draw the fixed lines H, K and L along the medians of the triangle. The reflections are about these lines and denote them by H, K and L. The rotations are about an axis through C and perpendicular to the plane of the triangle and there are three of them; R_1 through $120°$, R^2 through $240°$, and one through $360°$. This last one restores the triangle to its original position and we denote it by I, the identity bijection. Form the set G of elements I, R_1, R_2, H, K, L and define the product of any two of them as being map composition. These are the six symmetries of the equilateral triangle. In terms of the vertices we can write, for instance,

$R_1 = \begin{pmatrix} 123 \\ 231 \end{pmatrix}$ indicating that R_1 maps $1 \rightarrow 2$, $2 \rightarrow 3$ and $3 \rightarrow 1$. Thus

$L = \begin{pmatrix} 123 \\ 213 \end{pmatrix}$ and hence $R_1 L = \begin{pmatrix} 123 \\ 132 \end{pmatrix} = H$, while $LR_1 = \begin{pmatrix} 123 \\ 321 \end{pmatrix} = K$. These

six symmetries with map composition as the binary operation form the group of symmetries of the equilateral triangle as the following group table shows.

	I	R_1	R_2	H	K	L
I	I	R_1	R_2	H	K	L
R_1	R_1	R_2	I	K	L	H
R_2	R_2	I	R_1	L	H	K
H	H	L	K	I	R_2	R_1
K	K	H	L	R_1	I	R_2
L	L	K	H	R_2	R_1	I

G is not an abelian group. Its proper subgroups are $\{I, R_1, R_2\}$, $\{I, H\}$, $\{I, K\}$, and $\{I, L\}$. Note that any one of the three reflections, say H, and either one of the rotations R_1 and R_2, say R_1, will generate G. For all elements of G can be expressed in terms of H and R_1, since $I = R_1^3$, $R_2 = R_1^2$, $K = HR_1^2$, $L = HR_1$, as we see from the group table. Thus $G = [H, R_1]$.

Example 4. The **dihedral group** D_n is defined abstractly as the group of order $2n$ generated by the two elements x and y satisfying the three relations $x^n = e$, $y^2 = e$, $yx = x^{-1}y$. That the order of D_n is $2n$ follows from the fact that every element of D_n can be expressed in the form either of x^k or of $x^k y$, $1 \le k \le n$, and there are clearly $2n$ such elements. For by induction on k we can show that $yx^k = x^{n-k}y$, and hence $yx^i = x^{n-k}y$.

A realization of the dihedral group D_n is the group of symmetries of a regular polygon of n sides. If the vertices are numbered $1, 2, \ldots, n$ then all its symmetries are generated by one rotation x through an angle of $2\pi/n$ radians about a perpendicular axis through the center of the polygon and by one reflection y about a line through any vertex and the center of the polygon. Denoting the identity symmetry by e, it is easy to see that the three relations above are satisfied. The dihedral group is not an abelian group.

The group of symmetries of the equilateral triangle in Example 3 is the dihedral group D_3. It is generated by one rotation, say R, and one reflection, say L. For all its elements can be expressed in terms of products of these two elements and their inverses, $H = LR^2$ ($R^2 = R^{-1}$), $K = LR$.

Example 5. The **tetrahedral group.** Let $ABCD$ be a regular tetrahedron. There are three rotations about a perpendicular axis through A on BCD through angles of $0°$, $120°$, $240°$, which are all symmetries (distance-preserving bijections of the set of points of the tetrahedron to itself). Since any one of the 4 vertices can be moved into the position A, there are in all $4 \times 3 = 12$ such rotations. Under multiplication (map composition) they form a group of rotations of order 12 (that is a group of 12 elements). The other symmetries of the regular tetrahedron are the reflections (one for each edge) in the plane that passes through an edge and the midpoint of the opposite edge. Together with the rotations they form the complete group of symmetries of the regular tetrahedron of order 24. Actually it can be easily shown that the reflections generate this group.

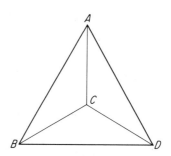

The order of the group of rotations of the regular tetrahedron is 12, but we claim it does not contain a subgroup of order 6. The substantiation of this claim is left to the reader.

Example 6. Let Z be the set of all integers and let n be a positive integer. An equivalence relation R is defined on Z by, xRy if and only if $n \mid x - y$, $x, y \in Z$. The equivalence classes of this relation are called **residue classes modulo** n. It is easy to see that there are n distinct residue classes modulo n. Denote them by Z_n. Let $\bar{x}$ stand for the residue class containing the integer x, then $\bar{x} = \bar{y}$ if and only if $n \mid x - y$. It is easy then to see that Z_n consists of the n distinct residue classes $\bar{0}, \bar{1}, \bar{2}, \ldots, \overline{n-1}$.

If we define $\bar{x} + \bar{y} = \overline{x + y}$ then it is apparent that this is a commutative and associative binary operation on Z_n. Its neutral element is $\bar{0}$ and the inverse $-\bar{x}$ of $\bar{x}$ is $-\bar{x} = (\overline{-x}) = \overline{n - x}$. Hence Z_n is an additive abelian group of order n.

EXERCISES

1. Find all the normal subgroups of S_3 and S_4.

2. Find all the normal subgroups of the group of symmetries of (a) the square and (b) the regular tetrahedron.

3. Find two nonabelian groups of order 8 and write out their group tables. Are there any more such groups?

4. Find all the subgroups and normal subgroups of the two groups of Exercise (3).

5. If the product of any two right cosets of a subgroup H in a group G is a right coset, prove that H is a normal subgroup of G.

6. Let x be an element of order m and y an element of order n of an abelian group A. If m and n are coprime prove A contains an element of order mn.

7. If x and y are elements of a finite group, prove that the elements xy and yx have the same order.

8. If a subgroup H of a group G contains the commutator subgroup $G^{(1)}$ of G, prove that H is a normal subgroup.

9. Show that the commutator subgroup of a quotient group G/H is generated by the set of elements $G^{(1)}/H$. If $H \subset G^{(1)}$ find the commutator subgroup of G/H.

10. Prove that the group of symmetries of a regular octahedron (6 vertices, 8 faces that are congruent equilateral triangles) has order 24.

2-7 HOMOMORPHISMS OF GROUPS

Let G and G' be two groups. There are many mappings from G to G' but the most important are those that preserve the group operation and hence the algebraic structure of the group. These are called **homomorphisms.**

Definition 9. A mapping f of $G \rightarrow G'$ is called a (group) *homomorphism* if and only if $(xy)f = (xf)(yf)$ for all $x, y \in G$. (For simplicity we are writing the two distinct group operations of G and G' as products.)

The homomorphism f is called an *epimorphism* if f is a surjective mapping, and it is called an *isomorphism* if f is a bijective mapping. It is not necessary for the group G' to be distinct from the group G. A homomorphism of a group G to itself is called an **endomorphism of G** and an isomorphism of a group G to itself is called an **automorphism of G.**

We shall use the notation $G \approx G'$ to denote that two groups G and G' are isomorphic.

The **kernel** of a homomorphism f of $G \to G'$, written ker f, is the subset of all elements of G that are mapped by f into the neutral element e' of G'. That is,

$$\ker f = \{x \in G \mid xf = e'\}.$$

THEOREM 9. If f is a homomorphism $G \to G'$ of two groups, then $ef = e'$ and $(x^{-1})f = (xf)^{-1}$ for all $x \in G$.

Proof: $(ef)^2 = (ef)(ef) = e^2f = ef$.   Hence ef is idempotent, and the only idempotent in a group is the neutral element.   Hence $ef = e'$. Also $(x^{-1}f)(xf) = ef = e'$, and similarly, $(xf)(x^{-1}f) = e'$. Hence $x^{-1}f$ is the inverse of xf.

Example 1. The mapping $e \to I$, $x \to R_1$, $x^2 \to R_2$, $y \to H$, $xy \to K$, $yx \to L$ is easily verified to be an isomorphism of the group G of Example 1, sec. 2-6, and the group of symmetries of the equilateral triangle.

Example 2. The mapping $e \to e$, $x \to e$, $x^2 \to e$, $y \to xy$, $xy \to xy$, $yx \to xy$ can be shown to be an endomorphism of the group G of Example 1, sec 2-6. Its kernel is the normal subgroup $\{e, x, x^2\}$.

Example 3. $e \to e$, $x \to x^2$, $x^2 \to x$, $y \to yx$, $xy \to xy$, $yx \to y$ is an automorphism of this same group.

Example 4. The mapping $n \to 2n$ is an endomorphism of the additive abelian group of integers.

Example 5. Two finite groups of the same order are not necessarily isomorphic.   The Klein 4-group $\{e, x, y, xy\}$ where $x^2 = y^2 = e$ and $xy = yx$ and the cyclic group $[t]$, $t^4 = e$, of order 4, are both abelian but they are not isomorphic.

Example 6. If f is an isomorphism $G \to G'$ of two groups, then if one is abelian, the other is.   For $x, y \in G$, $xy = yx$ if and only if $(xy)f = (yx)f$; that is, if and only if $(xf)(yf) = (yf)(xf)$. Since f is surjective, this is true in all of G'.

It should be observed that if a mapping $G \to G'$ is known to be a homomorphism, then it is completely determined by its values (images) on the generators of G. For if S is a subset of G that generates G, then every $x \in G$ can be expressed in the form

$$x = s_1 s_2 \cdots s_n,$$

where the s_i are either elements of S or inverses of elements of S. Hence if f is a homomorphism with G as its domain, then $xf = (s_1 \cdots s_n)f = (s_1 f) \cdots (s_n f)$. Further, if $s \in S$, since $(s^{-1})f = (sf)^{-1}$, we see that all factors in the product for xf can be expressed in terms of images under f of elements of S. Hence if f is a homomorphism and we are given its effect on the elements of S, the effect of f on any element of G is known.

THEOREM 10. A homomorphism f of two groups $G \to G'$ is injective if and only if ker $f = \{e\} = 1$, that is if the only element in ker f is the neutral element of G.

Proof: Let ker $f = 1$. Suppose $xf = yf$. Then $(xf)(yf)^{-1} = e'$, hence $(xf)(y^{-1}f) = (xy^{-1})f = e'$. Therefore $xy^{-1} = e$, and so $x = y$. This shows f is injective. Conversely, assume f is injective. Let $x \in$ ker f. Then $xf = e' = ef$ and hence $x = e$. Thus ker $f = 1$.

THEOREM 11. Let f be a homomorphism $G \to G'$ of two groups. If H is a subgroup of G then Hf is a subgroup of G', and if H' is a subgroup of G' then $f^{-1}H'$ is a subgroup of G.

Proof: Hf is the subset of G' comprising all elements of the form xf, $x \in H$. Assume H is a subgroup of G. Let $x', y' \in Hf$. Then $x' = xf$, $y' = yf$ for $x, y \in H$. We have $x'y' = (xf)(yf) = (xy)f$. Since H is a subgroup, $xy \in H$, hence $x'y' \in Hf$. Moreover, $(x')^{-1} = (xf)^{-1} = x^{-1}f$. Since $x \in H$ implies $x^{-1} \in H$, we see that $(x') \in Hf$. Hence Hf is a subgroup of G'.

Let H' be a subgroup of G'. $f^{-1}H'$ is the set of all $x \in G$ such that $xf \in H'$. Thus $x, y \in f^{-1}H'$ means $xf, yf \in H'$ and hence that $(xf)(yf) = (xy)f \in H'$. Therefore $xy \in f^{-1}H'$. Also if $x \in f^{-1}H'$, then $(xf)^{-1} = x^{-1}f \in H'$ and so $x^{-1} \in f^{-1}H'$. Thus $f^{-1}H'$ is a subgroup of G'.

Corollary. The kernel of a homomorphism $G \to G'$ is a normal subgroup of G.

Proof: ker $f = f^{-1}\{e'\}$. Since $\{e'\}$ is a subgroup of G', ker f is a subgroup of G. Let $x \in G$ and $y \in$ ker f. Form xyx^{-1}. Then $(xyx^{-1})f = (xf)(yf)(x^{-1}f) = (xf)e'(x^{-1}f) = (xf)(xf)^{-1} = e'$. Hence $xyx^{-1} \in$ ker f for all $x \in G$, $y \in$ ker f and therefore ker $f \vartriangle G$.

If G, G', G'' are groups and f is a homomorphism of $G \to G'$, and g is a homomorphism of $G' \to G''$, then the map composition fg is defined, and it is very easy to see that (1) fg is surjective if f and g are surjective, and (2) fg is injective if f and g are injective. Hence fg is an isomorphism if f and g are isomorphisms.

Looking back over these very desirable features of homomorphisms, it is not hard to see why such mappings may be expected to be of great importance. This will become even more apparent as we continue to study and use them. As one might expect there are homomorphisms of other algebraic systems besides groups and, in fact, homomorphisms will occupy our attention to the exclusion of other mappings.

2-8 EPIMORPHISMS OF GROUPS

THEOREM 12. If $H \triangle G$, then the mapping $xf = Hx$, $x \in G$, is an epimorphism f of $G \to G/H$.

Proof: $(xy)f = Hxy = Hx \cdot Hy = (xf)(yf)$, and hence f is a homomorphism. Clearly f is surjective.

THEOREM 13. If f is an epimorphism $G \to G'$ of two groups then $G/\ker f$ and G' are isomorphic groups, denoted by $G/\ker f \approx G'$.

Proof: Put $K = \ker f$. Define a mapping ϕ of $G/K \to G'$ by $(Kx)\phi = xf$, $x \in G$. (Note $xf \in G'$). Note there is some doubt attached to this mapping. Suppose coset Kx = coset Ky, what about their images under ϕ, does $xf = yf$? They had better be equal! Now $Kx = Ky$ implies $xy^{-1} \in K$. Hence $e' = (xy^{-1})f = (xf)(y^{-1}f) = (xf) \cdot (yf)^{-1}$, and therefore $xf = yf$, as would be required by a mapping. We say then that we have shown the mapping ϕ to be **well-defined.**

Now $(Kx \cdot Ky)\phi = (Kxy)\phi = (xy)f = (xf)(yf) = (Kx)\phi \cdot (Ky)\phi$ and hence ϕ is a homomorphism. Also ker ϕ is the set of all cosets Kx such that $xf = e'$, that is such that $x \in K$. Hence ker $\phi = \{K\}$, where K is the neutral element of G/K. This proves ϕ is injective, by Theorem 10. It is seen at once that since f is given to be surjective, this makes ϕ a surjective mapping. Hence ϕ is an isomorphism.

Definition 10. If $H \triangle G$, the epimorphism f of $G \to G/H$ defined by $xf = Hx$, $x \in G$, is called the **natural homomorphism of G** onto G/H.

Definition 11. A group G' is called a **homomorphic image** of a group G if there exists an epimorphism f of $G \to G'$.

Theorem 12 shows that every factor group of G is a homomorphic image of G. Theorem 13 shows that if G' is a homomorphic image of G, then G' is isomorphic to a factor group of G, in fact if f is the epimorphism of $G \to G'$, then $G' \approx G/\ker f$. Combining these two results, we see then that this essentially determines all homomorphic images of a group G. We express this important fact in a theorem.

THEOREM 14. A group G' is a homomorphic image of a group G if and only if it is isomorphic to a factor group of G.

If f is any homomorphism $G \to G'$ of two groups, then the mapping f of $G \to Gf$ is an epimorphism and Gf is a homomorphic image of G. Hence $G/\ker f \approx Gf$.

A homomorphic image G' of a group G inherits many of the properties of G itself. It is easy to verify for instance that if G is an abelian group, then so is its homomorphic image G'; if G is cyclic, then so is G'; if G is a torsion group, then so is G'. In fact, if G is a finitely generated group with the generators $x_1, x_2, \ldots, x_n$ then its homomorphic image G' is a finitely generated group with the generators $x_1 f, x_2 f, \ldots, x_n f$, where f is the epimorphism $G \to G'$. This is true of many other properties of G, and thus a homomorphic image is a sort of picture in miniature of the group itself. Theorem 14 above is fundamental, for it shows that many of the properties of a group are passed on to its factor groups; if a group is cyclic then every factor group is cyclic, and so on.

The natural epimorphism $x \to Hx$ of $G \to G/H$ is not of course the only epimorphism of $G \to G/H$. If $a \neq e$, is some fixed element of G, then $x \to Haxa^{-1}$ is an epimorphism of $G \to G/H$.

If f is a homomorphism $G \to G'$ of two groups and if H' is a subgroup of G', then $f^{-1}H'$ is always a subgroup of G that contains $\ker f$. For $e' \in H'$. If f is an epimorphism, this suggests that there might exist a bijection between the subgroups of G containing $\ker f$ and all the subgroups of G'.

THEOREM 15. Let f be an epimorphism $G \to G'$ of two groups. Then $K \to Kf = K'$ is a bijection of the set of all subgroups K of G containing $\ker f$ with the set of all subgroups K' of G'. Moreoever, $K \, \Delta \, G$ if and only if $K' \, \Delta \, G'$, whether K contains $\ker f$ or not.

Proof: Let Q be the mapping $K \to K' = Kf$. From the remark above we know Q is surjective. To show Q is injective we need to prove $K = f^{-1}K'$. We know $K \subset f^{-1}K'$. Let $x \in f^{-1}K'$, then $xf \in K'$. Since $K' = Kf$ we know there exists $k \in K$ such that $kf = xf$. Hence $e' = (xf)(kf)^{-1} = (xk^{-1})f$. Thus $xk^{-1} \in \ker f \subsetneq K$. Hence $x \in K$ and therefore $f^{-1}K' \subset K$. Hence $K = f^{-1}K'$. Thus Q is a bijection.

To prove the second part of the theorem, let $K \, \Delta \, G$. Let $x' \in G'$, then $x' = xf$ for some $x \in G$. Hence $x'K'x'^{-1} = (xf)(Kf)(x^{-1}f) = (xKx^{-1})f = Kf = K'$. Hence $K' \, \Delta \, G'$. Conversely, assume $K' \, \Delta \, G'$. Then $(xKx^{-1})f = (xf)K'(x^{-1}f) = K'$. Hence $xKx^{-1} = K$, since Q is injective. Therefore $K \, \Delta \, G$.

Corollary. Let $H \, \Delta \, G$. Then every subgroup of G/H has the form T/H, where T is a subgroup of G containing H. Moreover $T/H \, \Delta \, G/H$ if and only if $T \, \Delta \, G$.

Proof: Let f be the natural epimorphsim of $G \rightarrow G/H$; that is, $xf = Hx, x \in G$. Since ker $f = H$, it follows from the theorem that there is a bijection of all the subgroups T of G containing H with all subgroups of G/H. Thus every subgroup of G/H has the form T/H, where T is a subgroup of G containing H. The group T is mapped into the group Tf, which consists of all cosets of G/H of the form Ht, $t \in T$—that is, $Tf = T/H$. Incidentally this implies $H \, \Delta \, T$. The bijection is $T \rightarrow T/H$ and hence $T/H \, \Delta \, G/H$ if and only if $T \, \Delta \, G$.

This corollary thus determines the form of all subgroups of a factor group G/H and also the form of all normal subgroups of a factor group.

Example. Find all the homomorphic images of the group G of symmetries of the equilateral triangle.

To do this, first find all the nontrivial normal subgroups of G. Consulting the group table in Example 3 of Sec. 6, we see that the only such normal subgroup of G is the subgroup S of rotations I, R_2, R_3. Thus all nontrivial homomorphic images of G are isomorphic to the factor group G/S.

EXERCISE

1. Prove that the group of symmetries of the square (its order is 8) has two nontrivial normal subgroups, and therefore that it has essentially two nontrivial homomorphic images.

2-9 ISOMORPHISMS OF GROUPS

An isomorphism of two groups is tantamount to their identification with each other algebraically. For the isomorphism completely preserves the group structure. Isomorphic groups are equivalent algebraic systems, that is they are algebraically indistinguishable. For instance, if f is an injective homomorphism $G \rightarrow G'$ of two groups then $G \approx Gf \subset G'$. We can therefore identify G with its image Gf and hence regard G as being a subgroup of G'. This is frequently done for its considerable convenience. One can regard isomorphic groups as the same group symbolized in two different notations (the same man in different suits).

We next prove two fundamental isomorphism theorems about groups. These are also basic theorems in the applications of group theory to algebraic topology.

THEOREM 16 (First Isomorphism Theorem). Let f be an epimorphism $G \to G'$ of two groups. Let K be a normal subgroup of G containing ker f, and let $K' = Kf$. Then G/K and G'/K' are isomorphic groups —that is, $G/K \approx G'/K'$.

Proof: By Theorem 15 we know $K' \Delta G'$. We show that the mapping ϕ of $G/K \to G'/K'$, defined by $(xK)\phi = (xf)K'$, is an isomorphism. (As in Theorem 13, ϕ is well-defined, for $xK = yK$ implies $yf = xf$.) Now $(xK \cdot yK)\phi = (xy)K\phi = (xy)f \cdot K' = (xf) \cdot (yf)K' = (xf)K' \cdot (yf)K' = (xK)\phi \cdot (yK)\phi$, and hence ϕ is a homomorphism. Since f if surjective, the mapping ϕ is surjective. Moreover, ker ϕ is the set of all xK such that $xf \in K'$; that is, such that $x \in K$. If $x \in K$, $xK = K$, and ker ϕ consists of the single element K, the neutral element of G/K. Hence ϕ is injective, and therefore ϕ is an isomorphism.

Corollary. If $H \Delta G$, $K \Delta G$, and if $H \subset K$, then $G/K \approx G/H/K/H$.

Proof: Apply the theorem itself to the natural epimorphism f of $G \to G/H$—that is, take $G' = G/H$.

If A and B are subgroups of a group G, then the set AB of all elements ab, $a \in A$, $b \in B$, is not in general a subgroup of G. For the product of two such elements is not in AB. However if $A \Delta G$, then AB is a subgroup of G. For $(a_1 b_1)(a_2 b_2) = a_1(b_1 a_2 b_1^{-1})b_1 b_2$ and $(ab)^{-1} = b^{-1}a^{-1} = (b^{-1}a^{-1}b)b^{-1}$ are both in AB. Moreover, it is easy to show that $A \Delta AB$.

Let $H \Delta G$ and let f be the natural epimorphism of $G \to G/H$. If K is a subgroup of G, then Kf is a subgroup of G/H. This subgroup Kf corresponds to a subgroup of G containing H. Since KH is a subgroup of G and since $Kf = (KH)f$ it is seen that KH is the subgroup of G, containing H, and corresponding to the subgroup Kf in the bijection of Theorem 16.

THEOREM 17 (Second Isomorphism Theorem). If H and K are subgroups of a group G and if $H \Delta G$, then HK/H

$$K/H \cap K \approx hK/H.$$

Proof: A left coset of H in HK is hkH, $h \in H$, $k \in K$. Now $hkH = K(k^{-1}hk)H = kh'H = kH$, where $h' \in H$.

The mapping ϕ of $K \to HK/H$ defined by $k\phi = kH$, $k \in K$, is clearly an epimorphism. The kernel of ϕ is $H \cap K$, and hence by the fundamental homomorphism theorem we have the result. (Note that $H \cap K$ is a normal subgroup of K but not necessarily a normal subgroup of G. It is never an empty set, since $e \in H \cap K$.)

This theorem provides a very useful isomorphism fashioned out of two subgroups of a group, one of them being a normal subgroup.

In particular if $H \cap K = 1$, then $K \approx HK/H$.

EXERCISES

1. Find all the homomorphic images of S_3 and S_4.

2. G is an abelian group generated by the elements x and y, where $x^4 = 1$ and $y^3 = 1$. Find the order of G. Show that G has 6 automorphisms and 12 endomorphisms.

3. Under map composition as binary operation write out the group table for the group of automorphisms or order 6 of the group G in Exercise 2.

4. Prove that the group of plane rotations about the origin is isomorphic to the rotation group of 2×2 matrices of the form

$$\begin{pmatrix} \cos \alpha & -\sin \alpha \\ \sin \alpha & \cos \alpha \end{pmatrix},$$

which corresponds to a rotation through an angle α.

5. Let f be a homomorphism $G \rightarrow G'$ of two groups. If G' is abelian, is G abelian? Prove your answer.

6. A right translation T_a corresponding to an element a of a group G is the mapping of $G \rightarrow G$ defined by $xT_a = xa, x \in G$. (a) Prove T_a is bijective. Is T_a an isomorphism? (b) Prove $T_a \cdot T_b = T_{ab}$ and find the inverse of T_a. (c) Prove $a \rightarrow T_a$ is an isomorphism of G with the group of right translations. (d) A left translation $_aT$ is defined by $x_aT = ax, x \in G$. Explain what is meant by saying that $a \rightarrow {_aT}$ is an "antiisomorphism" of G with the group of left translations.

7. In Exercise 3, Sec. 2-5, prove $C/Z \approx Z$.

8. In Exercise 4, Sec. 2-5, prove $C/R \approx R$.

2-10 CYCLIC GROUPS

A group G is called **cyclic** if it contains an element x that generates the group. If we write the binary operation as multiplication, this means that if G is an infinite cyclic group, it consists of all the integral powers of x, positive, negative and zero, where we define $x^0 = e$, the neutral element of G. Thus a typical element of G has the form x^n with the inverse x^{-n}. We write $G = [x]$. If G is a finite cyclic group of order n, generated by an element x, then $x^n = e$ and the distinct elements of G can be written $e, x, x^2, \ldots, x^{n-1}$. We again use the notation $G = [x]$. Obviously a cyclic group is an abelian group.

Let H be a subgroup of a cyclic group $G = [x]$. If $x^i \in H$, then $x^{-i} \in H$. Let m be the least positive integer such that $x^m \in H$. Then $H = [x^m]$—that is, H is generated by x^m. For if $x^i \in H$, then the Euclidean algorithm gives $i = qm + r$, $0 \le r < m$. Since $(x^m)^q = x^{qm} \in H$, it follows that $x^r = x^{i-qm} \in H$. Since $r < m$ it follows that, to avoid a contradiction of the definition of m, $r = 0$ and hence $i = qm$. Thus $H = [x^m]$ and we see that every subgroup of a cyclic group is also a cyclic group.

Next let $H = [x^m]$ be a (cyclic) subgroup of the infinite cyclic group $G = [x]$. For any integer n we have $n = qm + r$, where $0 \le r < m$, and hence $r = 0, 1, 2, \ldots, m - 1$. It follows then that the distinct cosets of H in G are $H, Hx, Hx^2, \ldots, Hx^{m-1}$ (for every power of x is in one and only one of these cosets) and hence the index of H in G is m. Thus every subgroup of an infinite cyclic group has finite index, and in fact, if the index is m then the subgroup is generated by x^m. (This illustrates the interesting fact that an infinite group may have a subgroup of finite index.) Since for every positive integer m there is a subgroup $[x^m]$ of the infinite cyclic group $[x]$, it follows that there is a unique subgroup of index m in an infinite cyclic group.

Let $G = [x]$ be a finite cyclic group of order n. In this case we can write the elements of G as $e = x^0, x, x^2, \ldots, x^{n-1}$, with $x^n = e$. Let m be a divisor of $n, n = mr$. Then $e, x^m, x^{2m}, \ldots, x^{(r-1)m}$ is a subgroup of G of order r and index m. Thus there is a unique subgroup of each order r dividing the order n of a finite cyclic group. As we have observed earlier, this is not true of an arbitrary group.

An infinite cyclic group $[x]$ is isomorphic to the additive group Z of integers under the mapping ϕ of $[x] \rightarrow Z$ defined by

$$(x^n)\phi = n, \qquad n \text{ any integer.}$$

For if $(x^m)\phi = m$, then $(x^n \cdot x^m)\phi = (x^{n+m})\phi = n + m = x^n\phi + x^m\phi$ and ϕ is easily shown to be a bijection. Thus

$$[x] \approx Z.$$

On the other hand, a finite cyclic group $[x]$ of order n is isomorphic to the additive group Z_n of residue classes modulo n. The isomorphism is the mapping ϕ of $[x] \rightarrow Z_n$ given by $(x^n)\phi = \bar{n}$ where $\bar{n}$ is the residue class containing the integer n.

EXERCISES

1. G is a cyclic group of order m generated by the element x. G' is a cyclic group of order n. If f is a homomorphism of $G \rightarrow G'$ such that $x \notin \ker f$, show that m and n cannot be coprime.

2. If H is a subgroup of a cyclic group G, is the quotient group G/H always cyclic?

3. If G and G' are finite cyclic groups of the same order, prove that Hom (G, G') is a cyclic group of the same order. (Hom (G, G') is the set of all group homomorphisms of $G \rightarrow G'$.)

2-11 PERMUTATION GROUPS

A **permutation** of a set X is a bijection of $X \rightarrow X$. Under the associative binary operation of map composition, the set $P(X)$ of all permutations on X forms a group, the identity mapping I_X is the neutral element and every bijection has an inverse bijection. This group $P(X)$ is called the **group of permutations on X** or the **group of transformations of X.** Any subgroup Σ of $P(X)$ is called a *group of permutations* on the set X.

Let Σ be a group of permutations on X. We can define an equivalence relation R on X as follows: for $x, y \in X$, define xRy if and only if $y = x\sigma$ for some $\sigma \in \Sigma$. The reader can easily verify that R is an equivalence relationship, as claimed. The disjoint equivalence classes of this relation are called Σ-*orbits* in X. Two elements of X are in the same Σ-orbit if and only if one is mapped into the other by some permutation in Σ. We shall call a Σ-orbit *trivial* if it contains only one element. This would mean that the element is left unchanged by any permutation in Σ— that is, it is always mapped into itself by every permutation in Σ.

Our concern here will be with finite groups of permutations, and so we shall specialize the foregoing discussion to the case where X is a finite set. Let X be a set of n elements and let these elements be designated by the n natural numbers $1, 2, \ldots, n$. Write S_n for the group $P(X)$ of all permutations on this finite set X of n elements. S_n is called the **symmetric group of degree n.** The order of S_n is $n! = 1 \cdot 2 \cdot 3 \ldots n$. Any permutation $\alpha \in S_n$ can be written as

$$\alpha = \begin{pmatrix} 1 & 2 \ldots n \\ i_1 & i_2 \quad i_n \end{pmatrix}$$

where $i_1, i_2, \ldots, i_n$ is some arrangement of the natural numbers $1, 2, \ldots, n$ and the notation signifies that $1\alpha = i_1, 2\alpha = i_2, \ldots, n\alpha = i_n$.

Let $\alpha \in S_n$ and form the cyclic subgroup $[\alpha]$ of S_n that is generated by α. Let $\#[\alpha] = k$, so that $\alpha^k = e$, the identity permutation. Let $X_1, X_2, \ldots X_t$ denote the disjoint $[\alpha]$-orbits. An $[\alpha]$-orbit is trivial (contains only one element x) if and only if $x\alpha = x$. For this means $x\alpha^2 = x\alpha = x$, and so on, while if $y \neq x$, then, since α is injective, $y\alpha \neq x\alpha$ and hence $y\alpha \neq x$. A permutation α is called a **cycle** if one and only one of the $[\alpha]$-orbits is nontrivial. Let α be a cycle and let x be any element of the nontrivial $[\alpha]$-orbit, say X_i. Then $x\alpha, x\alpha^2, \ldots, x\alpha^{k-1}$ (where $\#[\alpha] = k$) are the other elements of X_i. These are all distinct. For if two of them are equal, then $x\alpha^m = x$ for some m where $0 < m < k - 1$, while if $y \notin X_i$ then y is in a trivial $[\alpha]$-orbit and therefore $y\alpha = y$, whence $y\alpha^m = y$. Thus $\alpha^m = e$ for $0 < m < k$, a contradiction of the definition of k.

We can write the cycle α as

$$\alpha = (x \quad x\alpha \quad x\alpha^2 \ldots x\alpha^{k-1}),$$

where the notation means that α cyclically permutes the elements in the one nontrivial $[\alpha]$-orbit (see the diagram) and leaves unchanged all other elements, which is signified in the notation above for α by the omission of any element that α maps into itself. Thus the elements appearing in the cycle are the same as those in its nontrivial orbit. α is called a *k-cycle*, since it contains k symbols. Its order (as an element of the group S_n) is k, $\alpha^k = e$.

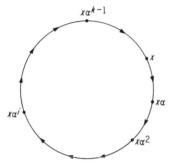

For instance, for $n = 5$ the permutation $\alpha = \begin{pmatrix} 1 & 2 & 3 & 4 & 5 \\ 3 & 2 & 5 & 4 & 1 \end{pmatrix}$ is a cycle and the single nontrivial $[\alpha]$-orbit is $\{1, 3, 5\}$ while the cycle itself is $\alpha = (135)$. Here $1\alpha = 3$, $1\alpha^2 = 3\alpha = 5$, and $1\alpha^3 = 5\alpha = 1$. Note the cycle can also be written (351) or (513), the cyclical order of the symbols being all that is important. The other $[\alpha]$-orbits are the trivial ones $\{2\}$ and $\{4\}$.

Definition. Two cycles are called **disjoint** if their nontrivial orbits are disjoint.

Of course a permutation may not be a cycle, however, it can be shown that every permutation can be expressed as the product (map composition) of disjoint cycles.

For example, let $\alpha = \begin{pmatrix} 1 & 2 & 3 & 4 & 5 \\ 3 & 5 & 4 & 1 & 2 \end{pmatrix}$ be permutations in S_5. The $[\alpha]$-orbits are $X_1 = \{1, 3, 4\}$ and $X_2 = \{2, 5\}$. Thus α is not a cycle. Let α_1 be the cycle for the orbit X_1, that is α_1 acts on X_1 just as α does, leaving unchanged all elements not in X_1. Let α_2 be the similar cycle for X_2. Then clearly $\alpha_1 = (134)$ and $\alpha_2 = (25)$, so that

$$\alpha = (134)(25) = (25)(134).$$

THEOREM 18. Every permutation $\alpha \in S_n$, $\alpha \neq e$, is a product (map composition) of disjoint cycles and this product is unique up to the order of its factors.

Proof: Let $\#[\alpha] = k$. Let $X_1, X_2, \ldots, X_t$ be the $[\alpha]$-orbits. For each trivial orbit X_i define $\alpha_i = (x_i)$ where $x_i \in X_i$ and call it a 1-cycle. Actually we can ignore α_i, since it leaves every element unchanged—that is, its effect is the same as that of the identity permutation. We shall elect to omit each such α_i. For each non-trivial orbit X_i let $x \in X_i$, then $x\alpha \neq x$. Let m_i be the smallest positive integer such that $x\alpha^{m_i} = x$. Now $m_i \leq k$ and $k = qm_i + r$, $0 \leq r < m_i$. Hence $\alpha' = \alpha^{k-qm_i}$ and therefore $x\alpha' = x$, which is a contradiction of the choice of m_i, unless $r = 0$. Therefore $r = 0$ and $m_i \mid k$. (In fact k is the least common multiple of all the m_i, for if all the m_i divide ω then $x\alpha\omega = x$ for every x. Hence $\alpha^\omega = e$ and therefore $k \mid \omega$.) For the nontrivial orbit X_i form the cycle $\alpha_i = (x \quad x\alpha \ldots x\alpha^{m_i-1})$. Continuing in this way through all the orbits we see that α is the product (map composition) of disjoint cycles

$$\alpha = \alpha_1 \alpha_2 \cdots \alpha_h$$

where we have omitted all α_i formed from the trivial orbits.

Since the set of $[\alpha]$-orbits is unique it follows that the expression for α as a product of cycles is unique up to the order of the factors.

Moreover we have shown that if m_i is the order (as an element of the group S_n) of the cycle α_i ($\alpha_i^{m_i} = e$), then the least common multiple k of the m_i is the order of their product α; that is, the order of a permutation is the least common multiple of the orders of its disjoint cycles.

Examples: Let $\alpha = \begin{pmatrix} 1 & 2 & 3 & 4 & 5 & 6 & 7 & 8 \\ 3 & 7 & 1 & 4 & 8 & 2 & 6 & 5 \end{pmatrix}$.

Then

$$\alpha = (13)(276)(58)$$

and the order of α is 6, $\alpha^6 = e$.

If $\alpha_1 = (x \quad x\alpha_1 \ldots x\alpha_1^{m_1-1})$ is a cycle then we can verify by map composition that its inverse is

$$\alpha_1^{-1} = (x\alpha_1^{m_1-1} \quad x\alpha_1^{m_1-2} \cdots x\alpha_1 x),$$

that is,

$$\alpha_1 \alpha_1^{-1} = e = \alpha_1^{-1}\alpha_1,$$

and that the inverse of $\alpha = \alpha_1 \alpha_2 \cdots \alpha_h$ is

$$\alpha^{-1} = \alpha_h^{-1} \alpha_{h-1}^{-1} \cdots \alpha_2^{-1} \alpha_1^{-1}.$$

A 2-cycle is called a **transposition,** and every n-cycle is a product of transpositions, for

$$(x_1 x_2 \cdots x_n) = (x_1 x_2)(x_2 x_3) \cdots (x_1 x_n).$$

This product is not unique, for

$$(x_1 x_2 \cdots x_n) = (x_2 x_3 \cdots x_n x_1) = (x_2 x_3)(x_2 x_4) \cdots (x_1 x_2).$$

It follows then that every permutation can be expressed as a product of transpositions.

Let us consider the effect of a transposition $(x_i x_j)$, $i < j$, on the product $P = \displaystyle\prod_{i<j} (x_i - x_j)$, $i, j = 1, 2, \ldots, n$. It is not hard to see that $(x_i x_j)$ causes changes of sign only in the $j - i$ factors $(x_i - x_{i+1}) \cdots (x_i - x_j)$ and in the $j - i - 1$ factors $(x_{i+1} - x_j) \cdots (x_{j-1} - x_j)$. Hence it effects a total of $2j - 2i - 1$ changes of sign in P, an odd number, and so a transposition $(x_i x_j)$ changes P to $-P$.

A permutation is called *even* if it does not change the sign of P and *odd* if it changes P to $-P$. Since every permutation is a product of transpositions, it follows that an even permutation must always be the product of an even number of transpositions, and an odd permutation always the product of an odd number of permutations. Thus an n-cycle is even if n is odd and odd if n is even. For example, $(123) = (12)(13)$ is an even permutation. Note $(123) = (12)(23)(23)(13)$, but the number of transpositions is still even.

Clearly the inverse of an even permutation is even and the inverse of an odd permutation is odd. The product of even permutations is even and the product of two odd permutations is even. The identity permutation e leaves the product P unchanged and so it is even.

These remarks prove that the set of all even permutations in the symmetric group S_n forms a subgroup A_n of S_n. A_n is called the **alternating group.**

Let us find the index of A_n in S_n. If α and β are two permutations then the cosets αA and βA are equal, $\alpha A = \beta A$, if and only if $\alpha \beta^{-1} \in A_n$ that is if and only if $\alpha^{-1}\beta$ is an even permutation. Since the product $\alpha^{-1}\beta$ of two odd permutations is always even, the index of A_n in S_n is 2. This proves that A_n is a normal subgroup of S_n of order $n!/2$.

THEOREM 19 (Cayley). Every abstract group is isomorphic to a group of permutations.

Proof: Let G be a group. To each element $a \in G$, assign a mapping f_a of $G \to G$ defined by $xf_a = xa$, $x \in G$. It is easy to see that f_a is a bijection—that is, a permutation on G. For $xa = ya$ implies $x = y$ and hence f_a is injective. Moreover, for any $z \in G$, $z =$

$(za^{-1})a$ and hence $(za^{-1})f_a = z$, thus f_a is surjective. Let T be the set of all such permutations formed for each element of G. Introduce the associative binary operation of map composition in T. For $x \in G$,

$$x(f_a f_b) = (xf_a)f_b = (xa)f_b = xab = xf_{ab}.$$

Hence $f_a f_b = f_{ab}$. Clearly f_e, where e is the neutral element of G, is the neutral element of the binary operation and it is easy to see that $f_{a^{-1}}$ is the inverse of f_a. Thus T is a group of permutations on G.

The mapping of $G \rightarrow T$ defined by $a \rightarrow f_a$ is a homomorphism, for if $b \rightarrow f_b$, then $ab \rightarrow f_{ab} = f_a f_b$. It is clearly a surjective mapping and its kernel contains only the neutral element e of G. Hence it is injective. Thus $G \simeq T$.

Every group G is therefore isomorphic to a group of permutations on G. This is our first example of what is called a **representation theorem.** It proves that every group G can be represented as a group of permutations. The isomorphism means that we can, if we choose, regard every group as a permutation group or transformation group. In particular, every finite group of order n is isomorphic to a subgroup of the symmetric group S_n of degree n.

Lemma 2. The alternating group A_n, $n \geq 3$, is generated by tricycles.

Proof: Every even permutation is a product of an even number of transpositions. A product of two transpositions is either a tricycle $(ab)(ac) = (abc)$, or a product of two tricycles $(ab)(cd) = (acd)(acb)$. Hence A_n is generated by tricycles.

Lemma 3. For $n \geq 5$, if a normal subgroup H of A_n contains a tricycle then $H = A_n$.

Proof: Let $(abc) \in H$. If x, y, z are all different from a, b, c (which is possible when $n > 5$) we have

$$(xyz) = (ax)(bzcy)(abc)(yczb)(xa)$$
$$= [(ax)(bzcy)](abc)[(ax)(bzcy)]^{-1}.$$

Since $H \Delta A_n$, $(xyz) \in H$.

Now consider the three cases where some of the x, y, z are the same as some of the a, b, c.
    Case I
$$(abx) = (cdx)(abc)(xdc)$$
$$= (cdx)(abc)(cdx)^{-1} \in H.$$

Here d is distinct from a, b, c, x (possible since $n \geq 5$).

Case II

$$(acx) = (bxc)(abc)(cxb)$$
$$= (bxc)(abc)(bxc)^{-1} \in H.$$

Case III

$$(axy) = (cy)(bx)(abc)(bx)(cy) \in H.$$

Hence H contains all tricycles and therefore $H = A_n$.

Corollary. A_4 contains no subgroup of order 6.

Proof: Let H be a subgroup of A_4 of order 6. The index of H in A_4 is 2 and hence H is a normal subgroup of A_4. Eight of the 12 elements of A_4 are tricycles. Hence H must contain a tricycle and therefore $H = A_4$.

This proves the converse of Lagrange's theorem is false. Thus if r is a divisor of the order of a finite group G, then G does not necessarily have a subgroup of order r.

We next prove a lemma that is most important in Galois theory.

Lemma 4. For $n \geq 5$, A_n is simple (that is, A_n has no proper normal subgroups).

Proof: Let $H \triangle A_n$, where $1 \subset H \subset A_n$. If H contains an element σ that has a cycle of length ≥ 4, $\sigma = (abcd \cdots x) \cdots (\cdots)$, then for $\sigma' = (cab)\sigma(cba) = (cabd \cdots x) \cdots (\cdots)$, which belongs to H, we have $\sigma^{-1}\sigma' = (cda) \in H$. Thus by Lemma 3, $H = A_n$.

We now examine the two remaining cases.
(a) Let $\sigma \in H$ contain cycles of length ≤ 3. If h has only one tricycle (the other factors being then transpositions), σ^2 is a tricycle and again we have $H = A_n$. If σ has at least two tricycles for factors, let $\sigma = (abc)(def) \cdots$. Then $\sigma' = (edc)\sigma(cde) \in H$ and hence

$$\sigma\sigma' = (adcbf \cdots) \in H.$$

This means H contains an element $\sigma\sigma'$ with a cycle of length ≥ 4. Hence $H = A_n$.
(b) Let $\sigma \in H$ have only transpositions (an even number, of course) as cycles. If $\sigma = (ab)(cd)$, then $\sigma' = (eba)\sigma(abe) = (be)(cd) \in H$, where e is distinct from a, b, c, d. (This is possible since $n \geq 5$.) Hence

$$\sigma'\sigma = (bea) \in H,$$

and by Lemma 3, $H = A_n$.

If σ contains more than two transpositions

$$\sigma = (ab)(cd)(ef)(gh) \cdots,$$

form $\sigma' = (cf)(eh)\sigma(eh)(cf)$, then $\sigma\sigma' = (cfg)(edh)$.   Hence by part (a), $H = A_n$.

Note that A_3 is simple, but A_4 has the normal subgroup 1, (12)(34), (13)(24), (14)(23).

EXERCISES

1. The generalized quaternion group Q_m, where m is a positive integer, is defined as the group generated by two elements x and y satisfying the relations $x^m = y^2$ and $yxy^{-1} = x^{-1}$. Prove that $x^{2m} = y^4 = 1$ and that the order of Q_m is $4m$. Show that every element of Q_m has a unique representation in the form $x^i y^j$, where $0 \leqq i \leqq 2m - 1, 0 \leqq j \leqq 1$.
Prove that the cyclic group $[x]$ is a normal subgroup of Q_m and find its index in Q_m.

2. Write out the group tables for Q_2 and Q_3.

3. Find a subgroup of order 8 of S_4.

4. Find the subgroup of S_4 that is generated by each of the following two sets (a) $\{(12),(13),(14)\}$ (b) $\{(23),(134)\}$.

5. Prove that the $n - 2$ 3-cycles $(123),(124),\ldots,(12n)$ for $n > 2$, generate the alternating group A_n. [HINT: $(abc) = (1ab)(1ca)$ and $(1xy) = (12y)(12x)(1y2)$.]

6. Prove by induction that the $n - 1$ transpositions $(12),(13),\ldots,(1n)$, for $n > 1$, form a set of generators of S_n.

7. If a normal subgroup N of S_n contains one 3-cycle, prove that N must contain all 3-cycles. Hence show $N = A_n$.

8. If the element x of a group G is of order mn, where m and n are coprime, prove that x is the product of elements of orders m and n.

9. Let p be a fixed prime. Prove that the set P of all rational numbers whose denominators are powers of p is a group under addition. Show that P contains as a subgroup the additive group Z of integers. Prove that the only subgroups of P/Z are the finite subgroups P_n of order p^n that are generated by $1/p^n, n = 1, 2, 3, \ldots$.

10. Let f and f' be homomorphisms of a group G into the groups H and H' respectively. If $\ker f \subset \ker f'$, prove there exists a unique homomorphism h of $Gf \to H$ such that $f' = fh$.

11. Let A, B, C be groups. Let f be an injective homomorphism of $B \to C$ and g a homomorphism of $A \to C$. Prove there exists a homomorphism h of $A \to B$ such that $g = fh$ if and only if $\operatorname{im} g \subset \operatorname{im} f$. Is h unique?

12. Determine the group of rotations of a cube. This group is isomorphic to the symmetric group S_4.

13. Find the groups of symmetries of the regular hexagon and of the regular octagon.

14. Find all the normal subgroups of the group G of symmetries of the regular hexagon. Find a homomorphic image of G.

15. Prove that the cube roots of 1 form a cyclic group.

16. If τ and σ are cycles in S_n prove that if $\sigma = (a_1 a_2 \cdots a_k)$ then $\tau^{-1}\sigma\tau = (a_1\tau \cdots a_k\tau)$—that is, the symbols in σ are replaced by their images under τ.

17. Prove that the tetrahedral group is isomorphic to A_4.

chapter 3

Vector Spaces and Linear Transformations

Often one's first acquaintance with vectors is in the guise of "quantities" having both magnitude and direction. They are represented as "arrows" or "directed line segments" and, when referred to a rectangular coordinate system, a vector **v** is an arrow drawn from the origin in a given direction and having a specified length.

The sum of two vectors **v** and **w** is defined by the parallelogram law which, in two dimensions, is illustrated in the accompanying diagram.

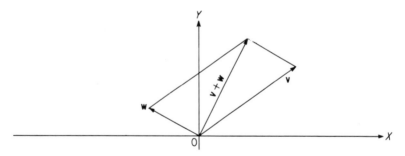

This sum can be proved geometrically to be both commutative and associative. The zero vector is one of zero length. If **v** is multiplied by a real number (scalar) c then $c\mathbf{v}$ is a vector whose magnitude is c times that of **v** and if $c > 0$, $c\mathbf{v}$ has the same direction as **v**, while if $c < 0$, $c\mathbf{v}$ has the opposite direction to **v**.

In algebra vectors are defined as elements of an algebraic system called a **vector space,** and their properties are enunciated by the set of axioms of this algebraic system. In this way great generality and power are realized. The dependency of vectors on a coordinate system is removed and their defining properties are abstracted from the intuitive but crude concept of a vector as an arrow or a boldface symbol. In this way a vector space emerges as a mathematical system occurring in a great many varieties. Needless to say, this greatly increases both our comprehension of the significance of vectors and the scope of their usefulness and their applications.

In order to define a vector space, we need the concept of a field. Fields and their properties are not formally introduced and discussed until later in the book. However the following definition should suffice for our present purposes.

Definition. A *field F* is an algebraic system with two commutative and associative binary operations, addition and multiplication, such that
  1. *F* is a group under addition.
  2. The nonzero elements of *F* form a group under multiplication.
  3. $a(b + c) = ab + ac$, for all $a,b,c \in F$.

Three familiar examples of fields are the rational numbers, the real numbers, and the complex numbers.

We shall denote by 1 the neutral element of the multiplicative group of nonzero elements of *F*, and by 0 the neutral element of *F* as an additive group.

Throughout this chapter the reader can, if he chooses to, safely identify *F* with the field of real numbers.

3-1 DEFINITION AND EXAMPLES

A *vector space* (*linear space*) *V* over a field *F* is an algebraic system that satisfies the following axioms:
  1. *V* is an additive abelian group.
  2. There is a mapping *f* of the Cartesian product $F \times V \to V$ which defines a "scalar multiplication" $x\alpha = (x,\alpha)f$ in *V* for all $x \in F$ and $\alpha \in V$.
  3. The scalar multiplication satisfies the following conditions:
     (a) $(x_1 + x_2)\alpha = x_1\alpha + x_2\alpha$, for all $x_1, x_2 \in F$ and all $\alpha \in V$.
     (b) $x(\alpha + \beta) = x\alpha + x\beta$, for all $x \in F$ and all $\alpha,\beta \in V$.
     (c) $x(y\alpha) = (xy)\alpha$, for all $x,y \in F$ and all $\alpha \in V$.
     (d) $1\alpha = \alpha$, for all $\alpha \in V$.
The elements of the field *F* are called **scalars,** and the elements of *V* are called **vectors.** We shall use Greek letters for vectors and Latin letters for scalars.

Denote the zero vector of *V* by $\bar{0}_V$. It is the neutral element of *V* as an additive abelian group. The zero scalar is of course simply the zero element 0 of the field *F*. Denote by $-\alpha$ the vector that is the unique (group) inverse of the vector α. Thus $\alpha + (-\alpha) = \bar{0}_V$.

Moreover, we write $\alpha - \beta$ for $\alpha + (-\beta)$, where $\alpha,\beta \in V$.

We next establish some elementary properties of operations in a vector space *V* with vectors and scalars.

Lemma 1. For every $\alpha \in V, 0\alpha = \bar{0}_V$.

Proof: $(x + 0)\alpha = x\alpha.$

Also $(x + 0)\alpha = x\alpha + 0\alpha$ by 3(a). Hence $x\alpha + 0\alpha = x\alpha$. Add to each side the inverse $-(x\alpha)$ of $x\alpha$ and we get $\bar{0}_V + 0\alpha = \bar{0}_V$ and hence $0\alpha = \bar{0}_V$.

Lemma 2. For every scalar x, $x\bar{0}_V = \bar{0}_V$.

Proof: $x(\alpha + \bar{0}_V) = x\alpha$.

Also $x(\alpha + \bar{0}_V) = x\alpha + x\bar{0}_V$ by 3(b). Hence $x\alpha + x\bar{0}_V = x\alpha$. Again adding $-(x\alpha)$ to each side we get

$$x\bar{0}_V = \bar{0}_V.$$

Lemma 3. For every $x \in F$ and every $\alpha \in V$,

$$(-x)\alpha = -(x\alpha) = x(-\alpha).$$

Proof: $(x + (-x))\alpha = 0\alpha = \bar{0}_V$ by Lemma 1.

Also $(x + (-x))\alpha = x\alpha + (-x)\alpha$ by 3(a). Hence $x\alpha + (-x)\alpha = \bar{0}_V$. Add the inverse $-(x\alpha)$ of $x\alpha$ to each side and we get

$$\bar{0}_V + (-x)\alpha = \bar{0}_V - (x\alpha)$$

and hence

$$(-x)\alpha = -(x\alpha).$$

Moreover, $x(\alpha + (-\alpha)) = x\bar{0}_V = \bar{0}_V$, by Lemma 2. Also $x(\alpha + (-\alpha)) = x\alpha + x(-\alpha)$ by 3(b). Hence $x\alpha + x(-\alpha) = \bar{0}_V$. Add $-(x\alpha)$ to both sides and we have

$$x(-\alpha) = -(x\alpha).$$

This completes the proof of the lemma.

Lemma 4. For every scalar x and for every finite set of vectors $\alpha_1, \ldots, \alpha_n$,

$$x(\alpha_1 + \alpha_2 + \cdots + \alpha_n) = x\alpha_1 + x\alpha_2 + \cdots + x\alpha_n.$$

Proof: The lemma is true for $n = 2$, by 3(b). Now use induction on n to prove it true for all $n \geq 2$.

Lemma 5. For every vector α and any finite number of scalars $x_1, \ldots, x_n$,

$$(x_1 + x_2 + \cdots + x_n)\alpha = x_1\alpha + x_2\alpha + \cdots + x_n\alpha.$$

Proof: By 3(a) the lemma holds for $n = 2$. Again induction on n yields an easy proof of this lemma.

We have defined what is called a "left" vector space over F; that is, in the scalar multiplication $x\alpha$ the scalar x is on the "left" of the vector α. Since multiplication in the field F is commutative we could just as well write αx for the product, defining $\alpha x = x\alpha$. We would then have a right vector space over F. It is easily verified that all the axioms are satisfied. For instance,

$$(\alpha x)y = y(\alpha x) = y(x\alpha) = (yx)\alpha = (xy)\alpha = \alpha(xy),$$

since $yx = xy$ in F.

Actually then, since F is a field, we have a "two-sided" vector space. It is very easy to show that $x\alpha \to \alpha x$ is an isomorphism (see Sec. 3-4) of the "left" with the "right" vector space over F.

Vector spaces can be defined over a division ring. This is an algebraic system which differs from a field only in that the multiplication is not required to be commutative. In other words, a division ring is a field if its multiplication is commutative. Since the multiplication is not assumed commutative in a division ring, there is a distinction to be made between left and right vector spaces over a division ring.

Example 1. Let F be any field. Let $V_n(F)$ be the set of all elements of the cartesian product $\prod_{i=1}^{n} F_i$ where each $F_i = F$. That is, an element of $V_n(F)$ has the form

$$(x_i) = (x_1, x_2, \ldots, x_n)$$

where $x_1, x_2, \ldots, x_n \in F$.

Define an equality in $V_n(F)$ by $(x_i) = (y_i)$ if and only if $x_i = y_i$, $i = 1, 2, \ldots, n$. It is easy to check that this equality is reflexive, symmetric, and transitive.

Define an addition in $V_n(F)$ by $(x_i) + (y_i) = (x_i + y_i)$. It is also easily proved that this addition is commutative and associative. Also the definition shows that $\bar{0} = (0, 0, \ldots, 0)$ behaves like a zero element.

Define a "scalar multiplication" in $V_n(F)$ by $c(x_i) = (cx_i)$, $c \in F$. Again it is quite easy to check that this definition does satisfy the conditions 3(a), 3(b), 3(c), 3(d) for scalar multiplication.

All the axioms are satisfied and $V_n(F)$ is a vector space over the given field F. We shall make frequent use of the vector space $V_n(F)$.

In particular, if F is the field of real numbers, $V_3(F)$ is the vector space over F whose elements are the triples (x_1, x_2, x_3) where the x_i are real numbers. This after all is a way of representing a point in three dimensions. We can relate it to the "arrows," mentioned earlier, by letting (x_1, x_2, x_3) be the coordinates of the tip of the arrow.

Example 2. Let S be any set and let F be any field. As usual denote by F^S the set of all mappings (functions) from S to F.

For $f, g \in F^S$, define $f + g$ by $s(f + g) = sf + sg$, for all $s \in S$. Let $\bar{0}$ be the mapping in F^S which maps every $s \in S$ into $0 \in F$. For $x \in F$, $f \in F^S$, define "scalar multiplication" in F^S by $s(xf) = x(sf)$. It is not hard to prove that with these definitions of vector addition and scalar multiplication F^S is a vector space over F. In this vector space, a mapping from $S \rightarrow F$ is a vector!

Example 3. In particular the set of all real functions whose domain is some interval, say $[0, 1]$, on the real line, is a vector space over the real field. In this vector space a real function becomes a vector!

Example 4. The set of real numbers is a vector space over itself as a field. Thus a real number is a vector!

Example 5. The "general solution" of a second-order homogeneous linear differential equation

$$\frac{d^2 y}{dx^2} + f(x) \frac{dy}{dx} + g(x)y = 0$$

has the form $y = c_1 y_1(x) + c_2 y_2(x)$, where c_1 and c_2 are arbitrary complex numbers and $y_1(x), y_2(x)$ are independent solutions. The set of all solutions $c_1 y_1(x) + c_2 y_2(x)$ is clearly a vector space over the field of complex numbers. What strange vectors these are!

Example 6. A polynomial $p(x)$ of degree n in a variable x over a field F is an expression of the form

$$a_0 + a_1 x + a_2 x^2 + \cdots + a_n x^n$$

where the $a_i \in F$ and $a_n \neq 0$.

Let $q(x)$ be a polynomial of degree m, $m \geq n$,

$$q(x) = b_0 + b_1 x + \cdots + b_m x^m, \quad b_i \in F, \quad b_m \neq 0.$$

Define $p(x) + q(x)$ by

$$p(x) + q(x) = (a_0 + b_0) + (a_1 + b_1)x + \cdots + (a_n + b_n)x^n \\ + b_{n+1} x^{n+1} + \cdots + b_m x^m$$

and for $c \in F$ define

$$cp(x) = ca_0 + ca_1 x + \cdots + ca_n x^n.$$

With these two definitions it is easy to check that the set of all polynomials over F is a vector space over F. The vectors here are polynomials.

EXERCISES

1. Find an algebraic system that satisfies all the axioms for a vector space except the axiom 3(d).

2. Prove that the set Z_p of all residue classes modulo a prime integer p is a field.

3. Find the total number of vectors in (a) the vector space $V_2(Z_5)$ and (b) the vector space $V_n(Z_p)$, where n is any positive integer and p is a prime integer.

4. Define a vector.

5. Write out complete proofs for Lemmas 4 and 5.

3-2 SUBSPACES

Definition. A nonempty subset U of a vector space V over F is called a **subspace** of V if U itself forms a vector space over F under the same operations as those of V.

THEOREM 1. A subgroup U of V is a subspace of V if U is closed under scalar multiplication.

Proof: Obvious.

THEOREM 2. A nonempty subset U of a vector space V over F is a subspace of V if (1) $\alpha, \beta \in U$ implies $\alpha + \beta \in U$, and (2) if $\alpha \in U$, $x \in F$, implies $x\alpha \in U$.

Proof: If $\alpha \in U$ then by (2), $(-1)\alpha = -\alpha \in U$. Thus U contains the inverse of α. This together with (1) proves U is a subgroup of V. From (2) it follows that U is closed under scalar multiplication. Hence by Theorem 1, U is a subspace of V.

A vector space V is trivially a subspace of itself. Also the subset of V consisting of the single element $\bar{0}_V$, the zero vector, is obviously a subspace of V. Moreover any subspace of a subspace of V is a subspace of V. A subspace $U \neq V$ and containing a nonzero vector is frequently called a **proper** subspace of V.

THEOREM 3. The intersection of a family V_i, $i \in I$, of subspaces of a vector space V over F is a subspace of V.

Proof: Let $\alpha, \beta \in \cap V_i$. Then α and β belong to each V_i and hence $\alpha + \beta$ belongs to each V_i and therefore $\alpha + \beta \in \cap V_i$. If $\alpha \in \cap V_i$ and $x \in F$, then $x\alpha$ belongs to each V_i and hence $x\alpha \in \cap V_i$. Thus by Theorem 2, $\cap V_i$ is a subspace of V.

Example 1. The following are five different subspaces of the vector space $V_n(F)$: all vectors of $V_n(F)$ of the forms

(1) $(0, x_2, x_3, \ldots, x_n)$. (2) $(x_1, x_1, x_1, \ldots, x_1)$.
(3) $(x_1, -x_1, x_2, x_3, \ldots, x_n)$. (4) $(x_1, 2x_1, x_3, \ldots, x_n)$.
(5) $(x_1, 0, 0, \ldots, 0)$.

Example 2. The following are three different subspaces of the vector space of all real functions on $[0, 1]$. See Example 3, Section 1. (1) all the real functions that are continuous (2) all those that have continuous first derivatives (3) all those that are constants, that is a function whose value $f(x) =$ some constant for all x of the interval.

Example 3. If m is a fixed positive integer, then the set of all polynomials of degree $\leq m$ is a subspace of the vector space of all polynomials. (See Example 6.) The set of all polynomials of degree m, however, is not a subspace.

If U and W are subspaces of a vector space V, denote by $U + W$ the set of all vectors of the form $\alpha + \beta$, $\alpha \in U$, $\beta \in W$. By Theorem 2 it follows that $U + W$ is a subspace of V. It is called the *linear sum* of U and W. Clearly $U + W$ contains both U and W. Moreover, any subspace of V containing U and W would obviously have to contain all the vectors of $U + W$. Hence $U + W$ is the smallest subspace of V that contains both U and W and so the linear sum $U + W$ can also be described as the intersection of all subspaces of V that contain both U and W.

Definition. If α_1, $\alpha_2, \ldots, \alpha_n$ is any finite set of vectors of a vector space V over F, an expression of the form

$$x_1 \alpha_1 + x_2 \alpha_2 + \cdots + x_n \alpha_n,$$

where the x_i are scalars, is called a **linear combination of the vectors** $\alpha_1, \alpha_2, \ldots, \alpha_n$. Note that a linear combination is always a finite sum.

If S is any subset (finite or infinite) of a vector space V then it is easy to prove that the set $L(S)$ of all linear combinations of vectors from S forms a substance of V. For the sum of two such linear combinations is a linear combination of vectors from S, and if a linear combination is multiplied by a scalar, then the product is a linear combination of vectors from S. Hence by Theorem 2 $L(S)$ is a vector space, a subspace of V.

Definition. If S is any subset of a vector space V, the subspace $L(S)$ of all linear combinations of vectors from S is called **the subspace spanned or generated by the set S.**

Clearly $S \subset L(S)$ and clearly $L(S)$ is the smallest subspace of V containing the subset S. The linear sum $U + W$ is the subspace generated by the set-union of the subspaces U and W.

For convenience, if ϕ is the empty set, we define $L(\phi) = \bar{0}_V$.

Lemma 6. If U is a subspace of a vector space V and if S is a subset of U then $L(S) \subseteq U$.

Proof: Any linear combination of vectors from S would have to be in U.

EXERCISES

1. Let V be an additive abelian group and let F be a field. Let scalar multiplication be defined in V by $x\alpha = \bar{0}_V$, for all $x \in F$ and all $\alpha \in V$, where $\bar{0}_V$ is the neutral element of the group. Check the axioms to determine if V is a vector space.

2. Show that in the axioms defining a vector space the field F can be replaced by a division ring, that is a "noncommutative" field.

3. Prove the uniqueness of the zero vector $\bar{0}_V$ and the uniqueness of the inverse $-\alpha$ of a vector $\alpha \in V$.

4. Define a vector and a scalar.

5. Prove $-(-\alpha) = \alpha$ for all $\alpha \in V$.

6. Prove

$$x(\alpha_1 + \alpha_2 + \cdots + \alpha_n) = x\alpha_1 + x\alpha_2 + \cdots + s\alpha_n,$$

$$(x_1 + x_2 + \cdots + x_n)\alpha = x_1\alpha + x_2\alpha + \cdots + x_n\alpha,$$

where the α_i are vectors and the x_i are scalars.

7. Show there are 5 ways of associating the four vectors $\alpha, \beta, \gamma, \delta$ to find their sum. Assuming the associative law for three vectors prove $\alpha + ((\beta + \gamma) + \delta) = \alpha + \beta + (\gamma + \delta)$.

8. State which of the following are subspaces of $V_3(F)$, where F is the real field, and describe them.
(a) The set of all vectors of the form $(x, x + y, y)$.
(b) The set of all vectors of the form $(x + 1, x, -x)$.
(c) The set of all vectors of the form $(x, x^2, 0)$.
(d) The set of all vectors of the form $(x + y + z, x - z, y + z)$.
(e) The set of all vectors (x, y, z) for which x, y, z are rational.

9. If S and T are subsets of a vector space V and if $S \subset T$ and $T \subset L(S)$, the subspace spanned by S, prove $L(S) = L(T)$.

10. Give the proofs for Examples 1, 2, 3, 6 of Sec. 3-1.

3-3 BASIS AND DIMENSION OF A VECTOR SPACE

Definition. A set S of vectors is called **dependent** if there are distinct vectors $\alpha_1, \alpha_2, \ldots, \alpha_k \in S$ and scalars $x_1, x_2, \ldots, x_k$, not all zero, such that

$$\sum_{i=1}^{k} x_i\alpha_i = x_1\alpha_1 + x_2\alpha_2 + \cdots + x_k\alpha_k = \bar{0}_V.$$

A set S of vectors is called **independent** if S is not dependent.

Thus the set S is dependent if a linear combination of vectors from S

can equal $\bar{0}_V$ without every scalar coefficient x_i having to be 0. Clearly any set S of vectors that contains the zero vector $\bar{0}_V$ is dependent.

If S is an independent set of vectors, then a linear combination of vectors from S can equal the zero vector $\bar{0}_V$ if and only if each of its scalar coefficients is zero.

Example 1. Let F be the real field. In the vector space $V_3(F)$ over F, the vectors $(1, 2, 3)$, $(-2, 1, 4)$ and $(-1, -\frac{1}{2}, 0)$ form a dependent set, since $4(1, 2, 3) - 3(-2, 1, 4) + 10(-1, -\frac{1}{2}, 0) = (0, 0, 0) = \bar{0}$.

On the other hand, the vectors $(1, -1, 0)$, $(2, 0, 1)$ and $(1, 1, -1)$ form an independent set in $V_3(F)$. For if $x, y, z \in F$, then

$$x(1, -1, 0) + y(2, 0, 1) + z(1, 1, -1) = (0, 0, 0) = \bar{0};$$

that is,

$$(x + 2y + z, -x + z, y - z) = (0, 0, 0)$$

if and only if $x + 2y + z = 0$, $-x + z = 0$ and $y - z = 0$. The only solution of these equations is $x = y = z = 0$.

Definition. A vector α is said to be **dependent** on a set S of vectors if $\alpha \in L(S)$.

Thus the zero vector is dependent on any nonempty set of vectors.

THEOREM 4. A set S of vectors is dependent if and only if there is a vector $\alpha \in S$ such that α is dependent on the other vectors of S.

Proof: Let $\alpha \in S$ and assume $\alpha \in L(S')$, where $S' = S \setminus \{\alpha\}$, the subset of S resulting from the removal of α from S. If $\alpha = \bar{0}$ then S is a dependent set. Assume $\alpha \neq \bar{0}$, then

$$\alpha = \sum_1^k x_i \beta_i,$$

where the x_i are scalars and $\beta_1, \ldots \beta_k$ are vectors in S'. Thus

$$\alpha - \sum_1^k x_i \beta_i = \bar{0}$$

and this is a linear combination of vectors of S whose scalar coefficients are not all zero. Hence by definition S is a dependent set of vectors.

Conversely, assume S is a dependent set. Then there exists a linear combination $\sum_1^k x_i \alpha_i$ of vectors of S such that

$$\sum_1^k x_i \alpha_i = \bar{0}, \qquad \alpha_i \in S,$$

and not all $x_i = 0$. Suppose $x_j \neq 0$, then solving for α_j,

$$\alpha_j = -x_j^{-1}(x_1\alpha_1 + \cdots + x_{j-1}\alpha_{j-1} + x_{j+1}\alpha_{j+1} + \cdots + x_k\alpha_k)$$

and hence α_j is dependent on the other vectors of S.

Lemma 7. Let S and T, where $S \subset T$, be subsets of a vector space V. If $S \subset T \subset L(S)$ then $L(T) = L(S)$.

Proof: Since $L(T)$ is the smallest subspace containing T, we have $L(T) \subsetneqq L(S)$. Let $\alpha \in L(S)$. Then α is a linear combination of vectors from S and hence α is a linear combination of vectors from T. Therefore $\alpha \in L(T)$ and so $L(S) \subsetneqq L(T)$. Thus $L(S) = L(T)$.

THEOREM 5. Every finite set S of vectors contains an independent subset S' such that $L(S) = L(S')$.

Proof: Use induction on the number of vectors in S. If S contains only one vector α, then if $\alpha = \bar{0}$, take S' to be the empty set (a subset of any set) and $L(S') = \bar{0}$. If $\alpha \neq \bar{0}$, take $S' = S$. Thus the theorem is true for a set S with only one element. Now assume it true for all sets containing less than k vectors. Let S be a set of k distinct vectors $\alpha_1, \alpha_2, \ldots, \alpha_k$. If S is independent, take $S' = S$, and we have $L(S) = L(S')$. If S is a dependent set, then for some j, α_j is dependent on the remaining vectors. By reindexing the α_i, if necessary, we can take $j = k$. Then α_k is dependent on the set S'' of vectors $\alpha_1, \alpha_2, \ldots, \alpha_{k-1}$. By the induction hypothesis, S'' contains an independent set S' such that $L(S') = L(S'')$. Since $\alpha_k \in L(S'')$, we have $S'' \subset S \subset L(S'')$. Therefore by Lemma 7, $L(S) = L(S'')$. Hence $L(S) = L(S')$. Thus the theorem is true for a set of k vectors. Hence we have shown by induction that the theorem is true for any finite set of vectors.

Lemma 8. If $\alpha_1, \ldots, \alpha_n$ is a sequence of vectors forming a dependent set, then for some j, $1 \leq j \leq n$, there exists a vector α_j such that α_j is dependent on (that is, is a linear combination of) the vectors $\alpha_1, \alpha_2, \ldots, \alpha_{j-1}$.

Proof: Being a dependent set of vectors, there exists a linear combination $\Sigma x_i\alpha_i$ of the vectors $\alpha_1, \ldots, \alpha_n$ such that $\Sigma x_i\alpha_i = \bar{0}$ and some of the scalar coefficients x_i are not 0. Let x_j be that coefficient with the largest subscript, which is not 0. Then solving for α_j,

$$\alpha_j = -x_j^{-1}(x_1\alpha_1 + \cdots + x_{j-1}\alpha_{j-1}).$$

THEOREM 6. In a vector space *spanned* (generated) by k vectors $\alpha_1, \ldots, \alpha_k$ there are at most k independent vectors. (That is, any independent set of vectors in the vector space is finite and contains at most k vectors.)

Proof: Note that this theorem differs from Theorem 5 which deals only in finite sets. If the field of scalars is infinite, this theorem asserts that the infinite set of vectors $L(\alpha_1, \alpha_2, \ldots, \alpha_k)$ contains an independent set of $k + 1$ vectors $\beta_1, \beta_2, \ldots, \beta_{k+1}$. Since

$$\beta_1 \in L, L = L(\beta_1, \alpha_1, \ldots, \alpha_k)$$

and the sequence of vectors $\beta_1, \alpha_1, \alpha_2, \ldots, \alpha_k$ is dependent. Hence, by Lemma 8, some vector of the sequence is dependent on its predecessors. This vector cannot be β_1, since β_1 belongs to an independent set and so $\beta_1 \neq \overline{0}$. If necessary we can reindex the α_i and assume α_k to be this vector. Then $\alpha_k \in L(\beta_1, \alpha_1, \ldots \alpha_{k-1})$ and $L = L(\beta_1, \alpha_1, \ldots, \alpha_{k-1})$.

Now $\beta_2 \in L$ and hence $L = L(\beta_2, \beta_1, \alpha_1, \ldots, \alpha_{k-1})$. Hence the sequence $\beta_2, \beta_1, \alpha_1, \ldots, \alpha_{k-1}$ is a dependent set. Again one of these vectors, and it cannot be β_2 or β_1 for they are independent vectors, is dependent on its predecessors. Again we can assume it to be α_{k-1}, and $\alpha_{k-1} \in L(\beta_2, \beta_1, \alpha_1, \ldots, \alpha_{k-2})$. Hence $L = L(\beta_2, \beta_1, \alpha_1, \ldots, \alpha_{k-2})$. Continuing in this way we add a β_i and simultaneously discard an α_j each time. Eventually, then, we reach $L = L(\beta_k, \beta_{k-1}, \ldots, \beta_1)$. Since $\beta_{k+1} \in L$, we have a contradiction of the assumption that the $k + 1$ vectors $\beta_1, \beta_2, \ldots, \beta_k$ constitute an independent set. Thus $L(\alpha_1, \ldots, \alpha_k)$ cannot contain an independent set of $k + 1$ vectors, which implies that any independent set of vectors in $L(\alpha_1, \ldots, \alpha_k)$ can contain at most k vectors.

Definition. A subset B of a vector space V over F is called a **basis** of V if B is an independent set and if $V = L(B)$.

Definition. A vector space V over F is said to be **finite-dimensional** if there is a finite subset S of vectors of V such that $V = L(S)$. A vector space that is not finite-dimensional is called **infinite-dimensional.**

A subspace U of a finite-dimensional vector space V is finite-dimensional. For if V is spanned by n vectors, then U can contain at most n independent vectors.

In Example 1, Sec. 3-1, it is not hard to see that the n vectors $(1, 0, 0, \ldots, 0), (0, 1, 0, \ldots, 0), \ldots, (0, 0, \ldots, 0, 1)$ span $V_n(F)$ and so $V_n(F)$ is a finite-dimensional space. In Example 5, Sec. 3-1 we see that the independent solutions $y_1(x)$ and $y_2(x)$ form a basis of two vectors for that vector space. However in Examples 3 and 4 it is not hard to show that the vector spaces are infinite-dimensional. In Example 2, F^S is an infinite dimensional vector space if and only if S is an infinite set.

It has been proved that every vector space, whether finite or infinite-dimensional, has a basis. However we shall prove here only that a finite dimensional vector space has a basis.

THEOREM 7. A finite-dimensional vector space V has a basis. Any basis of such a vector space is finite and all bases contain the same number of elements.

Proof: Since V is finite-dimensional, there is a finite set S' of vectors such that $V = L(S')$. By Theorem 5, S' contains an independent subset of vectors S such that $L(S) = L(S')$, and so S is a basis for V.

Let S'' be a second basis for V. By Theorem 6, S'' is a finite set. If the first basis S contains k vectors, then S'' contains $h \leq k$ vectors. But S is also an independent set in $L(S'')$ and therefore $k \leq h$. Thus $h = k$.

A direct and natural consequence of Theorem 7 is the following definition. The number of elements in a basis of a finite dimensional vector space is called the **dimension of the vector space.**

The next theorem demonstrates clearly the importance of a basis in a vector space.

THEOREM 8. If $\alpha_1, \ldots, \alpha_k$ is a basis for a vector space V over F, then every vector $\alpha \in V$ is a unique linear combination of the basis vectors.

Proof: Since $\alpha \in L(\alpha_1, \alpha_2, \ldots, \alpha_k)$, α is a linear combination of the α_i, $\alpha = \sum_{i=1}^{k} x_i \alpha_i$, $x_i \in F$. Suppose α has another such representa-tion, $\alpha = \sum_{i=1}^{k} y_i \alpha_i$, $y_i \in F$. Then $\sum_{1}^{k} (x_i - y_i) \alpha_i = \bar{0}_V$. Since the α_i are independent, each scalar coefficient must be 0. Hence $y_i = x_i$, $i = 1, 2, \ldots, k$. This proves the uniqueness.

Example 2. The vectors $\epsilon_1 = (1, 0, \ldots, 0)$, $\epsilon_2 = (0, 1, 0, \ldots, 0), \ldots$, $\epsilon_n = (0, 0, \ldots, 0, 1)$ form a basis called the **standard basis**, for the vector space $V_n(F)$, where F is the real field. Thus the dimension of $V_n(F)$ is n.

Example 3. The vector space of all polynomials in x over a field F of degrees $\leq m$ is a vector space of dimension $m + 1$. A basis of this vector space is the set of vectors $1, x, x^2, \ldots, x^m$.

Lemma 9. If U is a subspace of a finite-dimensional vector space V then any basis $\alpha_1, \alpha_2, \ldots, \alpha_k$ of U can be enlarged to a basis for V.

Proof: Let $\beta_1, \beta_2, \ldots, \beta_n$ be a basis for V. Write down the sequence of vectors $\alpha_1, \alpha_2, \ldots, \alpha_k, \beta_1, \beta_2, \ldots, \beta_n$. If β_1 depends on $\alpha_1, \ldots, \alpha_k$, cast it out and if not, retain it. Then on to β_2 and do the same thing. Continuing in this way, since every basis has the same number of vectors, we eventually add $n - k$ of the vectors from $\beta_1, \beta_2, \ldots, \beta_n$ to the vectors $\alpha_1, \ldots, \alpha_k$ to obtain a basis for V.

Note that a basis for a vector space V need not contain a basis for a subspace U of V.

EXERCISES

The notation $(x_1, x_2, \ldots, x_n)$ for a vector in $V_n(F)$ refers to the standard basis.

1. If U is a subspace of a finite-dimensional vector space, prove that U is finite-dimensional.

2. If U is a subspace of a vector space V, prove that dim U = dim V implies $U = V$.

3. Find the dimension of the subspace of $V_4(F)$, spanned by the four vectors $(1, 2, 1, 0)$, $(-1, 1, -4, 3)$, $(2, 3, 3, -1)$ and $(0, 1, -1, 1)$ when (a) F is the rational field, (b) F is the real field, and (c) F is the field Z_3 of integers modulo 3.

4. If dim $V = n$ and if S is a set of k vectors such that $L(S) = V$ prove $k \geq n$. If $k = n$ prove that the set S is independent.

5. Find the dimensions of the following vector spaces:

(a) The set of vectors $(x_1, x_2, \ldots, x_n)$ of $V_n(F)$, $x_i \in F$, such that
$$\sum_1^n x_i = 0.$$

(b) The set of vectors of $V_n(F)$ such that $\sum_1^k x_i = 0$, where $k < n$.

(c) The set of vectors of $V_n(F)$ such that $x_i = 0$ for $i < k$.

3-4 LINEAR TRANSFORMATIONS

Definition. Let V and W be vector spaces over the same field F. A mapping T of $V \to W$ is called a **linear transformation** if

(1) $(\alpha + \beta)T = \alpha T + \beta T$, for all $\alpha, \beta \in V$

(2) $(x\alpha)T = x(\alpha T)$, for all $x \in F$, and all $\alpha \in V$.

It follows by use of induction that if $\alpha = \Sigma x_i \alpha_i$ is a linear combination of vectors from V, then $\alpha T = (\Sigma x_i \alpha_i)T = \Sigma x_i (\alpha_i T)$.

The value αT of T for $\alpha \in V$ is called the **image** of α under T. A linear transformation of $V \to W$ is also called a **homomorphism**. It may be injective or surjective or neither. If T is bijective then it is an **isomorphism** and we know that there then exists a mapping T^{-1} from $W \to V$, called the inverse of T. We shall show soon that T^{-1} is an isomorphism of $W \to V$.

Definition. If T is a linear transformation of $V \rightarrow W$, then the **kernel** of T, ker T, is the set of all vectors $\alpha \in V$ such that $\alpha T = \overline{0}_W$, the zero vector of W. We denote by *im* T the set of all vectors $\alpha' \in W$ such that $\alpha' = \alpha T$ for at least one vector $\alpha \in V$. Clearly if T is surjective then *im* $T = W$. Another notation for *im* T is VT.

THEOREM 9. If T is a linear transformation of $V \rightarrow W$ then (1) $\overline{0}_V T = \overline{0}_W$ (2) if U is a subspace of V then UT is a subspace of W.

Proof: (1) For $\alpha \in V$, $(\alpha + \overline{0}_V)T = \alpha T$. Also $(\alpha + \overline{0}V)T = \alpha T + \overline{0}_V T$. Hence $\alpha T + \overline{0}_V T = \alpha T$. Add the inverse $-(\alpha T)$ of αT to both sides and we get $\overline{0}_V T = \overline{0}_W$.

(2) Let $\alpha', \beta' \in UT$. Then there exist $\alpha, \beta \in U$ such that $\alpha' = \alpha T$, $\beta' = \beta T$. Hence $\alpha' + \beta' = (\alpha + \beta)T$. U is a subspace and therefore $\alpha + \beta \in U$. Hence $\alpha' + \beta' \in UT$. Next for $x \in F$, $\alpha' \in UT$, we have $\alpha' = \alpha T$, where $\alpha \in U$. Hence $x\alpha' = x(\alpha T) = (x\alpha)T$. Since U is a subspace $x\alpha \in U$, hence $x\alpha' \in UT$. By Theorem 2, this proves UT is a subspace of W.

Corollary. *im* T is a subspace of W.

THEOREM 10. If T is a linear transformation from $V \rightarrow W$, then ker T is a subspace of V.

Proof: Let $\alpha, \beta \in$ ker T. Then $(\alpha + \beta)T = \alpha T + \beta T = \overline{0}_W$. Hence $\alpha + \beta \in$ ker T. Next if $x \in F$ and if $\alpha \in$ ker T, then $x(\alpha T) = x\overline{0}_W = \overline{0}_W$, that is $(x\alpha)T = \overline{0}_W$. Therefore $x\alpha \in$ ker T. Thus again by Theorem 2, ker T is a subspace of V.

THEOREM 11. A linear transformation T from $V \rightarrow W$ is injective if and only if ker $T = \overline{0}_V$.

Proof: Suppose T is injective. Let $\alpha \in$ ker T. Then $\alpha T = \overline{0}_W$. By Theorem 9, $\overline{0}_V T = \overline{0}_W$. Hence $\alpha = \overline{0}_V$ and so ker $T = \overline{0}_V$.

Conversely, let ker $T = \overline{0}_V$. Assume $\alpha T = \beta T$ for $\alpha, \beta \in V$. Then $(\alpha - \beta)T = \overline{0}_W$. Hence $\alpha - \beta = \overline{0}_V$ and therefore $\alpha = \beta$. This proves T is injective.

THEOREM 12. If T is an isomorphism $V \rightarrow W$ then T^{-1} is an isomorphism of $W \rightarrow V$.

Proof: First we show T^{-1} is a linear transformation. Let $\alpha', \beta' \in W$. Then there exist unique vectors $\alpha, \beta \in V$ such that $\alpha' = \alpha T$, $\beta' = \beta T$. Hence $\alpha' + \beta' = \alpha T + \beta T = (\alpha + \beta)T$. Thus

$$(\alpha' + \beta')T^{-1} = \alpha + \beta = \alpha' T^{-1} + \beta' T^{-1}.$$

Next if $\alpha' \in W$ and $x \in F$ then $\alpha' = \alpha T$ for $\alpha \in V$, and hence $x\alpha' = (x\alpha)T$. Therefore $x\alpha = (x\alpha')T^{-1}$. Thus $(x\alpha')T^{-1} = x\alpha = x(\alpha'T^{-1})$. Hence T^{-1} is a linear transformation.

Clearly T^{-1} is bijective, since T is an isomorphism. Hence T^{-1} is an isomorphism.

THEOREM 13. Let $\alpha_1, \ldots, \alpha_n$ be a basis for the finite-dimensional vector space V over F, and let $\beta_1, \ldots, \beta_n$ be arbitrary vectors in a vector space W over F. Then there exists a unique linear transformation T from $V \to W$ such that $\alpha_i T = \beta_i, i = 1, 2, 3, \ldots, n$.

Proof: For any vector $\alpha \in V, \alpha = \sum_{i=1}^{n} x_i\alpha_i, x_i \in F$. Define a mapping T of $V \to W$ by $\alpha T = \sum_{i=1}^{n} x_i\beta_i$. T can be shown easily to be a linear transformation and clearly $\alpha_i T = \beta_i, i = 1, 2, \ldots, n$.

Moreover T is unique. For let T' be a second linear transformation of $V \to W$, with the same requirement that $\alpha_i T' = \beta_i, i = 1, 2, \ldots, n$. If α is any vector of V, then $\alpha = \Sigma x_i\alpha_i, x_i \in F$, and we have $\alpha T' = \Sigma x_i\beta_i$. Hence $\alpha T' = \alpha T$ for every $\alpha \in V$, and therefore $T = T'$.

Lemma 10. If T is an injective linear transformation of $V \to W$, then if $\alpha_1, \ldots, \alpha_k$ are independent vectors of V, $\alpha_1 T, \alpha_2 T, \ldots, \alpha_k T$ are independent vectors of W.

Proof: If $x_1(\alpha_1 T) + \cdots + x_k(\alpha_k T) = \bar{0}_W$ where the x_i are scalars, then $(x_1\alpha_1 + \cdots + x_k\alpha_k)T = \bar{0}_W$ and, since T is injective, $x_1\alpha_1 + \cdots + x_k\alpha_k = \bar{0}_V$. The α_i are independent vectors and so $x_1 = x_2 = \cdots = x_k = 0$. This proves $\alpha_1 T, \ldots, \alpha_k T$ form an independent set of vectors.

THEOREM 14. If V is a finite-dimensional vector space and if T is a linear transformation $V \to W$, then $\dim V = \dim(\ker T) + \dim(im\, T)$.

Proof: Let $\dim V = k$. We know $\ker T$ is a subspace of V. Let $\alpha_1, \alpha_2, \ldots, \alpha_j$ be a basis for $\ker T$, so that $\dim \ker T = j$. By Lemma 9, this basis extends to a basis $\alpha_1, \ldots, \alpha_j, \alpha_{j+1}, \ldots, \alpha_k$ of V. We are going to show that $\alpha_j T, \ldots, \alpha_k T$ is a basis for $im\, T$.

Let $\alpha \in V$. Then

$$\alpha = \sum_{i=1}^{k} x_i\alpha_i$$

and hence

$$\alpha T = \sum_{i=1}^{k} x_i(\alpha_i T).$$

Now $\alpha_i T = \overline{0}_W$ for $i = 1, 2, \ldots, j$, and hence

$$\alpha T = \sum_{i=j+1}^{k} x_i(\alpha_i T).$$

This proves that the vectors $\alpha_j T, \ldots, \alpha_k T$ span the subspace *im T* of *W*. We next show they are independent vectors. Assume

$$\sum_{i=j+1}^{k} y_i(\alpha_i T) = \overline{0}_W, \quad y_i \in F.$$

Then

$$\sum_{i=j+1}^{k} (y_i \alpha_i) T = \overline{0}_W$$

and therefore

$$\sum_{i=j+1}^{k} y_i \alpha_i \in \ker T.$$

Since $\alpha_1, \ldots, \alpha_j$ is a basis for ker *T*, we have

$$\sum_{i=j+1}^{k} y_i \alpha_i = \sum_{i=1}^{j} x_i \alpha_i, \quad x_i \in F.$$

This yields

$$\sum_{1}^{j} x_i \alpha_i - \sum_{j+1}^{k} y_i \alpha_i = \overline{0}_V.$$

Since the $\alpha_1, \ldots, \alpha_j, \alpha_{j+1}, \ldots, \alpha_i$ are a basis for *V*, this linear combination can vanish if and only if all the scalar coefficients $x_i = 0$ and $y_i = 0$. Thus

$$\sum_{i=j+1}^{k} y_i(\alpha_i T) = \overline{0}_W$$

implies that all the $y_i = 0$. Hence the $\alpha_i T$, $i = j + 1, \ldots, k$ are independent vectors and, since they span *im T*, they are a basis for *im T*. Hence dim $(im\ T) = k - j$. This proves the theorem.

Corollary. If two finite-dimensional vector spaces *V* and *W* are isomorphic then dim *V* = dim *W*.

Proof: Let *T* be the isomorphism of $V \to W$. Then ker $T = \overline{0}_V$ and hence dim(ker *T*) = 0. Also *im T* = *W*. Hence dim *V* = dim *W*.

THEOREM 15. Let V be a vector space over F of dimension n. Then V is isomorphic to the vector space $V_n(F)$.

Proof: Let $\alpha_1, \ldots, \alpha_n$ be a basis for V. Then every vector $\alpha \in V$ has a unique representation in the form

$$\alpha = \sum_{i=1}^{n} x_i \alpha_i, \qquad x_i \in F.$$

Define a mapping T of $V \to V_n(F)$ by $\alpha T = (x_1, x_2, \ldots, x_n)$ where $\alpha = \sum_{i=1}^{n} x_i \alpha_i, \quad x_i \in F$.

It is a very simple exercise, left to the reader, to show that T is a linear transformation and that it is bijective.

Corollary. Two finite-dimensional vector spaces of the same dimension are isomorphic.

Proof: Isomorphism is a transitive property.

According, then, to Theorem 15, every finite-dimensional vector space V over F is isomorphic to the vector space $V_n(F)$, where $n = \dim V$. This theorem then determines, up to isomorphism, all finite-dimensional vector spaces. Every finite-dimensional vector space can be regarded as a vector space of the type $V_n(F)$.

EXERCISES

1. Prove that the mapping T defined in Theorem 15 is an isomorphism.

2. If T is a linear transformation $V \to W$ of two vector spaces where V is a finite dimensional vector space, prove that ker T and im T are finite dimensional spaces.

If in addition W is a finite dimensional space, show
(a) if T is injective, then $\dim V \leq \dim W$,
(b) if T is surjective, then $\dim V \geq \dim W$,
(c) if T is bijective, then $\dim V = \dim W$.

3-5 SETS OF LINEAR TRANSFORMATIONS

Let V and W be vector spaces over the same field F. We denote by Hom (V, W) the set of all linear transformations of $V \to W$. For $T_1, T_2 \in$ Hom (V, W), define $T_1 + T_2$ by $\alpha(T_1 + T_2) = \alpha T_1 + \alpha T_2$, for all $\alpha \in V$. It is easy to verify that, as defined, $T_1 + T_2$ is a linear transformation of $V \to W$. Hence $T_1 + T_2 \in$ Hom (V, W). It is also easily verified that this addition is commutative and associative. The constant linear transformation T_0 of $V \to W$ maps every α of V into $\overline{0}_W$. For $T \in$ Hom (V, W), $x \in F$, define xT by $\alpha(xT) = x(\alpha T)$, for all $\alpha \in V$ and all $x \in F$. It can be checked that this defines a scalar multiplication in Hom (V, W), which together with the definition above of vector addition makes Hom (V, W) a vector space over F. The reader should prove that all the axioms for a vector space are satisfied. We now have a vector space whose vectors are linear transformations. Its zero vector is T_0.

If in particular $W = V$ then we are able to define the product $T_1 T_2$, where now $T_1, T_2 \in$ Hom (V, V). For $\alpha \in V$, we define this product by $\alpha(T_1 T_2) = (\alpha T_1)T_2$. Now we show that $T_1 T_2 \in$ Hom (V, V). For $\alpha, \beta \in V$ we have, $(\alpha + \beta)T_1 T_2 = (\alpha T_1 + \beta T_1)T_2 = (\alpha T_1)T_2 + (\beta T_1)T_2 = \alpha(T_1 T_2) + \beta(T_1 T_2)$. Also $(x\alpha)T_1 T_2 = ((x\alpha)T_1)T_2 = (x(\alpha T_1))T_2 = x(\alpha T_1 T_2)$. Hence $T_1 T_2 \in$ Hom (V, V). We now have defined a multiplication in the vector space Hom (V, V). This multiplication is known to be associative, but it is not in general commutative. It is also readily proved that this multiplication obeys the following two distributive laws:

For $T_1, T_2, T_3 \in$ Hom (V, V),

$$T_1(T_2 + T_3) = T_1 T_2 + T_1 T_3,$$

$$(T_1 + T_2)T_3 = T_1 T_3 + T_2 T_3.$$

With these definitions of addition and multiplication of linear transformations the set Hom (V, V) forms an algebraic system known as a ring. Since, in addition to being a ring, Hom (V, V) is also a vector space, Hom (V, V) forms an algebraic system that is known as an algebra over the field F. We refer the reader to Chapter VIII for a discussion of algebras.

Definition. A linear transformation of a vector space V to itself is called an **endomorphism** or **linear operator** of V.

Thus Hom (V, V) is the set of all endomorphisms of V and Hom (V, V) is an algebra over F.

EXERCISES

1. Let T be the mapping of $V_3(F) \to V_3(F)$, F the real field, defined by

$$(x, y, z)T + (x + y + z, x - y, -z).$$

(a) Prove T is a linear transformation.
(b) Prove T is an automorphism.
(c) Find T^{-1} and T^2.

2. Let S and T be the endomorphisms of the vector space $V_2(F)$, where F is the real field, defined by

$$(x, y)S = (2x + y, x - y), \quad (x, y)T = (x, x + 3y).$$

Exhibit in a similar form the following endomorphisms of

$$V_2(F): \ S + T, ST, TS, S, -S \text{ and } S^2.$$

3. If two finite-dimensional vector spaces are isomorphic, prove they have the same dimension.

4. Find a linear transformation of $V_2(f) \rightarrow V_3(F)$ that is (a) injective, (b) not injective.

5. T is the linear transformation of $V_3(F) \rightarrow V_3(F)$ defined by $(x, y, z)T = (0, x + y, z)$. Find the kernel of T and the image of T and find their dimensions. If U is the subspace of $V_3(F)$ of all vectors of the form $(x, 0, z)$, find UT and $T^{-1}U$. What are their dimensions?

6. Prove that multiplication in Hom (V, V) is associative and that it does satisfy the two distributive laws.

7. T is a linear transformation $V \rightarrow W$ of two vector spaces and the subset B of V is a basis for V. What can be said about the subset BT of W if T is (a) injective, (b) surjective, (c) bijective.

3-6 THE MATRIX OF A LINEAR TRANSFORMATION

Suppose now that V and W are finite-dimensional vector spaces over the same field F, and let dim $V = m$, dim $W = n$.

Let T be a linear transformation of $V \rightarrow W$. If $\alpha_1, \alpha_2, \ldots, \alpha_m$ is a basis for V and $\beta_1, \beta_2, \ldots, \beta_n$ a basis for W, then for $i = 1, 2, \ldots, m$,

$$(1) \qquad \alpha_i T = \sum_{j=1}^{n} a_{ij}\beta_j, a_{ij} \in F.$$

The $m \times n$ rectangular array

$$(a_{ij}) = \begin{bmatrix} a_{11} & a_{12} & \cdots & a_{1n} \\ a_{21} & a_{22} & \cdots & a_{2n} \\ \cdots\cdots\cdots\cdots\cdots\cdots\cdots \\ \cdots\cdots\cdots\cdots\cdots\cdots\cdots \\ a_{m1} & a_{m2} & \cdots & a_{mn} \end{bmatrix}$$

of elements of F is called the **m $\times$ n matrix over F** of the linear transformation T with respect to the α-basis for V and the β-basis for W. Of course any change in these bases would result in a different matrix for T. If these bases are held fixed then each T determines a unique matrix and conversely to each $m \times n$ matrix (a_{ij}) over F (that is, its elements are in F) corresponds a unique linear transformation T determined by (1).

Thus a linear transformation has a matrix representation relative to a given choice of bases for V and W.

Let α be any vector in V. Then

$$\alpha = \sum_1^m x_i \alpha_i$$

where $x_i \in F$ and

$$\alpha T = \sum_1^m x_i (\alpha_i T) = \sum_{i=1}^m x_i \sum_{j=1}^n a_{ij} \beta_j$$

which we write in the form

(2)
$$\alpha T = \sum_{j=1}^n \sum_{i=1}^m x_i a_{ij} \beta_j \ .$$

Now $\alpha T \in W$, and therefore

(3)
$$\alpha T = \sum_{j=1}^n y_j \beta_j, \qquad y_j \in F.$$

Equating (2) and (3) we get a system of n equations

(4)
$$y_j = \sum_{i=1}^m x_i a_{ij}, \qquad j = 1, 2, \ldots, n$$

relating the scalars x_i of the vector α in V with the scalars y_i of the transformed vector αT in W. Thus a linear transformation is a function given by a system of linear equations.

If T is an isomorphism of two finite-dimensional vector spaces, T is called a **nonsingular** linear transformation, and the matrix of T relative to a given choice of bases is called a **nonsingular matrix.** Moreover T has an inverse T^{-1} and its matrix with respect to the same choice of bases is called the inverse of the matrix of T.

Note well that the rectangular array of the coefficients of the x_i in (4) is NOT the matrix (a_{ij}) of the linear transformation T but the *transpose* matrix of (a_{ij}). This means it is obtained from the matrix (a_{ij}) by interchanging rows and columns.

Let us now specialize our discussion to endomorphisms of a finite-dimensional vector space V.

If $\alpha_1, \alpha_2, \ldots, \alpha_n$ is a basis for V and if T is an endomorphism of V, then

$$\alpha_i T = \sum_{j=1}^n a_{ij} \alpha_j,$$

where (a_{ij}) is the $n \times n$ matrix of T relative to this particular choice of basis. Thus the endomorphisms of an n-dimensional vector space can be represented as $n \times n$ matrices with respect to any given basis of the space.

Example 1. From (4) we see that an endomorphism T of the vector space $V_n(F)$ always has the form

$$(x_1, x_2, \ldots, x_n)T = \left(\sum_1^n a_{1i}x_i, \sum_1^n a_{2i}x_i, \ldots, \sum_1^n a_{ni}x_i \right),$$

where all $a_{ij} \in F$.

Thus if T is an endomorphism of $V_2(F)$, then

$$(x_1, x_2)T = (a_{11}x_1 + a_{12}x_2, a_{21}x_1 + a_{22}x_2),$$

where the a_{ij} are scalars.

Definition. An endomorphism of a vector space V that is bijective is called an **automorphism of V.**

Let S and T be endomorphisms of the n-dimensional vector space V. With respect to some given basis of V, let (b_{ij}) be the matrix of S and (a_{ij}) the matrix of T. Then $S + T$ is an endomorphism with the matrix $(a_{ij} + b_{ij})$. If we define the sum of two $n \times n$ matrices by

$$(a_{ij}) + (b_{ij}) = (a_{ij} + b_{ij}),$$

then $S + T$ has a matrix which is the sum of the matrices of S and T.

Now TS, the product (map composition) of T and S, is an endomorphism of V. Let us find the matrix of TS.

$$\alpha_i(TS) = (\alpha_i T)S = \sum_{j=1}^n a_{ij}(\alpha_j S)$$

$$= \sum_{j=1}^n a_{ij} \sum_{k=1}^n b_{jk}\alpha_k = \sum_{k=1}^n \sum_{j=1}^n a_{ij}b_{jk}\alpha_k$$

and we see that (c_{ik}), where

$$c_{ik} = \sum_{j=1}^n a_{ij}b_{jk},$$

is the matrix of TS.

The $n \times n$ matrix (c_{ik}) is called the *product* $(a_{ij})(b_{ij})$ of the matrices (a_{ij}) and (b_{ij}) and gives the rule for the multiplication of matrices. Thus c_{ik} is itself the element in the ith row and kth column of the product matrix. We see that c_{ik} is obtained by multiplying the elements a_{i1}, $a_{i2}, \ldots, a_{in}$ of the ith row of (a_{ij}) by the elements $b_{1k}, b_{2k}, \ldots, b_{nk}$ of the kth column of (b_{ij}), so that

$$c_{ik} = a_{i1}b_{1k} + a_{i2}b_{2k} + \cdots + a_{in}b_{nk}.$$

If T is an automorphism of V, then T^{-1} is an automorphism of V. Now $TT^{-1} = T^{-1}T = 1_V$, the identity mapping of $V \to V$. Also $\alpha_i 1_V = \alpha_i$, $i = 1, 2, \ldots, n$, and hence the matrix of 1_V is

$$I = \begin{bmatrix} 1 & 0 & 0 & \ldots & 0 \\ 0 & 1 & 0 & \ldots & 0 \\ 0 & 0 & 0 & \ldots & 1 \end{bmatrix}$$

which is called the **identity matrix.**

It follows then that if (a_{ij}) is the matrix of T and if $(a_{ij})^{-1}$ denotes the matrix of T^{-1}, then $(a_{ij})(a_{ij})^{-1} = I = (a_{ij})^{-1}(a_{ij})$.

The matrix $(a_{ij})^{-1}$ is called the **inverse** of the matrix (a_{ij}). An endomorphism T of V has an inverse T^{-1} if and only if T is an automorphism. Hence a matrix (a_{ij}) has an inverse $(a_{ij})^{-1}$ if and only if it is the matrix of an automorphism with respect to some basis of V.

We have seen that Hom (V, V) is a ring and the $n \times n$ matrices over the field F form a ring. To each endomorphism $T \in$ Hom (V, V) corresponds a unique $n \times n$ matrix (a_{ij}), with respect to some fixed basis chosen for V. Since $T + S$ corresponds to the sum and TS to the product of the matrices of T and S, it is not hard to see that there is an isomorphism of the ring of endomorphisms of an n-dimensional vector space with the ring of $n \times n$ matrices over the scalar field of the vector space.

When T is an endomorphism of an n-dimensional vector space V over F, we defined, in Sec. 3-5, the endomorphism cT where $c \in F$. Clearly then if the matrix of T relative to some basis for V is (a_{ij}) then the matrix of cT is (ca_{ij}), that is every element of the matrix (a_{ij}) is multiplied by c. This serves to define $c(a_{ij})$, and we have

$$c(a_{ij}) = (ca_{ij}).$$

Furthermore, with this definition of a scalar multiplication of a matrix we see that the set of all $n \times n$ matrices over a field F is also a vector space, as well as a ring. Hence it is an algebra over F.

With respect to some given basis, every endomorphism of V can be represented as a matrix. However, whenever possible, it is always simpler (and much more elegant) to work with the endomorphisms themselves. Mathematicians with a passion for matrices may not admit this.

EXERCISES

1. Find the matrix of the endomorphism T of $V_3(F)$ with respect to the basis $(1, 0, 0), (0, 1, 0)$ and $(0, 0, 1)$, where T is defined by

$$(x, y, z)T = (2x - z, x + y + z, z - y).$$

2. Prove that T in Exercise (1) is an automorphism and find the matrix of T^{-1}. Next verify that the product of the matrices of T and T^{-1} is the identity matrix.

3. Let V be a 4-dimensional vector space with the basis $\alpha_1, \alpha_2, \alpha_3, \alpha_4$. An endomorphism T of V is defined by

$$(x_1\alpha_1 + x_2\alpha_2 + x_3\alpha_3 + x_4\alpha_4)T = (2x_1 + x_2)\alpha_1 - 3x_3\alpha_2 + x_4\alpha_3 + x_2\alpha_4.$$

Find the matrix of T with respect to this basis. Is T an automorphism?

3-7 CHANGE OF BASIS

Let V be a finite-dimensional vector space, dim $V = n$. Let T be an endomorphism of V and let (a_{ij}) be the matrix of T relative to the basis $\alpha_1, \alpha_2, \ldots, \alpha_n$ for V. If a change of basis is made in V from $\alpha_1, \alpha_2, \ldots, \alpha_n$ to a new basis $\beta_1, \beta_2, \ldots, \beta_n$, what is the matrix of T relative to this new basis?

By Theorem 13, there exists a unique endomorphism P of V such that $\alpha_i P = \beta_i$. Now

$$\beta_i = \sum_{j=1}^{n} p_{ij}\alpha_j, \qquad p_{ij} \in F, \quad i = 1, 2, \ldots, n.$$

Hence (p_{ij}) is the matrix of P relative to the basis $\alpha_1, \alpha_2, \ldots, \alpha_n$. It is easy to see that P is an automorphism of V, that is a bijective endomorphism of V. Hence P has an inverse P^{-1} which is also an automorphism of V.

We have $\alpha_i T = \Sigma a_{ij}\alpha_j$ and $\alpha_i P = \Sigma p_{ij}\alpha_j$, and hence

$$\alpha_i(PT) = \sum_{j=1}^{n}\left(\sum_{k=1}^{n} p_{ik}a_{kj}\right)\alpha_j.$$

Now $PTP^{-1} \in$ Hom (V, V). Let its matrix relative to the basis $\alpha_1, \ldots, \alpha_n$ be (c_{ij}). Then

$$(5) \qquad\qquad \alpha_i(PTP^{-1}) = \sum_{j=1}^{n} c_{ij}\alpha_j.$$

Applying P to (5), we get

$$\alpha_i(PT) = (\alpha_i P)T = \sum_{j=1}^{n} c_{ij}(\alpha_j P).$$

Since $\alpha_i P = \beta_i$ we have

$$\beta_i T = \sum_{j=1}^{n} c_{ij}\beta_j.$$

Thus (c_{ij}) is the matrix of T with respect to the new basis $\beta_1, \beta_2,$ $\ldots, \beta_n$, and

(6) $$(c_{ij}) = (p_{ij})(a_{ij})(p_{ij})^{-1}.$$

Two $n \times n$ matrices (a_{ij}) and (c_{ij}) are said to be **similar** if there exists a nonsingular matrix (p_{ij}) such that (6) is true. We have proved

THEOREM 16. If (a_{ij}) is the matrix of an endomorphism T of a finite-dimensional vector space and if (p_{ij}) is the nonsingular matrix defining a change of basis, then the similar matrix $(p_{ij})(a_{ij})(p_{ij})^{-1}$ is the matrix of T relative to the new basis.

Let T be an endomorphism of an n-dimensional vector space V. If $\alpha_1, \alpha_2, \ldots, \alpha_n$ and $\beta_1, \beta_2, \ldots, \beta_n$ are two bases for V, then the endomorphism P of V determined by $\beta_i = \alpha_i P$, $i = 1, 2, \ldots, n$, is an automorphism of V, and hence P has an inverse P^{-1}. Clearly P^{-1} is simply the automorphism given by $\beta_i P^{-1} = \alpha_i$. Now Theorem 16 simply states that the matrix of T with respect to the basis $\beta_1, \beta_2, \ldots, \beta_n$ is equal to the matrix of the endomorphism PTP^{-1} with respect to the basis $\alpha_1, \alpha_2,$ $\ldots, \alpha_n$.

In determining the matrix of an endomorphism of the vector space $V_n(F)$ with respect to some basis, it is important to bear in mind that when we write a vector in $V_n(F)$ as $(x_1, x_2, \ldots, x_n)$, the x_i are the coordinates of the vector with respect to the standard basis $\epsilon_1 = (1, 0, 0, \ldots, 0)$, $\epsilon_2 = (0, 1, 0, \ldots, 0), \ldots, \epsilon_n = (0, 0, \ldots, 0, 1)$. This means

$$(x_1, x_2, \ldots, x_n) = x_1 \epsilon_1 + x_2 \epsilon_2 + \cdots + x_n \epsilon_n.$$

Example. Consider the endomorphism T of $V_2(F)$, F the real field, defined by $(xy)T = (y - x, 2x + y)$. We find that $(1, 0)T = (-1, 2)$, $(0, 1)T = (1, 1)$. Hence the matrix of T with the respect to the standard basis is

$$\begin{bmatrix} -1 & 2 \\ 1 & 1 \end{bmatrix}.$$

Suppose now we choose another basis $(3, -1)$, $(-4, 5)$ for $V_2(F)$. Then

$$(3, -1)T = (-4, 5) = 17(3, -1) + 11(-5, 2)$$

$$(-4, 5)T = (7, -8) = -26(3, -1) - 17(-5, 2).$$

Thus the matrix of T with respect to this new basis is

$$\begin{bmatrix} -17 & 11 \\ -26 & -17 \end{bmatrix}.$$

To illustrate Theorem 16, let us calculate the matrices of the automorphism P and its inverse P^{-1} with respect to the standard basis.

Since P is given by $(1,0)P = (3,-1)$, $(0,1)P = (-5,2)$, the matrix of P with respect to the standard basis is

$$\begin{bmatrix} 3 & -1 \\ -5 & 2 \end{bmatrix}.$$

Now

$$(x,y)P = [x(1,0) + y(0,1)]P = x[(1,0)P] + y[(0,1)P]$$
$$= (3x,-x) + (-5y,2y) = (3x - 5y, 2y - x).$$

Hence

$$(3x - 5y, 2y - x)P^{-1} = (x,y).$$

For $x = 2, y = 1$, we get

$$(1,0)P^{-1} = (2,1).$$

For $x = 5, y = 3$, we get

$$(0,1)P^{-1} = (5,3).$$

Hence the matrix of P^{-1} with respect to the standard basis is $\begin{bmatrix} 2 & 1 \\ 5 & 3 \end{bmatrix}$, and we verify fast that

$$\begin{bmatrix} 3 & -1 \\ -5 & 2 \end{bmatrix}\begin{bmatrix} -1 & 2 \\ 1 & 1 \end{bmatrix}\begin{bmatrix} 2 & 1 \\ 5 & 3 \end{bmatrix} = \begin{bmatrix} 17 & 11 \\ -26 & -17 \end{bmatrix},$$

as Theorem 16 states it should.

3-8 QUOTIENT SPACES

Let U be a subspace of a vector space V over F. Then U is a subgroup of the additive abelian group V and hence V/U is an additive abelian quotient or factor group. Its elements have the form $\alpha + U$, $\alpha \in V$. For $x \in F$ let us define

$$x(\alpha + U) = x\alpha + U.$$

Note that if $\alpha + U = \beta + U$, then $\alpha - \beta \in U$, hence $x(\alpha - \beta) = x\alpha - x\beta \in U$ and therefore $x\alpha + U = x\beta + U$. Hence $x(\alpha + U) = x(\beta + U)$. This shows that the product $x(\alpha + U)$ is well-defined.

It is easily verified that this product satisfies the axioms for a scalar multiplication in the additive abelian group V/U. Hence with this definition V/U becomes a vector space over F. It is called the **quotient** or **factor space** of V with respect to the subspace U.

The mapping T of $V \to V/U$ defined by $\alpha T = \alpha + U$, $\alpha \in V$, is readily proved to be a linear transformation which is also surjective. T is called the **natural** or **canonical homomorphism** of $V \to V/U$.

THEOREM 17. If V is a finite-dimensional vector space and U is a subspace of V, then

$$\dim V = \dim U + \dim (V/U).$$

Proof: Consider the natural homomorphism T of $V \to V/U$. By Theorem 14 we have $\dim V = \dim(\ker T) + \dim(\operatorname{im} T)$. Here $\ker T = U$ and $\operatorname{im} T = V/U$. This proves the theorem.

Definition. A vector space V is said to be the **(internal) direct sum** $V = U \oplus W$ of two subspaces U and W if (1) $V = U + W$, and (2) $U \cap W = \bar{0}_V$.

Lemma 11. $V = U \oplus W$ if and only if each $\alpha \in V$ has a unique representation in the form $\alpha = \beta + \gamma$, where $\beta \in U$ and $\gamma \in W$.

Proof: Assume $V = U \oplus W$. Suppose $\alpha \in V$ has two such representations $\alpha = \beta + \gamma$, $\beta \in U$, $\gamma \in W$, and $\alpha = \beta' + \gamma'$, $\beta' \in U$, $\gamma' \in W$. Then $\beta + \gamma = \beta' + \gamma'$, and hence

$$\beta - \beta' = \gamma - \gamma' \in W.$$

However, $\beta - \beta' \in U$. Therefore $\beta - \beta' = \bar{0}_V$, that is $\beta = \beta'$, and hence $\gamma = \gamma'$. Conversely, assume each $\alpha \in V$ has a unique representation in the form $\alpha = \beta + \gamma$, $\beta \in U$, $\gamma \in W$. Then $V = U + W$. Let $\alpha \in U \cap W$. Then $\alpha = \bar{0}_V + \alpha = \alpha + \bar{0}_V$ would be two distinct representations of α in the given form. This is a contradiction unless $\alpha = \bar{0}_V$.

The concept of direct sum can be generalized to any finite number of subspaces.

Definition. A vector space V is said to be the *direct sum* $V = U_1 \oplus U_2 \oplus \cdots \oplus U_n$ of the n subspaces $U_1, U_2, \ldots, U_n$ if (1) $X = U_1 + U_2 + \cdots + U_n$ (2) the intersection of each U_i with the subspace generated by the remaining U_j is $\bar{0}_V$.

Lemma 12. If V is any vector space and if $V = U \oplus W$, then V/U and W are isomorphic vector spaces.

Proof: Now every vector $\alpha \in V$ has a unique representation in the form $\alpha = \beta + \gamma$, $\beta \in U$, $\gamma \in W$.

Define a mapping of $V/U \to W$ by $\alpha + U \to \gamma$, where $\alpha = \beta + \gamma$. This mapping is easily shown to be an isomorphism.

Lemma 13. If V is a finite-dimensional vector space and if

$$V = U \oplus W,$$

then

$$\dim V = \dim U + \dim W.$$

Proof: Since $W \approx V/U$, we obtain, by use of the corollary to Theorem 14 and by Theorem 17, a proof of the lemma.

Definition. An endomorphism P of a vector space V is called a **projection** if $PP = P$.

Lemma 14. If P is a projection of V, then

$$V = \ker P \oplus \text{im } P.$$

Proof: For any $\alpha \in V$, $\alpha = \alpha P + (\alpha - \alpha P)$. $\alpha P \in \text{im } P$. Since $(\alpha - \alpha P)P = \alpha P - \alpha PP = \overline{0}_V$, $\alpha - \alpha P \in \ker P$. Hence every vector of V is the sum of a vector of im P and a vector of ker P— that is, $V = \text{im } P + \ker P$. Let $\alpha \in \text{im } P \cap \ker P$. Then $\alpha = \beta P$ for some $\beta \in V$ and $\alpha P = \overline{0}_V$. Hence $\alpha = \beta P = \beta PP = \alpha P = \overline{0}_V$. Therefore im $P \cap \ker P = \overline{0}_V$. Thus $V = \text{im } P + \ker P$.

Example 1. If $V = U \oplus W$, then for each $\alpha \in V$ we have a unique representation of α in the form $\alpha = \beta + \gamma$, $\beta \in U$, $\gamma \in W$. It is easy to see that P_U defined by $\alpha P_U = \beta$ and P_W defined by $\alpha P_W = \gamma$ are projections. This can very simply be generalized to give n projections when V is the direct sum of n subspaces of V.

THEOREM 18. If U is a subspace of a finite-dimensional vector space V, then there exists a subspace W of V such that $V = U \oplus W$.

Proof: Let $\alpha_1, \ldots, \alpha_k$ be a basis for U. By Lemma 9 we can complete this to a basis $\alpha_1, \ldots, \alpha_k, \alpha_{k+1}, \ldots, \alpha_n$ for V. Let W be the subspace of V spanned by the vectors $\alpha_{k+1}, \ldots, \alpha_n$. Then clearly

$$V = U + W.$$

Furthermore $U \cap W = \overline{0}_V$. For if $\alpha \in U \cap W$, then

$$\alpha = a_1 \alpha_1 + \cdots + a_k \alpha_k \text{ and } \alpha = a_{k+1} \alpha_{k+1} + \cdots + a_n \alpha_n.$$

Hence $a_1 \alpha_1 + a_k \alpha_k - a_{k+1} \alpha_{k+1} - \cdots - a_n \alpha_n = \overline{0}_V$. Since the α_i are independent vectors, all the $a_i = 0$ and hence $\alpha = \overline{0}_V$. Hence

$$U \cap W = \overline{0}_V.$$

This completes the proof that $V = U \oplus W$.

EXERCISES

The notation $(x_1, \ldots, x_n)$ for a vector in $V_n(F)$ refers to the standard basis.

1. Let F be the real field. The matrix relative to the standard basis of an endomorphism T of $V_3(F)$ is

$$\begin{bmatrix} 2 & 0 & 0 \\ 0 & 1 & 2 \\ 0 & -1 & -2 \end{bmatrix}.$$

Find the kernel of T and its dimension. Find the codomain (range) of T and its dimension.

2. An endomorphism T of $V_3(F)$, F the real field, is defined by $(x, y, z)T = (2x - y, z, x + y)$.
   (a) Find the matrix A of T relative to the standard basis.
   (b) Find the matrix B of T relative to the basis

$$(1, 1, 0), \quad (1, 0, -1), \quad (1, 1, 1).$$

   (c) Find the matrix P for which $B = PAP^{-1}$.

3. Give an example of an endomorphism T of $V_3(F)$, F the real field, whose kernel is the plane spanned by the vectors $(1, 0, 1)$ and $(0, 1, 0)$ and whose image is the line spanned by the vector $(1, 0, -1)$.

4. Let T be the endomorphism of $V_3(F)$, F the real field, whose matrix with respect to the standard basis is

$$\begin{bmatrix} 0 & 2 & -2 \\ 1 & 1 & -4 \\ 0 & 1 & -1 \end{bmatrix}.$$

   (a) Find the kernel W_1 of T.
   (b) Find the matrix of the endomorphism T^2.
   (c) If I is the identity automorphism of $V_3(F)$ prove that the mapping $T^2 + I$ defined by $\alpha(T^2 + I) = \alpha T^2 + \alpha I = \alpha T^2 + \alpha$ is an endomorphism of $V_3(F)$.
   (d) Find the kernel W_2 of the endomorphism $T_2 + I$.
   (e) Prove $V = W_1 \oplus W_2$.

5. If U is a subspace of a vector space V, prove that the mapping T of $V \rightarrow V/U$ defined by $\alpha T = \alpha + U, \alpha \in V$, is a surjective linear transformation. Next prove that the mapping of $V/U \rightarrow VT$ defined by $\alpha + U \rightarrow \alpha T$ is an isomorphism. What is the kernel of T?

6. If V is a finite-dimensional vector space and if $V = U \oplus W$, prove that dim $V = $ dim $U + $ dim W.

3-9 THE INNER PRODUCT

Definition. An **inner product** in a vector space over the *real* field F is a mapping (function) f of the cartesian product $V \times V \to F$, denoted by $(\alpha, \beta)f = \alpha \cdot \beta \in F$ for all α and β of V, which has the following properties:

    I. (Symmetry) $\alpha \cdot \beta = \beta \cdot \alpha$ for all
      $\alpha, \beta \in V$.

    II. (Bilinearity)
      (a) $(x\alpha) \cdot \beta = x(\alpha \cdot \beta) = \alpha \cdot (x\beta)$ for all
         $x \in F$ and all $\alpha, \beta \in V$.
      (b) $(\alpha + \beta) \cdot \gamma = \alpha \cdot \gamma + \beta \cdot \gamma$ and
         $\gamma \cdot (\alpha + \beta) = \gamma \cdot \alpha + \gamma \cdot \beta$
         for all $\alpha, \beta, \gamma \in V$.

    III. (Positive Definiteness)
$$\alpha \cdot \alpha \geq 0 \text{ for all } \alpha \in V, \quad \text{and } \alpha \cdot \alpha = 0$$

if and only if $\alpha = \bar{0}_V$.

Thus an inner product in V is a function which assigns to each pair α and β of vectors of V a real number, denoted by $\alpha \cdot \beta$, and satisfying the above axioms. If an inner product is defined in a vector space V, then V is called an **inner product vector** space.

It follows at once from axiom II(b) that $\bar{0}_V \cdot \alpha = 0 = \alpha \cdot \bar{0}_V$ for all $\alpha \in V$.

It also follows easily by induction that the properties assumed in axiom II(b) can be generalized to

$$(\alpha_1 + \alpha_2 + \cdots + \alpha_n) \cdot \gamma = \alpha_1 \cdot \gamma + \alpha_2 \cdot \gamma + \cdots + \alpha_n \cdot \gamma$$

$$\gamma \cdot (\alpha_1 + \alpha_2 + \cdots + \alpha_n) = \gamma \cdot \alpha_1 + \gamma \cdot \alpha_2 + \cdots + \gamma \cdot \alpha_n.$$

Throughout this section, unless otherwise stated, the field F of scalars is taken to be the field of real numbers.

THEOREM 19. Let V be an n-dimensional vector space over F with a basis $\alpha_1, \alpha_2, \ldots, \alpha_n$. Then for $\alpha, \beta \in V$,

$$\alpha \cdot \beta = \sum_{i=1}^{n} a_i b_i$$

where $\alpha = \sum_{1}^{n} a_i \alpha_i$ and $\beta = \sum_{1}^{n} b_i \alpha_i$, is an inner product in V.

Proof: Axioms I and II are easily seen to be satisfied. Since $\alpha \cdot \alpha = \sum_{1}^{n} a_i^2$, and since the a_i are real numbers, axiom (III) is satisfied. For $\sum a_i^2 = 0$ if and only if each $a_i = 0$, that is if and only if $\alpha = \bar{0}_V$.

Example 1. In the vector space $V_3(F)$, F the real field, let $\alpha = (x_1, x_2, x_3)$ and $\beta = (y_1, y_2, y_3)$, then $\alpha \cdot \beta = x_1 y_1 + x_2 y_2 + x_3 y_3$ is an inner product in $V_3(F)$.

Example 2. A very interesting example of an inner product occurs in analysis. The set V of all continuous real functions defined on some real interval $[a, b]$, is an infinite-dimensional vector space over the real field. For $f, g \in V$, define

$$f \cdot g = \int_a^b f(x)g(x)d(x).$$

The axioms for an inner product can be shown to hold for this definition.

Definition. The *length* or *norm of a vector* α of an inner product vector space V is denoted by $|\alpha|$ and defined by

$$|\alpha| = + \sqrt{\alpha \cdot \alpha}.$$

Since $\alpha \cdot \alpha \geq 0$ this definition is meaningful. Thus $|\alpha| \geq 0$, and $|\alpha| = 0$ if and only if $\alpha = \bar{0}_V$.

Definition. The *distance* between two vectors α and β is defined to be $|\alpha - \beta| = \sqrt{(\alpha - \beta) \cdot (\alpha - \beta)} = \sqrt{\alpha \cdot \alpha - 2\alpha \cdot \beta + \beta \cdot \beta}$.

Example 3. In $V_3(F)$, F the real field, the vectors $\alpha = (2, -3, 6)$ and $\beta = (-1, 0, 4)$ have lengths $|\alpha| = 7$, $|\beta| = \sqrt{17}$. The distance between them is $|\alpha - \beta| = |(3, -3, 2)| = \sqrt{22}$.

If x is a real number, then $|x\alpha| = \sqrt{(xa) \cdot (x\alpha)} = \sqrt{x^2(\alpha \cdot \alpha)} = |x| \, |\alpha|$. In particular then $|-\alpha| = |\alpha|$. Here $|x|$ stands for the absolute value of the real number x.

THEOREM 20 (Schwarz Inequality). For any pair of vectors α and β,

$$|\alpha \cdot \beta| \leq |\alpha| \, |\beta|.$$

Proof: If either α or β is the zero vector, the statement is merely $\bar{0}_V \leq \bar{0}_V$, and so is true. Now assume α and β are nonzero vectors. For any scalars x, y,

$$0 \leq |x\alpha + y\beta|^2 = x^2|\alpha|^2 + 2xy(\alpha \cdot \beta) + y^2|\beta|^2.$$

Putting $x = |\beta|$, $y = |\alpha|$ and cancelling $2|\alpha| \, |\beta| \neq 0$, we get $-(\alpha \cdot \beta) \leq |\alpha| \, |\beta|$. Putting $x = |\beta|$, $y = -|\alpha|$ we find in the same way that

$$\alpha \cdot \beta \leq |\alpha| \, |\beta|.$$

Thus $|\alpha \cdot \beta| \leq |\alpha| \, |\beta|$.

Using the Schwarz inequality, we can prove the following inequalities:

(A) If α and β are any vectors, then

$$|\alpha + \beta| \leq |\alpha| + |\beta|.$$

(B) If α, β, γ are any three vectors, then

$$|\alpha - \beta| \leq |\alpha - \gamma| + |\gamma - \beta| \qquad \text{(Triangle Inequality)}.$$

To prove (A), we have

$$|\alpha + \beta|^2 = |\alpha|^2 + 2\alpha \cdot \beta + |\beta|^2 \leq$$
$$|\alpha|^2 + 2|\alpha||\beta| + |\beta|^2 = (|\alpha| + |\beta|)^2.$$

Hence

$$|\alpha + \beta| \leq |\alpha| + |\beta|.$$

(B) follows from (A) at once, for

$$\alpha - \beta = (\alpha - \gamma) + (\gamma - \beta)$$

and therefore

$$|\alpha - \beta| = |(\alpha - \gamma) + (\gamma - \beta)| \leq |\alpha - \gamma| + |\gamma - \beta|.$$

When V is a vector space with an inner product over the real field, we have defined the distance $d(\alpha, \beta)$ between two vectors to be

$$d(\alpha, \beta) = |\alpha - \beta|.$$

This distance function satisfies the four axioms for such a function: (1) $|\alpha - \beta| > 0$ if $\alpha \neq \beta$, (2) $|\alpha - \beta| = 0$ if and only if $\alpha = \beta$, (3) $|\alpha - \beta| = |\beta - \alpha|$, (4) the triangle inequality, $|\alpha - \beta| \leq |\alpha - \gamma| + |\gamma - \beta|$. Thus V becomes what is known as a **metric space.** The inner product led to a definition of the length $|\alpha|$ of a vector. $|\alpha|$ is also called the **norm** of α and a linear space with a norm is called a **normed linear space.**

Two very important normed linear spaces are Banach space and Hilbert space. They are *complete* normed linear spaces. A metric space is called *complete* if every Cauchy sequence in the spaces converges to a limit in the space. The real numbers, but not the rationals alone, form a complete linear space.

A great deal of the material developed in this chapter can be generalized to theorems in an infinite-dimensional linear space, of which one of the outstanding examples is Hilbert space.

EXERCISES

1. If α and β are vectors in an inner product vector space, prove that $|\alpha \cdot \beta| = |\alpha||\beta|$ if and only if α and β are linearly dependent.

2. If α and β are vectors in an inner product vector space, prove

$$2(\alpha \cdot \beta) = |\alpha + \beta|^2 - |\alpha|^2 - |\beta|^2.$$

3. If $\alpha = (x, y)$ and $\beta = (u, v)$ are vectors in $V_2(F)$, determine whether each of the following definitions of an inner product is acceptable:

(a) $\alpha \cdot \beta = (x + y)(u + v)$.
(b) $\alpha \cdot \beta = (x + y)(u + v) + vy$.
(c) $\alpha \cdot \beta = (x + u)(y + v)$.
(d) $\alpha \cdot \beta = (x + u)(y + v) + uv$.

3-10 ORTHONORMAL BASIS

Throughout this section we shall assume V to be an inner product vector space.

Definition. The **angle** θ, $0 \le \theta \le \pi$, between two nonzero vectors α and β is defined by

$$\cos \theta = \frac{\alpha \cdot \beta}{|\alpha| \, |\beta|}.$$

By the Schwarz inequality, $|\alpha \cdot \beta| \le |\alpha| \, |\beta|$, and hence $-|\alpha| \, |\beta| \le \alpha \cdot \beta \le |\alpha| \, |\beta|$, so that

$$-1 \le \frac{\alpha \cdot \beta}{|\alpha| \, |\beta|} \le 1.$$

This justifies our definition.

Definition. Two vectors α and β are said to be *orthogonal* if $\theta = 90°$.

Thus $\alpha \perp \beta$ if and only if $\alpha \cdot \beta = 0$.

Definition. A vector α is *normal* if $|\alpha| = 1$, that is if the length of α is 1.

Definition. A *basis* for a vector space V is called **orthogonal** if each pair of distinct vectors of the basis is orthogonal. An orthogonal basis is called **orthonormal** if each vector of the basis is normal.

Lemma 15. If U is a subspace of a vector space V over F, then the set $U^\perp$ of all vectors of V that are orthogonal to every vector of U, is a subspace of V.

Proof: For a, $\beta \in U^\perp$ and $\gamma \in U$, we have $\alpha \cdot \gamma = 0$, $\beta \cdot \gamma = 0$ and hence $(\alpha + \beta) \cdot \gamma = 0$. Thus $\alpha + \beta \in U^\perp$. Also if $x \in F$, then for $\alpha \in U^\perp$, $\gamma \in U$, $(x\alpha) \cdot \gamma = x(\alpha \cdot \gamma) = 0$. Hence $x\alpha \in U^\perp$ By Theorem 2 this proves $U^\perp$ is a subspace.

Definition. The subspace $U^\perp$ is called the **orthogonal complement** of U.

THEOREM 21. There exists an orthogonal basis for any finite-dimensional vector space.

Proof: Let $\alpha_1, \alpha_2, \ldots, \alpha_n$ be a basis for the finite-dimensional vector space V. We want to find an orthogonal basis $\beta_1, \beta_2, \ldots, \beta_n$ for V. For convenience, define

$$c_j^i = \frac{\beta_i \cdot \alpha_j}{|\beta_i|^2}, \quad i, j = 1, 2, \ldots, n.$$

We start by taking $\beta_1 = \alpha_1$. Hence $L(\beta_1) = L(\alpha_1)$. Next choose the scalar x so that $(\alpha_2 - x\beta_1) \cdot \beta_1 = 0$. We find $x = c_2^1$. Take $\beta_2 = \alpha_2 - c_2^1 \beta_1$. Then $\beta_2 \cdot \beta_1 = 0$ and $L(\beta_1, \beta_2) = L(\alpha_1, \alpha_2)$. Now choose scalars x_1 and x_2 so that the vector $\alpha_2 - x_1\beta_1 - x_2\beta_2$ is orthogonal to β_1 and to β_2. We find $x_1 = c_3^1$, $x_2 = c_3^2$. Next take $\beta_3 = \alpha_3 - c_3^1\beta_1 - c_3^2\beta_2$. Then $\beta_1, \beta_2, \beta_3$ are orthogonal vectors, and clearly $L(\beta_1, \beta_2, \beta_3) = L(\alpha_1, \alpha_2, \alpha_3)$.

Now use induction on k and assume $\beta_1, \beta_2, \ldots, \beta_k$ are orthogonal vectors given by

(7) $\beta_i = \alpha_i - c_i^1\beta_1 - c_i^2\beta_2 \cdots - c_i^{i-1}\beta_i, \, i = 1, 2, \ldots, k$

and such that $L(\alpha_1, \ldots, \alpha_k) = L(\beta_1, \ldots, \beta_k)$.

Determine scalars $x_1, x_2, \ldots, x_k$ such that

$$\gamma = \alpha_{k+1} - x_1\beta_1 - x_2\beta_2 - \cdots - x_k\beta_k$$

is orthogonal to each of the vectors $\beta_1, \beta_2, \ldots, \beta_k$. Then for each j from 1 to k, we get

$$\beta_j \cdot \gamma = \beta_j \cdot \alpha_{k+1} - x_j(\beta_j \cdot \beta_j) = 0;$$

that is,

$$x_j = c_{k+1}^j, j = 1, 2, \ldots, k.$$

Hence $\beta_{k+1} = \alpha_{k+1} - c_{k+1}^1\beta_1 - c_{k+1}^2\beta_2 - \cdots c_{k+1}^k\beta_k$ is a vector orthogonal to $\beta_1, \beta_2, \ldots, \beta_k$. Moreover, it is clear that $L(\beta_1, \beta_2, \ldots, \beta_{k+1}) = L(\alpha_1, \alpha_2, \ldots, \alpha_{k+1})$. Our induction is completed and proves that $\beta_1, \beta_2, \ldots, \beta_n$, where each β_i, $i = 1, 2, \ldots, n$, is given by (7), is an orthogonal set of vectors which span V, that is $L(\beta_1, \ldots, \beta_n) = L(\alpha_1, \ldots, \alpha_n)$. Hence $\beta_1, \beta_2, \ldots, \beta_n$ form an orthogonal basis for V.

Corollary 1. There exists an orthonormal basis for a finite-dimensional vector space V.

Proof: Simply normalize each vector $\beta_1, \beta_2, \ldots, \beta_n$ of an orthogonal basis for V and we obtain an orthonormal basis $\dfrac{\beta_1}{|\beta_1|}, \dfrac{\beta_2}{|\beta_2|}, \ldots, \dfrac{\beta_n}{|\beta_n|}$ for V.

Corollary 2. If $\alpha_1, \ldots, \alpha_k$ is an orthonormal basis for a subspace U of a vector space V of dimension n, then this basis can be enlarged to an orthonormal basis for V.

Proof: Let $U^\perp$ be the set of all vectors orthogonal to every vector of U. Now $U^\perp$ has an orthonormal basis $\beta_1, \beta_2, \ldots, \beta_h$. Then $\alpha_1, \ldots, \alpha_k, \beta_1, \ldots, \beta_h$ is an orthonormal set of vectors that is a basis for V and $h + k = n$.

If $\alpha_1, \alpha_2, \ldots, \alpha_n$ is an orthogonal basis for V then for every vector $\alpha \in V$ we have

$$\alpha = \sum_{i=1}^{n} x_i \alpha_i, \quad x_i \in F.$$

Hence

$$\alpha \cdot \alpha_j = \left(\sum_{i=1}^{n} x_i \alpha_i \right) \cdot \alpha_j = x_j(\alpha_j \cdot \alpha_j)$$

and so

$$x_j = \frac{\alpha \cdot \alpha_j}{|\alpha_j|^2}, \quad j = 1, 2, \ldots, n.$$

Hence with respect to an orthogonal basis $\alpha_1, \alpha_2, \ldots, \alpha_n$ every vector α can be expressed uniquely by

$$\alpha = \sum_{i=1}^{n} \frac{\alpha \cdot \alpha_i}{|\alpha_i|^2} \alpha_i.$$

If the basis is orthonormal, we have

(8) $$\alpha = \sum_{i=1}^{n} (\alpha \cdot \alpha_i) \alpha_i.$$

The reader should note well this last equation, which gives the form of a vector α with respect to an orthonormal basis. It will be used frequently in subsequent work.

Let $\alpha_1, \alpha_2, \ldots, \alpha_n$ be an orthonormal basis for a vector space V.

Let $\alpha \cdot \beta \in V$. Then $\alpha = \sum_{1}^{n} a_i \alpha_i, \beta = \sum_{1}^{n} b_i \alpha_i$. Therefore

$$\alpha \cdot \beta = \sum_{1}^{n} a_i b_i.$$

Thus we can see that, **relative to an orthonormal basis, the inner product has this unique form.**

THEOREM 22. If U is a subspace of a finite-dimensional vector space V, then $V = U \oplus U^{\perp}$.

Proof: Choose an orthonormal basis $\beta_1, \ldots, \beta_k$ for U. Let $\alpha \in V$. Then

$$\alpha = \sum_{1}^{k} (\alpha \cdot \beta_i)\beta_i + \left(\alpha - \sum_{1}^{k} (\alpha \cdot \beta_i)\beta_i \right).$$

Consider the vector $\gamma = \alpha - \sum_{i=1}^{k} (\alpha \cdot \beta_i)\beta_i$. Now

$$\gamma \cdot \beta_j = (\alpha \cdot \beta_j) - (\alpha \cdot \beta_j)(\beta_j \cdot \beta_j) = 0, \qquad \text{for } j = 1, 2, \cdots, k.$$

Hence γ is orthogonal to each vector of the basis $\beta_1, \beta_2, \ldots, \beta_k$ for U and hence $\gamma \in U^{\perp}$. Thus each vector $\alpha \in V$ is the sum of a vector from U and a vector from $U^{\perp}$. Moreover $U \cap U^{\perp} = \overline{0}_V$. Hence $V = U \oplus U^{\perp}$.

If $V = U \oplus U^{\perp}$ then every $\alpha \in V$ has a unique representation in the form $\alpha = \beta + \gamma$, $\beta \in U$, $\gamma \in U^{\perp}$. The linear transformation P of $V \to U$ defined by $\alpha P = \beta$ is called an **orthogonal projection** and β is called the *orthogonal projection of α on U.*

EXERCISES

1. If we propose in $V_2(F)$, F the real field, an inner product defined by $\alpha \cdot \beta = (x_1 + y_1)(x_2 + y_2)$ where $\alpha = (x_1, y_1)$ and $\beta = (x_2, y_2)$, test all the axioms to ascertain if this proposal is acceptable.

2. If U is a finite-dimensional subspace of an inner product vector space V, prove $(U^{\perp})^{\perp} = U$. (Recall that $V = U \oplus U^{\perp}$.)

3. Starting with the independent vectors $\alpha_1 = (0, 1, 1)$, $\alpha_2 = (-1, 0, 0)$, $\alpha_3 = (2, 1, 0)$ construct an orthogonal basis $\beta_1, \beta_2, \beta_3$ for $V_3(F)$, where F is the real field.

4. If V is an inner product vector space, when would $|\alpha + \beta| = |\alpha| + |\beta|$?

5. Let F be the real field and let $V_3(F)$ be the vector space with the standard or usual inner product. If U is the subspace of $V_3(F)$ spanned by the vector $(1, 2, 0)$, find the subspace $U^{\perp}$ and verify that $V_3(F) = U \oplus U^{\perp}$. Express the vector $(2, 3, 4)$ as the unique sum of a vector of U and a vector of $U^{\perp}$.

6. Let V be an inner product vector space. If $V = U \oplus W$, show that the mapping P defined by $\alpha P = \beta$ where $\alpha = \beta + \gamma \in V$, $\beta \in U$, $\gamma \in W$ is well-defined and is linear. The vector β is called the *projection of α on U along W.* If $W = U^{\perp}$, β is called the *orthogonal projection* of α on U. Prove that for any subspace of V, a vector of V has an orthogonal projection on the subspace.

7. If $\alpha_1, \ldots, \alpha_k$ are orthonormal vectors of an inner product vector space V, prove the orthogonal projection of a vector $\alpha \in V$ on

$$U = L(\alpha_1, \alpha_2, \ldots, \alpha_k)$$

is given by

$$\sum_1^k (\alpha \cdot \alpha_i)\alpha_i.$$

8. Prove that for every finite-dimensional real vector space an inner product exists.

9. $V_3(F)$ is a vector space over the real field with the usual inner product. If U is the subspace spanned by the vectors $(1, 0, 2)$ and $(-1, 1, 1)$, find the orthogonal projection of the vector $(2, 1, 0)$ on U.

10. Let V be the vector space of continuous real functions on the closed interval $[0, 1]$ with the inner product

$$f \cdot g = \int_0^1 f(x)g(x)\,dx, \qquad f, g \in V.$$

Find two orthonormal vectors of V that span the same subspace of V as the vectors f and g, where $xf = 1$, $xg = x$ for all $x \in [0, 1]$.

3-11 ISOMETRIES

Definition. Let V and W be vector spaces over the real field F with inner products. A linear transformation T of $V \to W$ is called an **isometry** if (1) im $T = W$ (2) $(\alpha T) \cdot (\beta T) = \alpha \cdot \beta$, for all $\alpha, \beta \in V$.

Thus an isometry is a surjective linear transformation that preserves the inner product. An isometry will therefore preserve all derivatives of the inner product, such as length, angle, distance, and hence, in particular orthogonality. An isometry of a vector space to itself is called an **orthogonal transformation**, and the matrix of an orthogonal transformation is called an **orthogonal matrix**.

If T is an isometry of $V \to W$ then the vector spaces V and W are said to be **isometric**.

THEOREM 23. An isomorphism T of $V \to W$ is an isometry if and only if T preserves the inner product.

Proof: If T is an isomorphism and preserves the inner product then from the definition, T is an isometry. Conversely, assume T is an isometry. Then T is surjective. Let α be any vector in ker T. Then $\alpha \cdot \alpha = (\alpha T) \cdot (\alpha T) = 0$. Hence $\alpha = \bar{0}_V$, and so by Theorem 11, T is injective. Therefore T is an isomorphism.

Corollary. If T is an isometry then T^{-1} is an isometry.

Proof: Since T is an isomorphism, we know T^{-1} is an isomorphism of $W \to V$. For $\alpha', \beta' \in W$, let $\alpha = \alpha' T^{-1}, \beta = \beta' T^{-1}$. Hence

$$(\alpha' T^{-1}) \cdot (\beta' T^{-1}) = \alpha \cdot \beta = (\alpha T) \cdot (\beta T) = \alpha' \cdot \beta'.$$

and therefore T^{-1} is an isometry.

THEOREM 24. If V and W are n-dimensional vector spaces with inner products, then there exists an isometry T of $V \to W$.

Proof: Let $\alpha_1, \ldots, \alpha_n$ and $\alpha_1', \ldots, \alpha_n'$ be orthonormal bases for V and W respectively. Then by Theorem 13 there exists an isomorphism T of $V \to W$ such that $\alpha_i T = \alpha_i', i = 1, 2, \ldots, n$.

If α and β are any vectors in V, then

$$\alpha = \sum_1^n a_i \alpha_i, \qquad \beta = \sum_1^n b_i \alpha_i,$$

and

$$\alpha T = \sum_1^n a_i \alpha_i', \qquad \beta T = \sum_1^n b_i \alpha_i'.$$

Now $\alpha \cdot \beta = \sum_1^n a_i b_i$ and $(\alpha T) \cdot (\beta T) = \sum_1^n a_i b_i$.   Hence

$$\alpha \cdot \beta = (\alpha T) \cdot (\beta T)$$

and therefore T is an isometry.

Theorem 24 demonstrates that all n-dimensional vector spaces with inner products are essentially the same—that is, such a vector space is unique up to isometry.

According to the Corollary to Theorem 23, orthogonal transformations have inverses and these inverses are orthogonal transformations. Consequently orthogonal matrices are nonsingular. It is quite easy to check that the product $T_1 T_2$ of two orthogonal transformations T_1 and T_2 preserves the inner product and hence is an orthogonal transformation. Moreover the identity map is an orthgonal transformation. We see then that the orthogonal matrices form a group under multiplication (map composition). It is called the **orthogonal group**.

EXERCISES

1. If T is a linear transformation $V \to W$ of two vector spaces which preserves lengths of vectors, that is $|\alpha T| = |\alpha|$ for every $\alpha \in V$, is T necessarily an isometry?

2. F is the real field. An endomorphism T of $V_3(F)$ is defined by

$$(1,0,0)T = (1,0,0), (0,1,0)T = \left(0, \frac{1}{2}, \frac{\sqrt{3}}{2}\right), (0,0,1)T = \left(0, \frac{-\sqrt{3}}{2}, \frac{1}{2}\right).$$

(a) Find $(x,y,z)T$, where (x,y,z) is an arbitrary vector.
(b) Prove T is an isometry.

3-12 DUAL VECTOR SPACES

The notion of a dual or conjugate space is a very important one in both algebra and analysis. We give here a brief introduction to the dual of a finite-dimensional vector space.

Definition. If V is any vector space over a field F, then a linear mapping f of $V \to F$ is called a **linear form** or **linear functional** on V.

This means $(\alpha + \beta)f = \alpha f + \beta f$ for all $\alpha, \beta \in V$ and $(x\alpha)f = x(\alpha f)$ for all $x \in F$ and $\alpha \in V$. The values of a linear form are the scalars of the field.

We can introduce an addition in the set of all linear forms on V by defining $f + g$ by

$$\alpha(f + g) = \alpha f + \alpha g, \quad \text{for all } \alpha \in V$$

and a scalar multiplication $xf, x \in F$, by

$$\alpha(xf) = x(\alpha f), \quad \alpha \in V.$$

The axioms for a vector space are actually satisfied and with these definitions the set of all linear forms on V becomes a vector space over F. It is called the **dual space** of V and is denoted by V^*.

From now on we deal exclusively with the dual of a finite-dimensional vector space.

First observe that if V is an n-dimensional vector space with the basis $\alpha_1, \alpha_2, \ldots, \alpha_n$ then any $f \in V^*$ is completely determined by its values $\alpha_i f, i = 1, 2, \ldots, n$, on the basis vectors. For if $\alpha \in V$, $\alpha = \sum_{1}^{n} a_i \alpha_i$ and $\alpha f = \sum_{1}^{n} a_i(\alpha_i f)$. For arbitrary values c_i of F, there is a unique $f \in V^*$ such that $\alpha_i f = c_i, i = 1, 2, \ldots, n$. The mapping f of $V \to F$ defined by

$$(a_1\alpha_1 + \cdots + a_n\alpha_n)f = a_1 c_1 + \cdots + a_n c_n$$

is easily shown to be linear and to be unique.

THEOREM 25. If dim $V = n$, then dim $V^* = n$.

Proof: Let $\alpha_1, \alpha_2, \ldots, \alpha_n$ be a basis for V. For each i from 1 to n, define $f_i \in V^*$ by

$$\alpha_j f_i = 0, \quad j \neq i,$$
$$= 1, \quad j = i.$$

We are going to show that $f_1, f_2, \ldots, f_n$ form a basis for V^*. Let $\alpha \in V$. Then $\alpha = \sum_1^n a_i \alpha_i$ and hence $\alpha f_i = a_i$, $i = 1, 2, \ldots, n$. Thus $\alpha = \sum_1^n (\alpha f_i) \alpha_i$. Let f be any linear form on V. Then

$$\alpha f = \sum_1^n (\alpha f_i)(\alpha_i f) = \alpha \sum_1^n (\alpha_i f) f_i \quad \text{for every } \alpha \in V.$$

Hence $f = \sum_1^n (\alpha_i f) f_i$, and so the f_i span V^*. Now suppose

$$\sum_1^n c_i f_i = \bar{0}_{V^*}, \quad c_i \in F,$$

where $\bar{0}_{V^*}$ is the zero linear form (the one that maps every $\alpha \in V$ into $0 \in F$). Then for each j from 1 to n, $\alpha_j \left(\sum_1^n c_i f_i \right) = \alpha_j \bar{0}_{V^*} = 0$. Now $\alpha_j \sum_1^n (c_i f_i) = c_j$. Hence $c_j = 0$ for all $j = 1, 2, \ldots, n$. This proves the linear forms f_i, $i = 1, 2, \ldots, n$ are independent. Hence they form a basis of n elements from V^*. Thus dim $V^* = n$.

Corollary. If V is a finite-dimensional vector space then $V \approx V^*$.

Proof: The isomorphism is given by the linear transformation T of $V \rightarrow V^*$ determined by $\alpha_i T = f_i$, $i = 1, 2, \ldots, n$.

Definition. The basis $f_1, \ldots, f_n$ is called the **dual basis** for V^*.

If $\alpha f = 0$ for every $f \in V^*$ then $\alpha = \bar{0}_V$. For $\alpha = \sum_1^n a_i \alpha_i$ and $\alpha f_i = a_i$, $i = 1, 2, \cdots, n$. Since $\alpha f_i = 0$ for each i, $a_i = 0$ for each i. Hence $\alpha = \bar{0}_V$.

For infinite-dimensional vector spaces it is not true in general that V and V^* are isomorphic spaces.

THEOREM 26. Let V be a finite-dimensional inner product vector space over the real field F. Then for every linear form $f \in V^*$ there exists a unique $\alpha \in V$ such that $\gamma f = \alpha \cdot \gamma$ for every $\gamma \in V$—that is, every $f \in V^*$ can be represented in this form.

Proof: f is a linear mapping of $V \to F$ and so im f is a subspace of F (Theorem 9, Corollary). Now the only subspaces of the vector space F are 0 and F. If im $f = 0$, take $\alpha = \bar{0}_V$, for we see that $\gamma f = \bar{0}_V \cdot \gamma = 0$ for all $\gamma \in F$. Also $\alpha = \bar{0}_V$ is unique.

Suppose then im $f = F$. Let dim $V = n$. Then dim ker $f = n - 1$ by Theorem 14, since dim im $f = 1$. Let $\alpha_1, \ldots, \alpha_{n-1}$ be an orthonormal basis for ker f and extend it to an orthonormal basis $\alpha_1, \ldots, \alpha_{n-1}, \alpha_n$ for V (Theorem 21, Corollary 2).

For $\gamma \in V$, $\gamma = \displaystyle\sum_1^n (\gamma \cdot \alpha_i)\alpha_i$ and hence $\gamma f = \displaystyle\sum_1^n (\gamma \cdot \alpha_i)(\alpha_i f)$.

For $i = 1, 2, \ldots, n - 1$, $\alpha_i \in$ ker f, hence $\gamma f = (\gamma \cdot \alpha_n)(\alpha_n f) = ((\alpha_n f)\alpha_n) \cdot \gamma$, since $\alpha_n f \in F$ and is therefore a scalar. Hence $\alpha = (\alpha_n f)\alpha_n$ is the promised vector for which $\gamma f = \alpha \cdot \gamma$, for all $\gamma \in V$.

Now this α is unique, for if $\alpha \cdot \gamma = \beta \cdot \gamma$ for all $\gamma \in V$, then $(\alpha - \beta) \cdot \gamma = 0$ for all $\gamma \in V$. If we put $\gamma = \alpha - \beta$ we get $(\alpha - \beta) \cdot (\alpha - \beta) = 0$ and this implies $\alpha = \beta$.

The linear forms of $V^* \to F$ form a vector space over F denoted by V^{**} and called the **bidual of V**. If dim $V = n$, then dim $V^* = n$ and so dim $V^{**} = n$.

Consider the mapping T of $V \to V^{**}$ defined by $\alpha T = \phi_\alpha$, where ϕ_α is the linear form on V^*, defined by $f\phi_\alpha = \alpha f$ for all $f \in V^*$.

If $\phi_\alpha = \phi_\beta$ then $f\phi_\alpha = f\phi_\beta$ for all $f \in V^*$, hence $\alpha f = \beta f$ and $(\alpha - \beta)f = 0$ for all $f \in V^*$. From the remark at the end of Theorem 26, we see $\alpha = \beta$. Hence T is injective. Since dim $V = $ dim V^{**}, it follows that T is an isomorphism and hence every linear form on V^* can be expressed in the form ϕ_α for some $\alpha \in V$.

If $\alpha_1, \ldots, \alpha_n$ is a basis for V, then it is quite straightforward to show that $\phi_{\alpha_1}, \phi_{\alpha_2}, \ldots, \phi_{\alpha_n}$ form a basis for V^{**}.

3-13 EIGENVECTORS AND THE SPECTRAL THEOREM

We give here a short introduction to the important notion of an eigenvector as well as a description of the spectral theorem for self-adjoint operators on a finite-dimensional inner product vector space.

Definitions. Let T be a linear operator (that is an endomorphism) on a vector space V over an arbitrary field F. A nonzero vector $\alpha \in V$ is called an **eigenvector (proper vector** or **characteristic vector)** of T if $\alpha T = \lambda \alpha$ for some $\lambda \in F$, and λ is called an **eigenvalue** of T.

The set of all eigenvalues of T is called the **spectrum of T.**

The eigenvalues and eigenvectors of T are defined by the equation $\alpha T = \lambda \alpha$, $\alpha \in V$, $\lambda \in F$. This equation means the vector αT is dependent on the vector α. In the language of "arrows," this equation states that the arrow for the vector αT must have either the same or opposite direction to the arrow for the vector α. The equation $\alpha T = \lambda \alpha$ is equivalent to $\alpha T = \lambda(\alpha I)$, that is to $\alpha(T - \lambda I) = \overline{0}_V$, where I is the identity operator on V. If α is an eigenvector of T, then $\alpha \neq \overline{0}_V$, and if λ is the eigenvalue associated with this α then the operator $T - \lambda I$ cannot be an automorphism of V, since $\alpha(T - \lambda I) = \overline{0}_V$. Thus the operator $T - \lambda I$ has no inverse; it is what is called a **singular operator.** Conversely, if for some $\lambda \in F$, $T - \lambda I$ is a singular operator then it is not an automorphism of V and hence ker $(T - \lambda I)$ must contain at least one nonzero vector; that is, there exists a nonzero vector $\alpha \in V$ such that $\alpha(T - \lambda I) = \overline{0}_V$. Hence $\alpha T = \lambda \alpha$ and λ is an eigenvalue of T.

Thus λ is an eigenvalue of T if and only if the operator $T - \lambda I$ is singular.

In practice the eigenvalues (if any exist) of an operator T on an n-dimensional vector space over the real field F are found as follows. With respect to some basis for V, $T - \lambda I$ has an $n \times n$ matrix of the form $(a_{ij}) - \lambda(e_{ij}) = (a_{ij} - \lambda e_{ij})$, where (a_{ij}) is the matrix of T and and where (e_{ij}) is the matrix of I. Then $e_{ij} = 0$, $i \neq j$, while $e_{ii} = 1$. The determinant of the $n \times n$ matrix $(a_{ij} - \lambda e_{ij})$ is a polynomial in λ of degree n. It is called the **characteristic polynomial** of T. It can be proved that the real roots of the characteristic polynomial constitute the eigenvalues of the operator. This condition is based on the fact that $T - \lambda I$ is singular if and only if the determinant of its matrix is 0, and this is equivalent to $\alpha(I - \lambda I) = \overline{0}_V$ for a nonzero vector α, which in turn is equivalent to $\alpha T = \lambda \alpha$.

Moreover, if some other basis is chosen we know T has a matrix with respect to this basis which is similar to the matrix (a_{ij}).

If we assume the following property of determinants: for $n \times n$ matrices A and B,

$$\det (AB) = (\det A)(\det B)$$

we can prove the following theorem.

THEOREM 27. Two similar matrices have the same characteristic polynomial.

Proof: The operator $T - \lambda I$ can be expressed as $A - \lambda I$ where A is the matrix of T relative to some basis for V and I is the identity matrix. If a change of basis is made then T has now a matrix B similar to A. Hence $B = PAP^{-1}$ where P is a nonsingular matrix.

Now

$$B - \lambda I = PAP^{-1} - \lambda PIP^{-1}$$
$$= P(A - \lambda I)P^{-1}.$$

Hence det $(B - \lambda I) = (\det P)(\det A - \lambda I)(\det P^{-1}) = \det (A - \lambda I)$.

Corollary. The eigenvalues of an operator T are independent of the particular choice of basis for V.

Example 1. Let V be the vector space $V_2(F)$ where F is the real field and choose $(1,0)$ and $(0,1)$ for a basis for $V_2(F)$. Let T be the operator on $V_2(F)$ defined by $(x_1, x_2)T = (3x_1 + 2x_2, 2x_1)$. Then

$$(1,0)T = 3(1,0) + 2(0,1)$$
$$(0,1)T = 2(1,0) + 0(0,1)$$

and the matrix of T is seen to be $\begin{pmatrix} 3 & 2 \\ 2 & 0 \end{pmatrix}$. The eigenvalues of T are given by

$$\det \begin{pmatrix} 3 - \lambda & 2 \\ 2 & -\lambda \end{pmatrix} = \lambda^2 - 3\lambda - 4 = 0.$$

Hence $\lambda = -1, 4$. $\lambda^2 - 3\lambda - 4$ is the characteristic polynomial.

Example 2. For the same basis and vector space as in the previous example, let T be the operator defined by $(x_1, x_2)T = (-x_2, x_1)$. We find its matrix is $\begin{pmatrix} 0 & 1 \\ -1 & 0 \end{pmatrix}$ and the characteristic polynomial is $\lambda^2 + 1$. Since this has no real roots, it follows that this operator has no eigenvalues in the real field F of scalars.

Lemma 16. If λ is an eigenvalue of T, then the set of all vectors $\alpha \in V$ such that $\alpha T = \lambda \alpha$ is a subspace S of V of dimension ≥ 1. (This subspace is called an **eigenspace of T**.)

Proof: If $\alpha T = \lambda \alpha$, $\beta T = \lambda \beta$, then $(\alpha + \beta)T = \lambda(\alpha + \beta)$. Thus if $\alpha, \beta \in S$, then $\alpha + \beta \in S$. If $x \in F$, $(x\alpha)T = x(\alpha T) = x\lambda \alpha = \lambda(x\alpha)$. Hence if $\alpha \in S$, then $x\alpha \in S$. Hence by Theorem 2, S is a subspace. If λ is an eigenvalue, then there exists $\alpha \neq 0_V$ such that $\alpha T = \lambda \alpha$. Hence dim $S \geq 1$.

Lemma 17. If $\alpha_1, \ldots, \alpha_k$ are eigenvectors of T with distinct eigenvalues $\lambda_1, \ldots, \lambda_k$, then $\alpha_1, \ldots, \alpha_k$ is an independent set of vectors.

Proof: Assume they are dependent. Then by Lemma 8 there is a least index m such that α_m depends on the independent vectors

$\alpha_1, \ldots, \alpha_{m-1}$. Then $\alpha_m = \sum_1^{m-1} x_i \alpha_i$, $x_i \in F$ and $\alpha_m T = \lambda_m \alpha_m = \sum_1^{m-1} x_i \lambda_i \alpha_i$. But

$$\lambda_m \alpha_m = \lambda_m \sum_1^{m-1} x_i \alpha_i = \sum_1^{m-1} x_i \lambda_m \alpha_i.$$

Hence
$$\sum_1^{m-1} x_i (\lambda_i - \lambda_m) \alpha_i = \bar{0}_V.$$

Since $\alpha_1, \ldots, \alpha_{m-1}$ are independent vectors, this forces $x_i(\lambda_i - \lambda_m) = 0$, $i = 1, 2, \ldots, m - 1$. Since $\lambda_i \neq \lambda_m$, we have $x_i = 0$, $i = 1, 2, \ldots, m - 1$. Hence $\alpha_m = \bar{0}_V$. This is a contradiction, since α_m is an eigenvector. Therefore the assumption that the α_i are dependent is impossible.

Corollary. If dim $V = n$, there can be at most n distinct eigenvalues of T.

We now want to define the *adjoint operator T^** of an operator T on a real inner product vector space V.

For each fixed $\beta \in V$ the function f of $V \to F$ defined by $\alpha f = \alpha T \cdot \beta$ is evidently a linear form on V, that is $f \in V^*$. Hence by Theorem 26 there exists a unique vector $\beta' \in V$ such that

(9) $\alpha T \cdot \beta = \alpha \cdot \beta'$, for all $\alpha \in V$.

Now define a mapping T^* of $V \to V$ by

(10) $\beta T^* = \beta'$, for all $\beta \in V$.

It is a straightforward exercise for the reader to prove that T^* is an operator (endomorphism) on V.

Definition. T^* is called the **adjoint operator of T.**

Moreover we have from (9) and (10) this important relationship between T and its adjoint T^*.

(11) $\alpha T \cdot \beta = \alpha \cdot \beta T^*$,   for all $\alpha, \beta \in V$.

Definition. If $T = T^*$, T is called a **self-adjoint** or **symmetric operator.**

Definition. If $TT^* = T^*T$ then T is called a **normal operator.** Obviously a self-adjoint operator is normal.

Definition. The **transpose** of a matrix $A = (a_{ij})$ is the matrix
$$A' = (a'_{ij}),$$
where
$$a'_{ij} = a_{ji}.$$

Lemma 18. If V is a real n-dimensional inner product vector space, then with respect to an orthonormal basis the matrix of an operator T is the transpose of the matrix of its adjoint operator T^*.

Proof: Let $\alpha_1, \alpha_2, \ldots, \alpha_n$ be an orthonormal basis for V. Let (a_{ij}) be the matrix of T relative to this basis. Then by (8),

$$\alpha_i T = \sum_1^n a_{ij} \alpha_j = \sum_1^n (\alpha_i T \cdot \alpha_j) \alpha_j.$$

Hence $a_{ij} = \alpha_i T \cdot \alpha_j$. Let (b_{ij}) be the matrix of T^* relative to this same basis. Then

$$\alpha_i T^* = \sum_1^n b_{ij} \alpha_j = \sum (\alpha_i T^* \cdot \alpha_j) \alpha_j.$$

Hence $b_{ij} = \alpha_i T^* \cdot \alpha_j$. By (11), $\alpha_i T^* \cdot \alpha_j = \alpha_j T \cdot \alpha_i = a_{ji}$. Therefore $b_{ij} = a_{ji}$.

Corollary. T is a self-adjoint operator if and only if its matrix (a_{ij}) relative to an orthonormal basis is symmetric, that is $a_{ij} = a_{ji}$ for all i and j.

We have not discussed inner product vector spaces over the complex field C, but it is to our advantage to discuss them briefly here, if only to indicate the important difference in the definition of the inner product. Let α and β be any two vectors in $V_n(C)$. Thus they have the form

$$\alpha = (x_1, x_2, \ldots, x_n), \quad \beta = (y_1, y_2, \ldots, y_n),$$

where the x_i and y_i are complex numbers.

The **inner product** in $V_n(C)$ is defined by

$$\alpha \cdot \beta = x_1 \bar{y}_1 + x_2 \bar{y}_2 \cdots + x_n \bar{y}_n,$$

where $\bar{y}_i$ denotes the complex conjugate of y_i.

It is easy to show that $\alpha \cdot \beta = \overline{\beta \cdot \alpha}$. Here again $\overline{\beta \cdot \alpha}$ denotes the conjugate of $\beta \cdot \alpha$. In particular then, $\alpha \cdot \alpha$ is real and vanishes if and only if $\alpha = \bar{0}_V$.

To a linear operator T on $V_n(C)$ there corresponds, as in the real case above, an adjoint operator T^*, related to T by

$$\alpha T \cdot \beta = \alpha \cdot \beta T^*, \text{ for all } \alpha, \beta \text{ in } V_n(C).$$

T is said to be *self-adjoint* or *hermitian* if $T = T^*$. It can be shown that, with respect to an orthonormal basis, the matrix A^* of T^* is the transpose of the conjugate matrix $\bar{A}$, where A is the matrix of T. If $A = (a_{ij})$, then $\bar{A} = (\bar{a}_{ij})$, and therefore $A^* = \bar{A}^T$.

If T is a hermitian operator, then for any $\alpha \in V_n(C)$,

$$\alpha \cdot \alpha T^* = \alpha \cdot \alpha T = \overline{\alpha T \cdot \alpha}.$$

Since $\alpha \cdot \alpha T^* = \alpha T \cdot \alpha$, it follows that $\overline{\alpha T \cdot \alpha} = \alpha T \cdot \alpha$. Hence if T is hermitian, then $\alpha T \cdot \alpha$ is real for all $\alpha \in V_n(C)$.

It follows at once from this last result that the eigenvalues of a hermitian operator are real; that is, all the roots of the equation

$$|A - \lambda I| = 0,$$

where A is a hermitian matrix, are real. For if $\alpha T = \lambda \alpha$, $\alpha \neq 0_V$, then

$$\alpha T \cdot \alpha = (\lambda \alpha) \cdot \alpha = \lambda(\alpha \cdot \alpha).$$

Since $\alpha T \cdot \alpha$ and $\alpha \cdot \alpha$ are both real, λ must be real. In particular then the eigenvalues of a real self-adjoint (symmetric) linear operator are real.

The "fundamental theorem of algebra" asserts that a polynomial equation over the complex field has all its roots in the complex field. A linear operator on $V_n(C)$ therefore always has eigenvalues.

One last remark about the inner product in $V_n(C)$. We find that for $x \in C$, $(x\alpha) \cdot \beta = x(\alpha \cdot \beta)$ but $\alpha \cdot (x\beta) = \bar{x}(\alpha \cdot \beta)$.

Example 3. Let T be the operator on the real inner product vector space $V_2(F)$ defined by $(x_1, x_2)T = (ax_1 + bx_2, bx_1 + cx_2)$, $a, b, c \in F$. The matrix of T relative to the standard basis $(1, 0)$, $(0, 1)$ is $\begin{pmatrix} a & b \\ b & c \end{pmatrix}$, hence T is self-adjoint.

On the other hand, the operator $(x_1, x_2)T = (ax_1 - bx_2, bx_1 + ax_2)$ has the matrix $\begin{pmatrix} a & b \\ -b & a \end{pmatrix}$ and is therefore not self-adjoint. The transpose of $\begin{pmatrix} a & b \\ -b & a \end{pmatrix}$ is $\begin{pmatrix} a & -b \\ b & a \end{pmatrix}$ and this yields the adjoint operator T^* where

$$(x_1, x_2)T^* = (ax_1 + bx_2, -bx_1 + ax_2).$$

It is easy to show, however, that T is normal.

Lemma 19. The eigenspaces S_1 and S_2 corresponding to two distinct eigenvalues λ_1 and λ_2 of a self-adjoint operator T on an inner product vector space V are orthogonal.

Proof: Since T is self-adjoint $\alpha T \cdot \beta = \alpha \cdot \beta T$ for all $\alpha, \beta \in V$. Let $\alpha_1 T = \lambda_1 \alpha_1$, $\alpha_2 T = \lambda_2 \alpha_2$.   Then $\alpha_1 T \cdot \alpha_2 = a_1 \cdot \alpha_2 T$, that is

$$\lambda_1(\alpha_1 \cdot \alpha_2) = \lambda_2(\alpha_1 - \alpha_2).$$

Since $\lambda_1 \neq \lambda_2$, $\alpha_1 \cdot \alpha_2 = 0$. Hence every eigenvector α_1 of S_1 is orthogonal to every vector α_2 of S_2.

Example 4. Let T be an operator on $V_2(F)$ defined by

$$(x_1, x_2)T = (4x_1 + 2x_2, 2x_1 + x_2).$$

It is easily verified that $\alpha T \cdot \beta = \alpha \cdot \beta T$ for all vectors α and β of $V_2(F)$ so that T is self-adjoint. Alternatively, the matrix $\begin{pmatrix} 4 & 2 \\ 2 & 1 \end{pmatrix}$ of T is symmetric and so $T = T^*$. The eigenvalues of T are found to be $\lambda_1 = 0$ and $\lambda_2 = 5$, and $\alpha_1 = (1, -2)$, $\alpha_2 = (2, 1)$ are the corresponding eigenvectors. Thus the eigenspace S_1 is spanned $(1, -2)$ and S_2 is spanned by $(2, 1)$. Clearly $(1, -2) \cdot (2, 1) = 0$ and $S_2 = S_1^\perp$. The eigenvectors span $V_2(F)$ and hence $V_2(F) = S_1 \oplus S_2$. If E_1 and E_2 are the projections of $V_2(F)$ on S_1 and S_2 respectively, then $E_1 + E_2 = I_V$ and $E_1 E_2 = 0$, where here 0 stands for the operator mapping every element of $V_2(F)$ into $\overline{0}_V$. Thus E_1 and E_2 are orthogonal. Now $\alpha_1 T = \overline{0}_V$ and $\alpha_2 T = 5\alpha_2$. $S_1 = \ker T$, and hence $S_2 = \operatorname{im} T$.

If $\alpha \in V$, then $\alpha = \beta_1 + \beta_2$, $\beta_1 \in S_1$ and $\beta_2 \in S_2$. Hence

$$\alpha T = \beta_1 T + \beta_2 T = 0 + 5\beta_2 = 5(\alpha E_2) = \alpha(5E_2).$$

Hence $T = 5E_2$. We note here in anticipation of the spectral theorem that we can write $T = \lambda_1 E_1 + \lambda_2 E_2$, where $\lambda_1 = 0$, $\lambda_2 = 5$ are the eigenvalues. This is a decomposition of T into orthogonal projections for which $E_1 + E_2 = I_V$.

Example 5. Let T be the operator on the inner product vector space $V_3(F)$, F the real field, defined by

$$(x_1, x_2, x_3)T = (2x_1 - x_2, -x_1 + 2x_2 - x_3, -x_2 + 2x_3).$$

Relative to the basis $(1, 0, 0)$, $(0, 1, 0)$, $(0, 0, 1)$ the matrix of T is $\begin{bmatrix} 2 & -1 & 0 \\ -1 & 2 & -1 \\ 0 & -1 & 2 \end{bmatrix}$, and hence T is self-adjoint. The characteristic equation is $(2 - \lambda)^3 - 2(2 - \lambda) = 0$, and the eigenvalues are found to be $\lambda_1 = 2$, $\lambda_2 = 2 + \sqrt{2}$, $\lambda_3 = 2 - \sqrt{2}$. The eigenspaces S_1, S_2, S_3 corresponding to these eigenvalues are spanned by the eigenvectors $(1, 0, -1)$, $(1, -\sqrt{2}, 1)$ and $(1, \sqrt{2}, 1)$ respectively. $V_3(F) = S_1 \oplus S_2 \oplus S_3$, by Lemma 19. Let E_1, E_2, E_3 be the projections of $V_3(F)$ on S_1, S_2, S_3 respectively. This means that if $\alpha \in V_3(F)$ and $\alpha = \alpha_1 + \alpha_2 + \alpha_3$, where $\alpha_i \in S_i$, $i = 1, 2, 3$, then $\alpha E_i = \alpha_i$, $i = 1, 2, 3$.

It can easily be verified that $E_i E_j$, $i \neq j$, is the zero operator and that $E_1 + E_2 + E_3$ is the identity operator. Moreover it follows readily that $T = \lambda_1 E_1 + \lambda_2 E_2 + \lambda_3 E_3$.

The two previous examples illustrate the **spectral theorem for self-adjoint operators** on a **real finite dimensional inner product vector space,** which we now state without proof.

THEOREM 28 (Spectral Theorem). Let $\lambda_1, \lambda_2, \ldots, \lambda_p$ be the distinct eigenvalues of T and let $S_1, S_2, \ldots, S_p$ be the corresponding eigen-

EXERCISES

1. Find the eigenvectors and the eigenvalues of the endomorphism T of $V_2(F)$ defined by

$$(x, y)T = (2x + 3y, 4x - 2y).$$

F is the real field.

2. If U and W are subspaces of a vector space V, prove

$$U + W/U \approx W/U \cap W.$$

3. If V is a finite-dimensional vector space and if U and W are subspaces of V, prove

$$\dim (U + W) = \dim U + \dim W/U \cap W.$$

4. Let T be an endomorphism of an n-dimensional vector space V. If the eigenvectors $\alpha_1, \ldots, \alpha_n$ of T form a basis for V, that is $\alpha_i T = c_i \alpha_i,$ $i = 1, 2, \ldots, n$ and the c_i are scalars, then the matrix of T relative to this basis is diagonal.

Conversely, if the matrix of T is diagonal with $c_1, c_2, \ldots, c_n$ along the principal diagonal, then the α_i are eigenvectors of T and the c_i are the corresponding eigenvalues.

Fill in the details for these statements and prove that a matrix A of a linear transformation T is similar to a diagonal matrix if and only if its eigenvectors span V.

5. Show that similar matrices have the same eigenvalues and eigenvectors.

6. Let

$$A = \begin{bmatrix} 2 & 0 & 0 \\ -3 & 1 & 0 \\ 0 & 1 & 0 \end{bmatrix}$$

be the matrix of an endomorphism of $V_3(F)$, F the real field, relative to the standard basis. Find the eigenvalues of A and show that the eigenvectors of A span $V_3(F)$. Prove that A is similar to the matrix

$$B = \begin{bmatrix} 1 & 0 & 0 \\ 0 & 2 & 0 \\ 0 & 0 & 0 \end{bmatrix}$$

and find the matrix P for which $B = PAP^{-1}$.

7. Prove that under traditional definitions the set Hom (V, W) of linear transformations of $V \to W$, where V and W are vector spaces over the same field F, is a vector space over F. If dim $V = n$, prove that dim Hom $(V, V) = n^2$.

8. If dim $V = n$ and if $T \in$ Hom (V, V), prove that $1_V, T, T^2, \ldots, T^{n^2}$ form a dependent set of vectors in Hom (V, V). Hence show that T satisfies a polynomial equation with coefficients in the field F of scalars.

Let B be the set of all polynomials $g(x)$ in the variable x with coefficients in F for which $g(T) = 0$. Prove there is a unique polynominal $m(x)$ of least degree with leading coefficient equal to 1 for which $m(T) = 0$. $m(x)$ is called the **minimal polynomial** of T.

9. Prove that the minimal polynomial $m(x)$ of T divides the characteristic polynomial $f(x) = |x1_V - T|$ of T.

10. Let T be a surjective linear transformation from a vector space V to a vector space W. If U is a subspace of V such that ker $T \subset U$, prove

$$V/U \approx W/UT.$$

11. If T is a linear transformation $V \to W$ of two finite-dimensional inner product vector spaces that preserves lengths of vectors, is T an isometry? Prove your answers.

12. An endomorphism S of an inner product vector space V over a field F for which $|\alpha S| = c|\alpha|$ for every $\alpha \in V$ where c is a fixed scalar, is called a **similarity transformation.**

(a) Prove the set of all similarity transformations on V form a group G.

(b) Prove that the orthogonal transformations H on V form a normal subgroup of G.

(c) Prove that the quotient group G/H is isomorphic to the multiplication group of all positive real numbers.

13. T is the endomorphism of $V_3(F)$, F the real field, whose matrix with respect to the standard basis is

$$A = \begin{bmatrix} -1 & 0 & 2 \\ -3 & -2 & 6 \\ -1 & 0 & 2 \end{bmatrix}.$$

(a) Find the characteristic and minimal polynomials of T.

(b) Find the eigenvalues and eigenvectors of T.

(c) Show A is similar to the diagonal matrix

$$\begin{bmatrix} 0 & 0 & 0 \\ 0 & 1 & 0 \\ 0 & 0 & -2 \end{bmatrix}.$$

14. The endomorphism T of the vector space $V_3(F)$, with the usual inner product, has the matrix

$$\begin{bmatrix} 1 & 0 & 0 \\ 0 & 1/2 & \sqrt{3/2} \\ 0 & -\sqrt{3/2} & 1/2 \end{bmatrix}$$

relative to the standard basis. Prove that T is an isometry.

15. Let T be a given endomorphism of a finite-dimensional inner product vector space V over the real field F. Prove that T', defined by $\gamma T' = \gamma'$, $\gamma \in V$

and where γ' is given by $\alpha T \cdot \gamma = \alpha \cdot \gamma'$ for all $\alpha \in V$, is an endomorphism of V. (Choose an orthonormal basis in V and relate the matrices of T and T' with respect to this basis.) T' is called the **transpose of T.**

16. Let $V_n(F)$, where F is the real field, have the usual inner product. If T is an endomorphism of $V_n(F)$ with the matrix (a_{ij}), and T' an endomorphism with the matrix (b_{ij}), both relative to the standard basis, prove that if $b_{ij} = a_{ji}$ for all i and j, $(\alpha T) \cdot \beta = \alpha \cdot (\beta T')$ for all $\alpha, \beta \in V_n(F)$.

17. Let V be an inner product finite-dimensional vector space and let T be an orthogonal endomorphism of V—that is, $\alpha \cdot \beta = (T\alpha) \cdot (T\beta)$ for all $\alpha, \beta \in V$. If T' is the transpose of T prove $TT' = T'T = I$, the identity endomorphism of V.

chapter 4

Structure of Groups

In this chapter we resume the study of groups. We shall be principally concerned with theorems on the structure of groups. The type and the number of subgroups that a group may possess determine to a great extent its structure. We shall first consider finite groups, and for these the Sylow theorems are the most important, because of their rich yield of information on the structure of such groups. For this purpose we introduce the notion of **conjugacy**, both for the elements of the group and for its subgroups. Two special types of subgroups, **center** and **normalizer**, are defined and used to derive both quantitative and qualitative properties of the group. This will enable us to establish the important class equation, which is a means of counting the number of elements in a group and classifying them in conjugate classes. With this equipment we can prove the Sylow theorems.

The rest of the chapter is mainly given over to the structure of finitely generated abelian groups. It is possible to determine precisely the structure of a finitely generated abelian group, and for this purpose the concepts of direct product and free abelian group are introduced. The universal mapping property, which occurs first in the definition of a free abelian group, will prove to be of great importance in later chapters.

4-1 CONJUGATE CLASSES AND SUBGROUPS

Let $C(G)$ be the subset of all elements of a group G that commute with every element of G. Hence $a \in C(G)$ if and only if $ax = xa$, for every $x \in G$. $C(G)$ is not empty, for the neutral element e of G is in $C(G)$.

THEOREM 1. $C(G)$ is an abelian normal subgroup of G.

Proof: Let $a, b \in C(G)$. Then $ax = xa$, $bx = xb$ for every x in G. Hence $abx = axb = xab$, and therefore $ab \in C(G)$. Also if $a \in C(G)$, $ax = xa$ and so $a^{-1}(ax)a^{-1} = a^{-1}(xa)a^{-1}$. Hence $xa^{-1} = a^{-1}x$, and we see $a^{-1} \in C(G)$. Thus $C(G)$ is a subgroup of G. Obviously $C(G)$ is an abelian subgroup of G. Since $xC(G) = C(G)x$ for every $x \in G$, it follows that $C(G)$ is a normal subgroup of G.

Definition. The subgroup $C(G)$ is called the **center of G.**

$C(G)$ may consist of the neutral element only, that is $C(G) = 1$. On the other hand $C(G) = G$ if and only if G is an abelian group.

Definition. Two elements x and y of a group G are said to be *conjugate* if and only if there exists an element $z \in G$ such that $zxz^{-1} = y$.

Conjugacy is an equivalence relation. Let $x \sim y$ denote that x and y are conjugate elements. If e is the neutral element of G, then $exe^{-1} = x$ and so $x \sim x$, for all $x \in G$. The relation is therefore reflexive. Also $x \sim y$ means $zxz^{-1} = y$ for some $z \in G$. Hence $z^{-1}yz = x$ and we have $y \sim x$. This is symmetry. Finally, $x \sim y$ and $y \sim z$ mean $uxu^{-1} = y$ and $wyw^{-1} = z$, for some u and some w of G, and therefore $wuxu^{-1}w^{-1} = z$, that is $(wu)x(wu)^{-1} = z$. Hence the relation is transitive.

Conjugacy partitions the group G into what are called **conjugate classes.** The conjugate class containing $x \in G$ consists of all elements of G that are conjugate to x. It is readily seen that an element $x \in G$ is the only member of its conjugate class if and only if $x \in C(G)$, the center of G. The order of $C(G)$ is thus equal to the number of conjugate classes that contain only one element each.

THEOREM 2. If S is any nonempty subset of a group G, the set $N(S)$ of all elements x of G for which $xSx^{-1} = S$ is a subgroup of G.

Proof: The condition $xSx^{-1} = S$ is equivalent to $xS = Sx$. This latter condition means that the element x commutes with S setwise, but not necessarily elementwise. That is to say, $xs = s'x$, $s, s' \in S$, but s and s' are not necessarily the same element.

Now we prove that $N(S)$ is a subgroup of G. Let $x, y \in N(S)$. Then $xS = Sx$ and $yS = Sy$ imply that $xyS = xSy = Sxy$ and therefore $xy \in N(S)$. Also if $xS = Sx$, then clearly $Sx^{-1} = x^{-1}S$. Thus $x \in N(S)$ implies $x^{-1} \in N(S)$. All this means that $N(S)$ is a subgroup.

Definition. $N(S)$ is called the **normalizer** of the subset S.

Observe that the subset S is not necessarily contained in its normalizer.

In particular if $S = \{a\}$, then $N(a)$ is called the normalizer of the element $a \in G$. It is the subgroup of G consisting of all $x \in G$ for which $xax^{-1} = a$; that is, for which $xa = ax$. $N(a)$ is thus the subgroup of those elements of G that commute with the element a. (For this reason it is often called the **centralizer** of the element a.) Of course $N(a)$ always contains the cyclic group $[a]$ generated by the element a, since an element always commutes with itself. Also, of course, $C(G) \subset N(a)$. In fact we see that $C(G) = \bigcap_{a \in G} N(a)$. If $a \in C(G)$ then $N(a) = G$.

Since not all the conjugates xax^{-1}, $x \in G$, of an element a are distinct (for example, $eae^{-1} = aaa^{-1} = a$) we need a means of determining the number of distinct elements in the conjugate class containing the element a. The normalizer of the element a will enable us to do this, as the next theorem shows.

THEOREM 3. If $a \in G$, then the number of distinct elements in the conjugate class containing the element a, is equal to the index of the normalizer $N(a)$ in G.

Proof: First we observe that the left coset $xN(a)$, $x \in G$, consists of those elements $z \in G$ for which $zaz^{-1} = xax^{-1}$. For if $z \in xN(a)$ then $x^{-1}z \in N(a)$, and therefore $x^{-1}za = ax^{-1}z$. Hence $zaz^{-1} = xax^{-1}$. Conversely, if $zaz^{-1} = xax^{-1}$, then $x^{-1}z \in N(a)$, and hence $z \in xN(a)$.

Two left cosets are equal (identical), $xN(a) = yN(a)$, if and only if $y^{-1}x \in N(a)$. Now $y^{-1}5 \in N(a)$ if and only if $xax^{-1} = yay^{-1}$. It therefore follows that the number of distinct cosets is the same as the number of distinct conjugates of the element a. This is the statement of the theorem.

Corollary. The number of distinct elements in the conjugate class containing the element a is a divisior of the order of G.

Let H be a subgroup of G. We can form the normalizer $N(H)$ of H. Clearly $H \subset N(H)$, and in fact we see that H is a normal subgroup of $N(H)$. In fact $N(H)$ is seen to be the largest subgroup of G in which H is normal. Since any group is a normal subgroup of itself, it is possible that $N(H) = H$. Of course $N(H) = G$ if and only if H is a normal subgroup of G.

Definition. Two subgroups H and K of a group G are called **conjugate subgroups** in G if $K = x^{-1}Hx$ for some $x \in G$.

As noted before the relationship is a mutual one, since $H = xKx^{-1}$. Moreover, as before, we see that an equivalence relation R is set up on the set of all subgroups of G by defining HRK if and only if H and K are conjugate subgroups. The subgroups conjugate to a given subgroup H have the form $x^{-1}Hx$ where x runs through all the elements of G. Not all of these subgroups conjugate to H are necessarily distinct and the number of distinct ones is specified by the following theorem.

THEOREM 4. The number of distinct subgroups conjugate to a given subgroup H of G is equal to the index of the normalizer $N(H)$ in G.

Proof: $x^{-1}Hx = y^{-1}Hy$ if and only if $Hxy^{-1} = xy^{-1}H$ that is if and only if $xy^{-1} \in N(H)$, whence if and only if the cosets $N(H)x$ and

$N(H)y$ of $N(H)$ in G are equal. Thus the mapping $x^{-1}Hx \rightarrow N(H)x$ of the set of subgroups of G conjugate to H into the set of cosets of $N(H)$ in G is well-defined and injective. Clearly it is surjective, and hence is bijective. This proves the theorem.

Because all the subgroups conjugate to a given subgroup H are identical with H if and only if H is a normal subgroup of G, normal subgroups are sometimes called **self-conjugate subgroups.**

The mapping $h \rightarrow x^{-1}hx$ of $H \rightarrow x^{-1}Hx$ is a bijection since $x^{-1}h_1x = x^{-1}h_2x$ if and only if $h_1 = h_2$. Thus conjugate subgroups contain the same number of elements, that is have the same order.

THEOREM 5. If x and y are conjugates, $y = z^{-1}xz$, then $N(x)$ and $N(y)$ are conjugate subgroups, in fact $N(y) = z^{-1}N(x)z$.

Proof: Let $u \in N(x)$, then $ux = xu$. Then from $y = z^{-1}xz$ it follows that $y(z^{-1}uz) = z^{-1}xuz = z^{-1}uxz = (z^{-1}uz)y$. Hence $z^{-1}uz \in N(y)$ and therefore $z^{-1}N(x)z \subset N(y)$. In the same way, it follows from $x = zyz^{-1}$ that $zN(y)z^{-1} \subset N(x)$—that is, $N(y) \subset z^{-1}N(x)z$. Hence $N(y) = z^{-1}N(x)z$.

Corollary. If x and y are conjugate elements then $N(x)$ and $N(y)$ contain the same number of elements.

Proof: $N(x)$ and $N(y)$ are conjugate subgroups.

The conjugate class containing an element $x \in G$ contains only x (a singleton set) if and only if $x \in C(G)$, and hence it contains more than this one element (a multiple set) if and only if $x \notin C(G)$. Choose one (any one) representative element x from each of the distinct multiple conjugate classes, then the index $|G:N(x)|$ of $N(x)$ in G is the number of distinct conjugate elements in this conjugate class. By the above corollary, if y comes from the same conjugate class as x then $|G:N(y)| = |G:N(x)|$. Let $\sum_{x \notin C(G)} |G:N(x)|$ denote the summation taken over all these representative elements, one from each distinct multiple conjugate class. This sum is the total of all elements of G in the multiple conjugate classes. Since $\#C(G)$ is the total number of elements of G in the singleton conjugate classes and since G is the union of all these disjoint conjugate classes, we have for the order $\#G$ of G,

$$\#G = \#C(G) + \sum_{x \notin C(G)} |G:N(x)|$$

This equation in quantities is known as the **class equation** and it merely counts up the number of elements in G. It is very useful for computation purposes with finite groups.

Illustrative Example. Let us illustrate use of the class equation by proving that if G is a group of order p^n, where p is a prime, then $C(G) \neq 1$, that is $C(G)$ contains an element $x \neq e$. For $x \neq e$, $N(x)$ is a subgroup of order p^α, $0 < \alpha \leq n$. If $\#N(x) = p^n$ then $N(x) = G$ and hence $x \in C(G)$. Thus $C(G) \neq 1$. If for every $x \neq e$, $\#N(x) < p^n$ then $p \mid \mid G:N(x) \mid$. In the class equation we see then that $p \mid \sum_{x \notin C(G)} \mid G:N(x) \mid$ and since $p \mid \#G$, it follows that being a prime, p must divide $\#C(G)$. Hence $C(G) \neq 1$.

Illustrative Example. A deduction from this result is that a group of order p^2, p a prime, must be abelian. For $C(G) \neq 1$ and therefore $\#C(G) = p$ or p^2. Suppose $\#C(G) = p$. Then there exists $x \notin C(G)$. Since $C(G) \subset N(x)$, then $\#N(x) = p^2$ and $N(x) = G$. But this means $x \in C(G)$, which is a contradiction. Hence $\#C(G) = p^2$, and therefore $C(G) = G$, that is G is abelian.

Lemma 1. Let p be a prime dividing the order of a finite abelian group G, then G contains at least one element of order p.

Proof: Assume $\#G = np$. We use finite induction on n. The lemma is obviously true for $n = 1$, and assume it true for all abelian groups of order kp, $k < n$. Let $x \in G$, $x \neq e$. If the order of x is divisible by p, $x^{mp} = e$, then x^m is an element of order p. If the order of x is not divisible by p, then the order of the cyclic group $[x]$ is not divisible by p and hence, by Lagrange's theorem, p divides $\mid G:[x] \mid$, the index of $[x]$ in G. Now $G/[x]$ is an abelian group of order kp, $k < n$. Hence by the induction hypothesis it contains an element, that is a coset $y[x]$, of order p, that is $(y[x])^p = y^p[x] = [x]$. ($[x]$ is the neutral element of $G/[x]$). Hence $y^p \in [x]$ and therefore $y^p = x^r$ for some positive integer r. Since x^r is of some finite order m, $y^{mp} = e$, and hence y^m is an element of order p in G. This proves the lemma for a group of order np. Hence it is true of all finite abelian groups whose orders are divisible by p.

We now use the class equation to generalize this result to Cauchy's theorem.

THEOREM 6. Let p be a prime dividing the order of a finite group G. Then G contains at least one element of order p.

Proof: Use induction on the order of G and assume the theorem true for all groups of orders less than the order of G. The theorem is certainly true if $\#G = p$. The theorem is true if G is abelian. If G is not an abelian group, then $C(G) \neq G$. Let $x \in G$, $x \notin C(G)$. Now $\#N(x) < \#G$. For if $N(x) = G$, then $x \in C(G)$. Hence if $p \mid \#N(x)$ then $N(x)$, and therefore G, would contain an element of spaces (S_i is the subspace of all eigenvectors of T for the same eigenvalue λ_i). Let $E_1, E_2, \ldots, E_p$ be the projections of V on $S_1, S_2, \ldots,$ S_p respectively.

Then $T = \displaystyle\sum_{j=1}^{p} \lambda_j E_j$ where

1. $E_i E_j$ is the zero operator, $i \neq j$.
2. $\displaystyle\sum_{i=1}^{p} E_i$ is the identity operator.
3. $TE_j = E_j T, j = 1, 2, \ldots, p$.
4. dim S_j (the range of E_j) $\leq$ the multiplicity of λ_j as a root of the characteristic equation.
5. An operator A on V commutes with T if and only if it commutes with each E_j.

Theorem 28 can also be shown to be true for normal operators on V.

The spectral theorem accomplishes the decomposition of a self-adjoint operator into a linear combination of projections whose coefficients are the scalars of the spectrum of T, that is the eigenvalues of T. Such a decomposition greatly simplifies the study of the operator.

We make one more comment. A projection E that commutes with an operator T, $TE = ET$, is said to *reduce* T. Thus the projections in the spectral theorem reduce T. This follows from (3). The reducibility of an operator T signifies the existence of a proper T-invariant subspace U of V—that is, a subspace U such that $(U)T \subseteq U$. An operator T is said to be *fully reducible* if V is the direct sum of proper T-invariant subspaces.

It is a theorem that an operator T on V is fully reducible if and only if it is reduced by each projection E_i of V on the subspaces U_i, where

$$V = \sum_{1}^{n} \oplus U_i,$$

that is each U_i is T-invariant if and only if $TE_i = E_i T, i = 1, 2, \ldots, n$.

Examples 4 and 5 of this section illustrate this theorem. In Example 4, $(1, -2)T = (0, 0) \in S_1$, hence $S_1 T \subset S_1$. Also $(2, 1)T = (10, 5) \in S_1$ and in this case $S_2 T = S_2$. Thus S_1 and S_2 are T-invariant subspaces. Similarly in Example 5 we see that $S_i T \subseteq S_i, i = 1, 2, 3$.

These are very important considerations in the decomposition of an operator and the resulting decomposition of the vector space into a direct sum of invariant subspaces.

All this is said with the deliberate intention of whetting the reader's appetite for further study of a fascinating branch of mathematics, the theory of linear operators on Hilbert and Banach spaces.

order p. If $p \mid \#N(x)$ for all $x \in C(G)$, then p divides $\mid G:N(x) \mid$ for all such x. Since $p \mid \#G$ it follows then from the class equation that $p \nmid \#C(G)$. Now $C(G)$ is an abelian group and hence, by the previous lemma, contains an element of order p, and so G contains this element of order p. The theorem is therefore true for all groups whose orders are divisible by p.

(Thus a group of order 6 would contains elements of orders 2 and 3.)

Lemma 2. Let H and K be subgroups of a group G. Then the number of distinct subgroups conjugate to H in K is

$$\mid K:K \cap N(H) \mid .$$

Proof: By the conjugates of H in K, we mean the subgroups of the form $kHk^{-1}, k \in K$.

Now $K \cap N(H)$ is a subgroup of K. Consider the mapping $kHk^{-1} \to k(K \cap N(H))$ of the set of conjugates of H in K into the set of cosets of $K \cap N(H)$ in K.

Two cosets are equal, $k_1(K \cap N(H)) = k_2(K \cap N(H))$, if and only if $k_1^{-1}k_2 \in K \cap N(H)$, that is if and only if $k_1^{-1}k_2 \in N(H)$, hence if and only if $k_1Hk_1^{-1} = k_2Hk_2^{-1}$. This proves the mapping is injective. Since it is obviously surjective, the mapping is a bijection and hence the lemma is true.

EXERCISES

1. Find the subgroups that are conjugate to the subgroup $H = \{1, (1234), (13)(24), (1432)\}$ of S_4 and find the order of the normalizer $N(H)$.

2. Find the centers of the alternating group A_4 and the group of symmetries of the regular pentagon.

3. Find the normalizers of (a) the subset $\{(12)\}$ of S_3, (b) the subset $\{(123)\}$ of S_3, (c) the subgroup $\{1, (123), (132)\}$ of S_4.

4. G is the nonabelian group of order 8 generated by the elements x and y, satisfying the relations $x^4 = 1$, $y^2 = 1$, $xyx = y$. Find the center of G and write out the class equation for G.

5. Show that a set of conjugate subgroups never contains all the elements of the group.

4-2 THE SYLOW THEOREMS

With the aid of the notions of conjugacy, normalizer, and center of a group, we are now ready to derive the three Sylow theorems, which are fundamental in the study of finite groups. The first, Theorem 10, is an existence theorem. It states that there are such things as Sylow subgroups of a group. The last two Sylow theorems, Theorems 11 and 12, yield both qualitative and quantitative properties about Sylow subgroups. In fact, Theorem 10 is almost able to tell us just how many such subgroups there are in any given finite group.

Definition. A **p-group,** p a prime, is a group in which the order of every element is some power of p (not necessarily the same power).

A p-group can be infinite or finite. Clearly any subgroup of a p-group is a p-group.

THEOREM 7. A finite group G is a p-group if and only if its order is a power of p.

Proof: If G is a p-group then its order cannot be divisible by a prime $q \neq p$, for then G would have to contain an element of order q. Hence the order of G is a power of p. Conversely, if the order of G is a power of p, then since the orders of all its elements must divide the order of G, these orders must be powers of p.

Corollary. The center of a finite p-group always contains more than one element.

Proof: See the first illustrative example.

THEOREM 8. If G is a finite abelian group whose order is divisible by some power of a prime p, then the set H of all elements of G whose orders are powers of p, is a p-subgroup of G. Moreover H is a maximal p-subgroup—that is, it is not contained in any other proper p-subgroup.

Proof: Let x and y be elements of H and let their orders be p^α and p^β respectively. Then the product xy has the order p^γ where γ is the least common multiple of α and β. Moreover x^{-1} has the order p^α. Hence H is a p-group. It is maximal since it contains all elements whose orders are powers of p.

Note this theorem is not true for nonabelian groups—for example, the symmetric group S_3 where all the elements of order 2 do not form a group.

Definition. If p is a prime, then a p-subgroup P of a finite group G is called a **Sylow p-subgroup** of G if p does not divide $|G:P|$; that is, if and only if the order of P is the highest power of p that divides the order of G.

In particular if G is itself a p-group, then $P = G$. Note also that the center of a Sylow p-subgroup always contains more than one element. Moreover all Sylow p-subgroups contain the same number of elements.

THEOREM 9. If H is a p-subgroup of G and P is a Sylow p-subgroup of G, then $H \cap N(P) = H \cap P$.

Proof: Clearly $H \cap P \subset H \cap N(P)$ since $P \subset N(P)$. Put

$$H_1 = H \cap N(P).$$

Because it is a subgroup of the p-group H, it follows that H_1 is a p-group. Now P is a normal subgroup of $N(P)$ and H_1 is a subgroup of $N(P)$, hence by the second isomorphism theorem for groups it follows that

$$H_1/H_1 \cap P \cong H_1P/P.$$

Since the order of $H_1/H_1 \cap P$ is a power of p, then so is the order of H_1P/P. Thus H_1P is a p-subgroup containing P, and since P is a maximal p-subgroup, it follows that $H_1P = P$ or $H_1P = G$. Now $H_1P = G$ is a p-group, and hence $G = P$. Thus $H_1P = P$ under all circumstances. Hence $H_1 \subset P$. Since $H_1 \subset H$ it follows that $H_1 \subset H \cap P$. Since $H \cap P \subset H_1$, we have $H \cap P = H \cap N(P)$.

Corollary. If H is a p-subgroup of G and P is a Sylow p-subgroup of G, then the number of distinct conjugates of P in H is $|H : H \cap P|$.

Proof: This follows at once from Lemma 2 and the theorem.

Definition. The elements of H for which $hPh^{-1} = P$ form a subgroup of H called the **normalizer of P in H**, written $N_H(P)$. [We have so far dealt only with normalizers in G, and we have written simply $N(P)$ for $N_G(P)$.] Clearly $N_H(P) = H \cap N(P)$, and so the last theorem states that $N_H(P) = H \cap P$. Moreover, the number of distinct conjugates of P in H is $|H : N_H(P)| = |H : H \cap P|$.

THEOREM 10 (Sylow). Let G be a group of order $p^k q$, where p is a prime and $p \nmid q$. Then G has subgroups of order p^i, $1 \leq i \leq k$, and each subgroup of order p^i is normal in some subgroup of order p^{i+1} for $1 \leq i \leq k$.

Proof: We use induction on n where $\#G = np$. The theorem is true for $n = 1$. Assume it is true for all groups whose orders are less than np and divisible by p. The proof is treated in two cases.

Case I. Suppose $p \nmid \#C(G)$. Then from the class equation we see that there exists an $N(x)$ such that $p \mid |G:N(x)|$ and therefore $\#N(x) = p^k s$, where $s < q$. By the induction hypothesis the theorem is true of $N(x)$, whose order contains the same power of p as the orders of G, and hence it is true of G. Thus it is true of all such groups, provided p does not divide the order of their centers.

Case II. Suppose $p \mid \#C(G)$. Then $C(G)$ must contain an element of order p, and hence a cyclic subgroup P of order p. Being a subgroup of $C(G)$, P is a normal subgroup of G. Form the factor group G/P. Then by the induction hypothesis the theorem is true of G/P, since $\#G/P = p^{k-1}q < \#G$. In the natural epimorphism f of $G \to G/P$ the subgroups of G containing P are in one-to-one corcorrespondence with the set of all subgroups of G/P.

Let $\overline{H}_{i-1}$ be a subgroup of G/P of order p^{i-1}. Then $\overline{H}_{i-1}$ is a normal subgroup of some subgroup $\overline{H}_i$ of P of order p^i. Now $\overline{H}_{i-1} = H_i/P$, where H_i is a subgroup of G of order p^i; that is, $H_i f = \overline{H}_{i-1}$ and $f^{-1}\overline{H}_{i-1} = H_i$. Put $H_{i+1} = f^{-1}\overline{H}_i$. Then H_i is a normal subgroup of H_{i+1}. (For $H_{i+1} \to \overline{H}_i = H_{i+1}f$ is an epimorphism, and hence if $\overline{H}_{i-1} \, \Delta \, \overline{H}_i$ then $f^{-1}\overline{H}_{i-1} = H_i \, \Delta \, H_{i+1}$.)

This defines a family of subgroups H_i in G for $i = 2, 3, \ldots, k$. Define $H_1 = P$. Since P is a normal subgroup of G, it is a normal subgroup of any subgroup of G containing it. We see then that the subgroups $H_1, H_2, \ldots, H_k$ of G have the properties stated in the theorem and so the theorem is true of G. Hence the theorem is true of all groups specified in its statement.

Corollary 1. Any finite group whose order is divisible by a prime p contains a Sylow p-subgroup.

Corollary 2 (Cauchy's Theorem). A finite group G whose order is divisible by a prime p contains an element of order p.

Proof: For G contains a Sylow p-subgroup, and the center of this subgroup contains more than one element. This center is abelian and so contains an element of order p, which element is therefore in G.

Corollary 3. A p-subgroup P of a finite group G is a Sylow p-subgroup of G if and only if it is a maximal p-subgroup of G. (Thus a Sylow p-subgroup of G can be defined as a maximal p-subgroup of G.)

Proof: If P is a Sylow p-subgroup, then clearly it must be maximal, since any proper p-subgroup containing it would have to be of a higher order of p then appears in the order of G. Conversely, let P be a maximal p-subgroup of G. If $\#G = p^k q$, $p \mid q$, let $\#P = p^\alpha$, $0 < \alpha \leqq k$. If $\alpha < k$ then P would be a normal subgroup of a p-subgroup P' of order $p^{\alpha+1}$, by the first Sylow theorem. This is impossible since P is a maximal p-subgroup. Hence $\alpha = k$ and therefore P is a Sylow p-subgroup.

Corollary 4. All the conjugates of a Sylow p-group P are Sylow p-groups and there are $\mid G:N(P)\mid$ of them.

Proof: Since conjugate subgroups have the same order, it follows from Corollary 3, that the conjugates $P_x = xPx^{-1}$, $x \in G$, of a Sylow p-subgroup P of G are also Sylow p-subgroups of G, and we know by Theorem 2 that there are $|G:N(P)|$ of such distinct conjugates. Note that $|G:N(P)|$ is not divisible by p, in fact $|G:P|$ is not divisible by p.

Lemma 3. If P is a Sylow p-subgroup of G then every p-subgroup of G is contained in some subgroup conjugate to P.

Proof: Let G be a finite group whose order is divisible by the prime p. Let H be a fixed p-subgroup of G. Subdivide the set of all distinct conjugates P_x, $x \in G$, of P into classes of H-conjugates—that is, P_x and P_y are in the same H-conjugate class if $P_y = hP_xh^{-1}$ for some $h \in H$. The number of H-conjugates of P_x is then

$$|H:H \cap N(P_x)| = |H:H \cap P_x|,$$

by Corollary, Theorem 9. Hence

(1) $$|G:N(P)| = \Sigma |H:H \cap P_x|$$

where the sum is a set of representatives P_x, one from each H-conjugate class. The P_x run through the distinct H-conjugates of P. Each number in the sum is either 1 or a power of p. Since $p \nmid |G:N(P)|$, it follows that at least one of these numbers is 1. Hence $H \cap P_x = H$ for some P_x and therefore $H \subseteq P_x = xPx^{-1}$.

THEOREM 11 (Sylow Theorem). Any two Sylow p-subgroups of G are conjugate. Their total number is $|G:N(P)|$ where P is any Sylow p-subgroup.

Proof: If the p-subgroup H in the above lemma is taken to be an arbitrary Sylow p-subgroup then $H \subset P_x$. But $\#H = \#P_x$, since they are both Sylow p-subgroups. Hence $H = P_x = xPx^{-1}$.

If G is a finite abelian group, then it follows at once from this theorem that G has one and only one Sylow p-subgroup for each prime p dividing the order of G.

THEOREM 12 (Sylow Theorem). The number of distinct Sylow p-subgroups of a finite group G is $1 + mp$, where m is a nonnegative integer, and $1 + mp$ is a divisor of the order of G.

Proof: Put $H = P$ in the sum (1). Then if $x \notin N(P)$, that is $P_x \neq P$, then $|P:P \cap P_x|$ is a positive power of p. If $x \in N(P)$ then $P_x = P$ and $|P:P \cap P_x| = 1$. Now $x \in N(P)$ occurs only once in the summation and this happens when $x = e$. Hence $|G:N(P)| \equiv 1 \pmod{p}$; that is, $|G:N(P)| = 1 + mp$, where m is a nonnegative integer. Moreover $1 + mp \mid \#G$. Thus if $\#G = p^k q$, $p \nmid q$, then $1 + mp \mid q$, for $\mid G : N(P) \mid$ is a divisor of $\#G$, by Lagrange's theorem.

EXERCISES

1. If P is a Sylow p-subgroup of a group G and $N(P)$ is the normalizer of P, prove that $N(P)$ is the normalizer of $N(P)$.

2. Let G be an abelian group of order 45 generated by the elements x, y, z where $x^3 = y^3 = z^5 = 1$. Find the Sylow subgroups of G. Find a cyclic subgroup of G of order 15.

3. Construct a group of order 12 that contains an element of order 6, and prove that it contains a normal Sylow 3-subgroup.

4. If P is a Sylow p-subgroup of a finite group G and H is a normal subgroup of G containing P, prove that if P is normal in H then P is normal in G.

5. Prove that a subgroup H of a p-group G, for which the index of H in G is p, is a normal subgroup.

6. Determine all groups of order 15.

7. Determine the structure of all groups of order pq where p and q are distinct primes.

4-3 AUTOMORPHISMS OF GROUPS

Let G be a group and let $a \in G$. Consider the mapping f of $G \to G$ defined by $xf = a^{-1}xa$, for all $x \in G$. Clearly $(xy)f = a^{-1}xya = (a^{-1}xa)(a^{-1}ya) = (xf)(yf)$, and hence f is an endomorphism of G. It is surjective, for if $z \in G$, then $(aza^{-1})f = a^{-1}(aza^{-1})a = z$. It is injective, for $a^{-1}xa = a^{-1}ya$ implies $x = y$. Hence f is an automorphism of G. An automorphism of G of this type is called an **inner automorphism** of G, and we can form such an inner automorphism with each element $a \in G$. Not all automorphisms of G necessarily have this form. An automorphism of G that is not an inner automorphism is called an **outer automorphism.**

If a and b are two distinct elements of G, it does not follow that they give rise to two distinct inner automorphisms. For instance, in an abelian group it is easy to see that there is only one inner automorphism of the group, namely the identity mapping I_G of G. Thus any other automorphisms which an abelian group may have would have to be outer automorphisms.

A normal subgroup H of a group G is invariant under an inner automorphism f of G. This means $Hf \subset H$. This is the reason why some writers call normal subgroups **invariant subgroups.**

The automorphisms (inner and outer) of a group G form a group

$A(G)$ under map composition. For if $\alpha, \beta \in A(G)$ then $\alpha\beta$ is an automorphism and α^{-1} is an automorphism. For $(xy)(\alpha\beta) = ((xy)\alpha)\beta = ((x\alpha)(y\alpha))\beta = (x(\alpha\beta))(y(\alpha\beta))$ which proves $\alpha\beta$ is an endomorphism of G. Also $x(\alpha\beta) = y(\alpha\beta)$ implies $(x\alpha)\beta = (y\alpha)\beta$. Hence $x\alpha = y\alpha$, and this implies $x = y$. Thus $\alpha\beta$ is injective. Moreover $\alpha\beta$ is surjective, since both α and β are surjective. Therefore $\alpha\beta \in A(G)$. Moreover, since α is bijective it has a bijective inverse α^{-1}. We next show $\alpha^{-1} \in A(G)$. For $x, y \in G$, put $x\alpha^{-1} = u$, $y\alpha^{-1} = v$. Then $x = u\alpha$, $y = v\alpha$ and hence $xy = (uv)\alpha$. Therefore $uv = (xy)\alpha^{-1}$ and hence $(xy)\alpha^{-1} = (x\alpha^{-1})(y\alpha^{-1})$.

The group $A(G)$ is called the **group of automorphisms of G.** It is a multiplicative group. In general, there is no exchange of properties between a group and its group of automorphisms. Examples exist, for instance, where G is abelian and $A(G)$ is not, and where G is nonabelian and yet $A(G)$ is abelian.

The inner automorphisms $J(G)$ of G form a subgroup of $A(G)$. For if $\alpha, \beta \in J(G)$ then for some $a \in G$, $x\alpha = a^{-1}xa$, for all $x \in G$, and for some $b \in G$, $x\beta = b^{-1}xb$ for all $x \in G$. Hence $x(\alpha\beta) = (a^{-1}xa)\beta = (ab)^{-1}x\,(ab)$, and we see that $\alpha\beta \in J(G)$. It follows readily that $J(G)$ is a subgroup of $A(G)$. We can say more than this for we next show $J(G)$ is a normal subgroup of $A(G)$. Let $\phi \in A(G)$, $\alpha \in J(G)$, where $x\alpha = a^{-1}xa$, $x \in G$. Then $x(\phi^{-1}\alpha\phi) = x\phi^{-1}(\alpha\phi) = [a^{-1}(x\phi^{-1})a]\phi = (a^{-1}\phi)x(a\phi) = (a\phi)^{-1}x(a\phi)$. Thus $\phi^{-1}\alpha\phi$ is the inner automorphism of G corresponding to the element $a\phi$ of G. Hence $J(G) \mathrel{\Delta} A(G)$.

We define a mapping f of $G \to J(G)$ by $af = \alpha$, where α is the inner automorphism of G corresponding to the element $a \in G$. Thus $x\alpha = a^{-1}xa$ for all $x \in G$. If $bf = \beta$, $b \in G$, then $x\beta = b^{-1}xb$ for all $x \in G$. Hence $(af)(bf) = \alpha\beta$ where, as we saw above, $x(\alpha\beta) = (ab)^{-1}x(ab)$ for all $x \in G$. Thus $(af)(bf) = (ab)f$. This proves f is a group homomorphism. Clearly it is surjective, hence by the fundamental theorem of group homomorphisms $G/\ker f \approx J$. What is $\ker f$? Of course we know it is a normal subgroup of G, but which normal subgroup? Now $a \in \ker f$ if and only if $af = \alpha$ is the neutral element of J—that is, the identity mapping I_G of G. Hence $a \in \ker f$ if and only if $a^{-1}xa = x$ for every $x \in G$—that is, if and only if $xa = ax$ for every $x \in G$. Hence $\ker f$ is the center $C(G)$ of G. Our final result is then

$$G/C(G) \approx J(G).$$

For an abelian group $C(G) = G$ and $J(G)$ consists then of the one automorphism of G, the identity mapping I_G of G. The other extreme is if $C(G) = 1$, in which case G and $J(G)$ are isomorphic groups.

EXERCISES

1. Find all the automorphisms of the group S_3. Find the center of S_3 and the order of the group of inner automorphisms of S_3.

2. Do the same for S_4.

3. Find all the inner automorphisms of S_4 and find the order of the center of S_4.

4. If A and B are subgroups of a group G, define $[A, B]$ as the subgroup of G generated by all elements of the form $aba^{-1}b^{-1}$, $a \in A$, $b \in B$. Thus $[G, G]$ is the commutator subgroup of G. If H is a normal subgroup of G and K/H is the center of G/H prove that $[K, G] \subset H$.

5. Prove the converse of Exercise 4.

6. If a subset S of a group G is mapped into itself by every inner automorphism of G, prove that the subgroup of G generated by S is normal.

7. Find the two automorphisms of the additive group Z of integers. Are they inner automorphisms?

8. If the center of a group G is 1, the unit group, prove the center of the group of automorphisms of G is 1.

4-4 DIRECT PRODUCTS OF GROUPS

Direct Products

If A and B are subgroups of a group G and if at least one of them, say A, is a normal subgroup of G, then the set AB of all elements ab, $a \in A$, $b \in B$, is a subgroup of G. For the product of two such elements $(ab)(a'b') = a(ba'b^{-1})bb' \in AB$, since $ba'b^{-1} \in A$. Moreover $(ab)^{-1} = b^{-1}a^{-1} = (b^{-1}a^{-1}b)b^{-1} \in AB$. In fact it is easy to see that $AB = BA$, but not elementwise. That is, every $ab = b(b^{-1}ab) = ba'$, $a' \in A$. Furthermore, it readily follows that AB is the subgroup of G generated by the set $A \cup B$—that is, $AB = A \vee B$, the join of A and B.

If A and B are both normal subgroups of G then AB is also a normal subgroup of G. For if $x \in G$, then $x(ab)x^{-1} = (xax^{-1})(xbx^{-1}) \in AB$. Of course AB is still the subgroup of G generated by $A \cup B$, but now $AB = BA$ elementwise, that is $ab = ba$ for all $a \in A$, $b \in B$.

If the group G is abelian then all its subgroups are normal subgroups. If A and B are subgroups of the additive abelian group G, then the set $A + B$ of all $a + b$, $a \in A$, $b \in B$, is the subgroup generated by $A \cup B$.

Definition. A group G is said to be the *internal direct product* of two normal subgroups A and B of G if $G = A \vee B$ and $A \cap B = 1$ (the unit subgroup of G). This direct product is denoted by $G = A \times B$.

Thus G is the direct product $G = A \times B$ of two normal subgroups A and B of G if and only if $G = AB$ and $A \cap B = 1$.

If G is an additive abelian group then G is called the **direct sum** of two subgroups A and B if $G = A + B$ (that is, G is generated by $A \cup B$), and if $A \cap B = 0$, the unit subgroup of G. We write the direct sum as $G = A \oplus B$.

THEOREM 13. If $G = A \times B$ then $B \approx G/A$.

Proof: The mapping ϕ of $B \rightarrow G/A$ defined by $b\phi = bA$, $b \in B$, is clearly a homomorphism. Since ker $\phi = 1$, ϕ is injective. Moreover, if $x \in G$ then $x = ab$ (unique), so that a coset $xA = abA = b(b^{-1}ab)A = bA$ and hence ϕ is surjective.

Corollary: If G is a finite group and if $G = A \times B$, then the order of G is the product of the orders of A and B.

Proof: By Lagrange's theorem, $\#G$ is the product of the order of A and the index of A in G. By the theorem this index is equal to the order of B.

We now generalize these ideas to any finite number of subgroups.

Definition. Let $H_1, H_2, \ldots, H_n$ be n distinct normal subgroups of a group G. G is called the **internal direct product** of the normal subgroups H_i, $i = 1, 2, \ldots, n$, $G = H_1 \times H_2 \times \cdots \times H_n$ if (1) G is generated by $\bigcup_{i=1}^{n} H_i$, (2) the intersection of every H_i with the subgroup generated by all $H_j, j \neq i$, is the unit subgroup 1.

THEOREM 14. G is the direct product of the distinct subgroups H_i, $i = 1, 2, \ldots, n$ if and only if (1') $H_i H_j = H_j H_i$, $i \neq j$, elementwise, and (2') every element $x \in G$ has a unique representation in the form of a product, $x = h_1 h_2 \ldots h_n, h_i \in H_i, i = 1, 2, \ldots, n$.

Proof: Assume G is the direct product of the normal subgroups H_i, $i = 1, 2, \ldots, n$. We first prove (1'). Let $h_i \in H_i$, $h_j \in H_j$, $i \neq j$. Then

$$h_i h_j h_i^{-1} h_j^{-1} = h_i(h_j h_i^{-1} h_j^{-1}) \in H_i$$
$$= (h_i h_j h_i^{-1})h_j^{-1} \in H_j.$$

By part (2) of the definition of a direct product we know

$$H_i \cap H_j = 1.$$

Hence, $h_i h_j h_i^{-1} h_j^{-1} = e$, and therefore $h_i h_j = h_j h_i$.

Next we prove (2'). Let x be any element of G. Since G is generated by the H_i, x is a finite product of elements from the H_i. Since for $h_i \in H_i, h_j \in H_j, i \neq j, h_i h_j = h_i h_i$, we can write this product in the form $x = h_1 h_2 \cdots h_n$. Assume $x = h_1' h_2' \cdots h_n', h_i' \in H_i, i = 1, 2, \ldots, n$

is a second such form. Then $h_1h_2 \cdots h_n = h_1'h_2' \cdots h_n'$ and therefore $h_1'^{-1}h_1 = h_2'h_3' \cdots h_n' (h_2 \cdots h_n)^{-1} = h_2'h_2'^{-1} \cdots h_n'h_n^{-1}$. Thus $h_1'^{-1}h_1$ belongs to the subgroup generated by $H_2 \cup H_3 \cdots \cup H_n$. But $h_1'^{-1}h_1 \in H_1$. Hence, $h_1'^{-1}h_1 = e$, that is $h_1' = h_1$. Similarly

$$h_i' = h_i, i = 2, 3, \ldots, n.$$

This proves (2').

Conversely, assume H_i, $i = 1, 2, \cdots, n$ are subgroups of G that satisfy (1') and (2'). We first show that each $H_i \vartriangle G$. Let $x \in G$, $y \in H_i$. Then by (2'), $x = h_1h_2 \cdots h_n$, $h_i \in H_i$ for all i. Hence $xyx^{-1} = h_1h_2 \ldots h_n yh_n^{-1}h_{n-1}^{-1} \ldots h_1^{-1} = h_iyh_i^{-1} \in H_i$, by (1'). Therefore $H_i \vartriangle G$.

From (2') we see that G is generated by the H_i, $i = 1, 2, \ldots, n$. Next, let $y \in H_i \cap [\bigcup_{j \neq 1} H_j]$. Then $y = h_i$, $h_i \in H_i$ and

$$y = h_1h_2 \ldots h_{i-1}h_{i+1} \ldots h_n \in [\bigcup_{j \neq 1} H_j].$$

Hence by (2'), $h_j = e$, $j = 1, 2, \ldots, n$, and therefore $y = e$. Hence, by definition, G is the direct product of the H_i.

Lemma 4. If G is a finite group and if $G = H_1 \times H_2 \times \cdots \times H_n$ then the order of G is the product of the orders of the H_i.

Proof: This follows from the corollary of Theorem 13 and by finite induction.

THEOREM 15. A finite abelian group is the direct product of its Sylow groups.

Proof: Let A be a finite abelian group. If p is a prime dividing the order of A, denote by A_p the set of all elements of A whose orders are powers of p. It is easy to prove that A_p forms a subgroup of A. In fact, if $\#A = p^kq$, where $p \nmid q$, then $\#A_p = p^k$, that is A_p is the Sylow p-subgroup of A (the only one, since A is abelian). To see this, assume $\#A_p < p^k$. We know A_p is a finite p-group and so its order is a power of p. Hence p divides the order of the factor group A/A_p. Hence A/A_p must contain an element aA_p of order p, that is $(aA_p)^p = a^pA_p = A_p$. Thus $a^p \in A_p$. Hence a^p, and therefore a itself, must have orders which are a power of p. Hence $a \in A_p$ which would contradict aA_p being an element of order p. Therefore $\#A_p = p^k$. (This result also follows at once from theorems 8 and 10).

Suppose then that $\#A = p_1^{k_1}p_2^{k_2}\ldots p_r^{k_r}$, where the p_i are distinct primes. Form the r Sylow groups A_{p_i}, $i = 1, 2, \ldots, r$. Their product $A_{p_1}A_{p_2}\ldots A_{p_r}$ is a subgroup of A whose order is clearly equal to the order of A. Hence $A = A_{p_1}A_{p_2}\ldots A_{p_r}$. Moreover, since the p_i are distinct primes, the intersection of any A_{p_i} with the subgroup generated by the remaining ones is always 1. Thus A is the direct product of the A_{p_i},

$$A = A_{p_1} \times A_{p_2} \times \cdots \times A_{p_r}.$$

This direct product is absolutely unique, that is the terms themselves are unique.

We can generalize the definition of internal direct product to an arbitrary family (H_i), $i \in I$, of normal subgroups of a group G (where I is an infinite set).

If (1) G is generated by this family (that is, every element of G is a product of a finite number of elements from the H_i), and (2) if the intersection of each H_i with the subgroup generated by all H_j, $j \neq i$, is the unit subgroup 1, then G is called the **internal direct product of the family** (H_i), $i \in I$, **of normal subgroups.** It can be shown as before that (2) is equivalent to: (2') the representation of an element of G as a product of elements from the H_i is unique. We write $G = \underset{i \in I}{\times} H_i$ for this internal direct product.

For additive abelian groups the internal direct product is called the **direct sum.** If the additive abelian group G is the direct sum of a family of subgroups (H_i), $i \in I$, then we write

$$G = \sum_{i \in I} \oplus \; H_i.$$

Note that each element of $\Sigma \; \oplus \; H_i$ is a finite sum. The terms H_i in a direct sum are called **direct summands.**

Definition. A group is said to be **indecomposable** if it is not the direct product of proper subgroups.

4-5 THE EXTERNAL DIRECT PRODUCT

The External Direct Product

Let (G_i), $i \in I$ be a family of groups and let $\displaystyle\prod_{i \in I} G_i$ be the cartesian product of this family. We can create a new group out of this family by defining a binary operation in ΠG_i and turning it into a group, as follows.

For $f, g \in \Pi G_i$ define a product fg by $i(fg) = (if)(ig)$, $i \in I$. It is verified at once that this is an associative binary operation with the neutral element e, defined by $ie = e_i$, $i \in I$, where e_i is the neutral element of G_i, and with the inverse f^{-1} of $f \in \Pi G_i$ given by if^{-1}, $= (if)^{-1}$, $i \in I$.

This new group will still be designated by ΠG_i and is called the **external direct product** of the family of groups (G_i), $i \in I$. As we have observed in Chapter 1, its elements can be expressed in the "vector" notation (x_i), $i \in I$, $x_i \in G_i$. For each mapping $f \in \Pi G_i$ determines a unique (x_i), in which $if = x_i$, $x_i \in G_i$. Conversely, each (x_i) determines a unique function f by $if = x_i$, $x_i \in G_i$. In this notation the binary operation takes the form $(x_i)(y_i) = (x_i y_i)$, $x_i, y_i \in G$. The neutral element is (e_i), $i \in I$, and the inverse $(x_i)^{-1} = (x_i^{-1})$, $i \in I$.

For each $j \in I$ the group ΠG_i contains a subgroup G'_j isomorphic to G_j. G'_j is the subgroup of all elements (x_i) where $x_i = e_i$, $i \neq j$, $x_j \in G_j$. The isomorphism is given by $x_j \rightarrow (x_i)$ where $x_i = e_i$, $i \neq j$, $x_j \in G_j$. Moreover, it is easy to show that each $G'_j \, \Delta \, \Pi G_i$.

The subset W of the direct product ΠG_i consisting of all elements (x_i) for which almost all (all but a finite number) $x_i = e_i$, $i \in I$, clearly forms a subgroup of the direct product. W is called the **weak direct product** of the family (G_i), $i \in I$. It is a straightforward exercise to show that if $(x_i) \in W_i$, $(y_i) \in \Pi G_i$, then $(y_i)(x_i)(y_i)^{-1} = (y_i x_i y_i^{-1}) \in W$, that is $W \, \Delta \, \Pi G_i$.

Obviously if I is a finite set then $W = \Pi G_i$, that is the external direct product of a finite number of groups is the same as the weak direct product of these groups. If I is the set $\{1, 2, \ldots, n\}$ then $(x_i) = (x_1, x_2, \ldots, x_n)$.

If the (G_i), $i \in I$, are additive abelian groups, then the binary operation in ΠG_i is defined by $i(f + g) = if + ig$, $i \in I$, $f, g \in \Pi G_i$, or in the "vector" notation $(x_i) + (y_i) = (x_i + y_i)$, $x_i, y_i \in G_i$. The symbol 0 is now used for the neutral element and $-(x_i) = (-x_i)$.

If the (G_i), $i \in I$, are all subgroups of a group G and if G is the internal direct product of the G_i then G is isomorphic to the weak direct product W of the G_i. In particular if the G_i are subgroups of an additive abelian group G and if G is the direct sum of the G_i, then G is isomorphic to the weak direct product of the G_i. To see this, let $f \in W$. Then for $i \in I$, $if = 0$ except at a finite number of points $i_1, i_2, \ldots, i_n$ of I. The isomorphism is then the mapping $f \rightarrow x_{i_1} + x_{i_2} + \cdots + x_{i_n}$, where $x_{i_r} \in G_{i_r}$, $r = 1, 2, \cdots, n$. This is easily checked, recalling that addition is the binary operation in both W and G.

This isomorphism illuminates the connection between the external and internal direct products and justifies the common term *direct product* in both names. We shall identify the direct sum with the weak direct product, since they represent essentially the same group in two different notations.

THEOREM 16. Let C be a cyclic group of order $n = p_1^{\alpha_1} p_2^{\alpha_2} \cdots p_s^{\alpha_s}$, where the p_i are distinct primes. Then

$$C = P_1 \oplus P_2 \oplus \cdots \oplus P_s$$

where the P_i are cyclic groups of orders p^{α_i}, $i = 1, 2, \cdots, s$.

Proof: Let P_i be the subgroup of all elements of C whose orders are powers of the prime p_i. Being a subgroup of a cyclic group, P_i is itself a cyclic group. Since the p_i are distinct primes, the intersection of each P_i with the subgroup generated by the remaining P_j, $j \neq i$, is the unit subgroup 0. The order of the direct sum $P_1 \oplus P_2 \oplus \cdots \oplus P_s$ is n and, since it is a subgroup of C, it must be C.

Lemma 5. A finite cyclic group $C = [x]$ is indecomposable if and only if its order is a power p^α of a prime p.

Proof: Assume the order of G is p^α. If $C = A \oplus B$, then $\#A = p^\beta$ and $\#B = p^\gamma$, where $\beta = \gamma = \alpha$. If $\beta < \gamma$ then $p^\gamma x \in A \cap B$, and if $\beta = \gamma$ then $A = B$, both of which imply C is not the direct sum of A and B and hence is indecomposable. Conversely, assume C is indecomposable. If $\#C = rp^\beta$, where $p \nmid r$, then $C = A \oplus B$, where (by the previous theorem) $\#A = r$ and $\#B = p^\beta$. This is a contradiction and hence the order of C must be a power of some prime.

Lemma 6. An infinite cyclic group $C = [x]$ is indecomposable.

Proof: Suppose $C = A \oplus B$, then A and B are cyclic groups and $A = [rx]$, $B = [sx]$ for $r \neq s$. But $rsx \in C_1 \cap C_2$ which contradicts $C = C_1 \oplus C_2$. Hence C is indecomposable.

Example. The additive group of integers is an infinite cyclic group and hence indecomposable.

We combine the last two lemmas into the theorem.

THEOREM 17. A cyclic group is indecomposable if and only if it is an infinite group or a finite group of prime power order.

EXERCISES

1. Prove that the direct product $H_1 \times H_2 \times \cdots \times H_r$ of cyclic groups H_i whose orders are powers $p^{\alpha i}$, $1 \leq i \leq r$, of distinct primes is a cyclic group.

2. A and B are normal subgroups of orders m and n respectively of a finite group G. If m and n are coprime prove that $AB \approx A \times B$.

3. If $G = H_1 \times H_2 \times \cdots \times H_n$, prove
(a) $G_i = H_{i+1} \times \cdots \times H_n$ is a subgroup of G.
(b) G_i is a normal subgroup of G_{i-1}.
(c) G_i is a normal subgroup of G.

4. A and B are normal subgroups of G and $G = A \times B$. If H is a subgroup of G containing A, prove $H = A \times (H \cap B)$.

4-6 FREE ABELIAN GROUPS

Let S be any nonempty set. The pair (F, f) consisting of an abelian group F and a mapping f of $S \to F$ is called a **free abelian group on S** if it has the following property: For each choice of an abelian group X and of a mapping g of $S \to X$, there exists a unique (group) homomorphism h of $F \to X$ such that $fh = g$ on S.

This property is illustrated in the accompanying diagram.

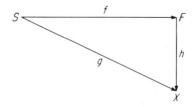

It is called a **commutative** diagram by reason of the equation $g = fh$. The property is known as the **universal factorization property** (U.F.P.) and the free abelian group (F,f) is said to have the U.F.P. on S. It means that for any choice of the pair (X,g) there exists a homomorphism h of $F \to X$, unique for this choice of (X,g), such that the mapping g factors into $g = fh$.

As yet we do not know whether a free abelian group (F,f) exists on a set S. Before proving its existence, we derive some properties of a free abelian group. The group F itself is frequently spoken of as the free abelian group on S.

THEOREM 18. If (F,f) is a free abelian group on a set S then the mapping f is injective.

> **Proof:** The theorem is clearly true if S contains only one element. Let x and y be any two distinct elements of the given set S and let X be a group of more than one element. Choose a mapping g of $S \to X$ such that $xg \neq yg$. Then $xfh \neq yfh$ and hence $xf \neq yf$. This proves f is an injective mapping.

THEOREM 19. The image Sf of f is a set of generators of the free group F.

> **Proof:** Let F' be the subgroup of F generated by the set Sf. Apply the U.F.P. to the free abelian group (F,f), taking the arbitrary group X to be F' and the mapping g of $S \to X = F'$ to be f. This is possible since the mapping f of $S \to F$ can be regarded as a mapping of $S \to F'$. Then there exists a unique group homomorphism h

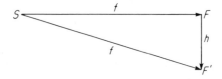

of $F \to F'$ such that $f = fh$. Again using the U.F.P. of (F,f), but this time with $X = F$ and $g = f$, we get a unique group homomorphism ϕ of $F \to F$ such that $f = f\phi$. Since $f = fI_F$, where I_F is

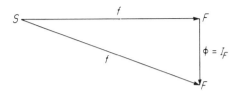

the identity mapping of F on itself, $\phi = I_F$, for I_F is a homomorphism. Since F' is a subgroup of F, the homomorphism h can be regarded as a homomorphism of $F \to F$ such that $f = fh$. Since $\phi = I_F$ is unique, $h = I_F$, that is $F = F'$ and therefore Sf generates F.

THEOREM 20. If (F,f) and (F',f') are free abelian groups on the same set S then F and F' are isomorphic groups and there exists a unique isomorphism h of $F \to F'$ such that $f' = fh$.

Proof: The U.F.P. of (F,f) with $X = F'$ and $g = f'$, yields the unique homomorphism h of $F \to F'$ for which $f' = fh$. The U.F.P. of (F',f') with $X = F$ and $g = f$ yields the unique homomorphism h' of $F' \to F$ for which $f = f'h'$. Thus $f = fhh'$, where hh' is a homomorphism of $F \to F$. The U.F.P. of (F,f) with $X = F$ and $g = f$ yields the unique homomorphism I_F (the identity mapping of F on itself) of $F \to F$ for which $f = fI_F$. Hence $hh' = I_F$. Similarly $h'h = I_{F'}$. Hence both h and h' are isomorphisms. Since h is unique, it is the unique isomorphism of $F \to F'$ for which $f' = fh$.

This last theorem shows that if the free abelian group (F,f) on S exists then the U.F.P. characterizes it up to an isomorphism. Thus the free abelian group on a given set S is essentially unique.

Next we prove the existence of a *free* abelian group (F,f) on any set S by actually constructing it.

THEOREM 21. There exists a free abelian group (F,f) on any set S.

Proof: Let Z be the additive group of integers. Let F be the set of functions α of $S \to Z$, such that almost all (that is all but a finite number) $s\alpha = 0$, $s \in S$. This means then that $s\alpha$ is a nonzero integer at only a finite number of points of S. For $\alpha, \beta \in F$ define $\alpha + \beta$ by $s(\alpha + \beta) = s\alpha + s\beta$, for all $s \in S$. This defines a binary operation in F which is clearly both commutative and associative.

Define $\alpha_0 \in F$ as the function which maps every $s \in S$ into the integer 0. Then F becomes an abelian group with α_0 as the neutral element and $-\alpha$, defined by $s(-\alpha) = -(s\alpha)$ for all $s \in S$, as the inverse of α.

Next define a mapping f of $S \rightarrow F$ by $sf = f_s$, $s \in S$, where f_s is defined as follows:

For $x \in S$, $xf_s = 0$ if $x \neq s$ and $sf_s = 1$. Then (F,f) is a free abelian group on S. For let X be any abelian group and let g be any mapping of $S \rightarrow X$. Define a mapping h of $F \rightarrow X$ by

$$\alpha h = \sum_{s \in S} (s\alpha)(sg), \ \alpha \in F.$$

Note that this is a finite sum since almost all $s\alpha = 0$. For $\alpha, \beta \in F$, $s\alpha$ and $s\beta$ are integers and it is easy to see that $(\alpha + \beta)h = \alpha h + \beta h$ which proves h is a homomorphism. From the definition of h it is seen that $f_s h = sg$, that is $(sf)h = sg$ and therefore $s(fh) = sg$ for every $s \in S$. Hence $fh = g$. We have thus shown that for any abelian group X and for any mapping g of $S \rightarrow X$, there exists a homomorphism h of $F \rightarrow X$ for which $fh = g$.

To complete the proof that (F,f) has the U.F.P. on S, that is that (F,f) is the free abelian group on S, we need to prove that the homomorphism h is unique. Suppose then that h' is a homomorphism of $F \rightarrow X$ with the property that $g = fh'$. Let $\alpha \in F$ and let $s_1, s_2, \ldots, s_n$ be the elements of S at which $s\alpha \neq 0$. Let $\alpha \in F$, then $s\alpha = 0$ except at a finite number of points $s_1, \ldots, s_n$ of S. Consider the element ϕ of F defined by $\phi = \sum_{1}^{n} (s_i \alpha) f_{s_i}$. $s\phi = \Sigma (s_i \alpha)(sf_{s_i})$. Clearly $s\phi = 0$ if s is not one of the s_i, $i = 1, \ldots, n$, while if $s = s_i$, then $s_i \phi = s_i \alpha$. Therefore $\phi = \alpha$.

Hence $\alpha = \sum_{i=1}^{n} (s_i \alpha) f_{s_i}$. For $s\alpha = \sum_{i=1}^{n} (s_i \alpha)(sf_{s_i})$ and all $sf_{s_i} = 0$ if s is not equal to any s_i, whence $s\alpha = 0$. However if $s = s_i$ then $s\alpha = s_i \alpha$, since $s_j f_{s_i} = 0$, $i \neq j$, while $s_i f_{s_i} = 1$. Hence, since h' is a homomorphism

$$\alpha h' = (\Sigma (s_i \alpha) f_{s_i}) h' = \Sigma (s_i \alpha)(f_{s_i} h') = \Sigma (s_i \alpha)[s_i(fh')]$$

$$= \Sigma (s_i \alpha)(s_i g) = \alpha h$$

for all $\alpha \in F$. Hence $h' = h$ and therefore h is unique. This completes the proof that the pair (F,f) we have constructed is the free abelian group on S.

Corollary. The elements f_s, $s \in S$, constitute a set of generators of the free abelian group (F,f) constructed in the theorem.

Proof: This follows at once from Theorem 19, since $f_s = sf$.

4-7 THE BASIS OF A FREE ABELIAN GROUP

Let $\sum_{i=1}^{k} n_i f_{s_i}$, where the n_i are integers, be any finite linear sum of the generators of the free abelian group (F, f) constructed in the last theorem. Suppose $\Sigma n_i f_{s_i} = \alpha_0$, where α_0 is the neutral or zero element of F. Then $\Sigma n_i(s f_{s_i}) = s\alpha_0 = 0$, for every $s \in S$. Putting $s = s_1, s_2, \ldots, s_n$ in succession, we see that each $n_i = 0, i = 1, 2, \ldots, k$.

We have proved that if any finite linear sum of the generators f_s of F is zero, then each integral coefficient n_i of this sum has to be zero. This property is called **linear independence,** and the f_s are therefore a linearly independent set of generators of F. A linearly independent set of generators of a free abelian group is called a *basis* of the group. Thus F has a basis $f_s, s \in S$.

The mapping f of a free abelian group (F, f) on S is injective and hence S and Sf are bijective sets. We can regard therefore S as being identified with Sf. If this is done, S becomes a subset of F and we speak of F as the free abelian group generated by S. The mapping f becomes then the inclusion mapping $S \to F$.

THEOREM 22. A set S of generators of an abelian group A is a basis if and only if each element of A has a unique representation in the form of a finite sum of elements of S with integral coefficients.

Proof: Let S be a basis. Assume $a \in A$ has two such representations, $a = \sum_{i=1}^{k} n_i s_i = \sum_{i=1}^{k} m_i s_i, s_i \in S$. (Note there is no loss of generality in taking the same s_i in each representation, since we can always fill in terms by using zero coefficients.) Hence

$$\Sigma(n_i - m_i)s_i = 0,$$

and therefore $n_i = m_i$ for each i. Hence we have uniqueness. Conversely, assume uniqueness of representation. Let $\Sigma n_i s_i = 0$. Since $\Sigma 0 \cdot s_i = 0$, the uniqueness implies $n_i = 0$ for each i. Hence S is a linearly independent set of generators and therefore is a basis.

THEOREM 23. An abelian group A and an injective mapping γ of $S \to A$ is a free abelian group on S if and only if the image $S\gamma$ of γ is a basis of A.

Proof: If (A, γ) is free on S then, by Theorem 20, A is isomorphic to F, where (F, f) is the free abelian group constructed on S. Let j be the isomorphism of $F \to A$. The f_s, $s \in S$, form a basis for F. We prove that the $f_s j$, $s \in S$, form a basis for A. Let x be any element of A. Then there is a unique $\alpha \in F$ such that $x = \alpha j$, and $\alpha = \Sigma(s_i\alpha)f_{s_i}$. Hence $x = \Sigma(s_i\alpha)(f_{s_i}j)$. This shows that the $f_s j, s \in S$, are a set of generators of A.

We next prove they are linearly independent. Assume $\Sigma n_i(f_{s_i}j) = 0$, where the n_i are integers. Hence $(\Sigma n_i f_{s_i})j = 0$. Since j is an isomorphism, this implies $\Sigma n_i f_{s_i} = \alpha_0$, the zero element of F. Since the f_s are a basis of F, each $n_i = 0$, and hence the $f_s j$ are linearly independent and therefore a basis of A.

Conversely, let $S\gamma$ be a basis of A. We prove (A, γ) is a free abelian group on S. Let X be any abelian group and let g be any mapping of $S \rightarrow X$. Let a be any element of A. Then $a = \Sigma n_i(s_i\gamma)$, a unique representation of a in terms of the basis. Define a mapping h of $A \rightarrow X$ by $ah = \Sigma n_i(s_i g)$, where g is the given mapping of $S \rightarrow X$. It is easy to verify that h is a homomorphism. Let s be any element of S. Then $(s\gamma)h = sg$ and hence $s(\gamma h) = sg$ for all $s \in S$. Therefore $\gamma h = g$. All we have left to show is that h is unique. Assume h' to be a homomorphism of $A \rightarrow X$ for which $\gamma h' = g$. Then $ah' = (\Sigma n_i(s_i\gamma))h' = \Sigma n_i(s_i\gamma h') = \Sigma n_i(s_i g) = ah$. Thus $ah = ah'$ for all $a \in A$ and therefore $h' = h$. Hence the pair (A, γ) has the U.F.P. and therefore (A, γ) is a free abelian group on S.

Corollary 1. A subset S of an abelian group A is a basis of A if and only if A is the free abelian group generated by S.

Corollary 2. An abelian group is free if and only if it has a basis.

Proof: This follows from the Corollary and Theorem 1.   ??

Let S be any set. Consider the family of sets X_s, $s \in S$, where each $X_s = Z$, the additive group of integers. The direct sum (weak direct product) of this family of abelian groups is by definition the set of all functions from $S \rightarrow \bigcup_{s \in S} X_s = Z$ whose values are almost all 0. But this is precisely the free abelian group F constructed on S. Hence

$$F = \sum_{s \in S} \oplus X_s.$$

Since any free abelian group A on S is isomorphic to F (Theorem 20) we have

$$A \cong \sum_{s \in S} \oplus X_s.$$

Now any infinite cyclic group is isomorphic to Z and each $X_s = Z$. Hence a free abelian group on S is isomorphic to a direct sum of any family of infinite cyclic groups indexed by the set S. The converse of this is true. We now prove it by showing that if an abelian group A is isomorphic to the free abelian group F constructed on S, then A is a free abelian group on S.

Let j be the isomorphism of $F \to A$. Then $S \xrightarrow{f} F \xrightarrow{j} A$. Define $\gamma = fj$. Then (A, γ) has the U.F.P. on S. Let X be an abelian group and g any mapping of $S \to X$. The U.F.P. of (F, f) yields a unique homomorphism k of $F \to X$ such that $fk = g$. Define a mapping h of $A \to X$ by $h = j^{-1}k$, where j^{-1} is the inverse isomorphism of j. As a product of homomorphisms, h is clearly a homomorphism. Moreover $\gamma h = fjh = fk = g$. Let h' be a homomorphism of $A \to X$ such that $\gamma h' = g$. If a is any element of A, $a = \alpha j$, $\alpha \in F$ and $\alpha = \Sigma(s_i \alpha)f_{s_i} = \Sigma(s_i \alpha)(s_i f)$. Hence $\alpha = \Sigma(s_i \alpha)(s_i fj) = \Sigma(s_i \alpha)(s_i \gamma)$. Therefore, since h' is a homomorphism,

$$ah' = \Sigma(s_i \alpha)(s_i \gamma h') = \Sigma(s_i \alpha)(s_i g) = \Sigma(s_i \alpha)(s_i \gamma h) = [\Sigma(s_i \alpha)(s_i \gamma)]h = ah.$$

Hence $h' = h$. Therefore (A, γ) is a free abelian group on S.

We can summarize these results in the following theorem.

THEOREM 24. An abelian group is free if and only if it is isomorphic to a direct sum of infinite cyclic groups.

Corollary 1. If an abelian group is the direct sum of infinite cyclic subgroups, then the generators of these subgroups are a basis of the abelian group.

THEOREM 25. All bases of the same free abelian group contain the same number of elements (that is, have the same cardinal number).

Proof: Let S be a subset of A that is a basis of A. Let p be a prime. The set $pA = \{pa \mid a \in A\}$ is a (normal) subgroup of A. Then A/pA is an abelian group whose elements are the cosets $a + pA$, $a \in A$. Let F_p be the field of residue classes modulo p. If $\bar{0}, \bar{1}, \bar{2}$, $\ldots, \overline{p-1}$ are the elements of F_p, then we can define a scalar multiplication in A/pA by

$$\bar{x}(a + pA) = xa + pA, \qquad \bar{x} \in F_p.$$

It is easy to verify that this satisfies the axioms for scalar multiplication and A/pA becomes a vector space over F_p. If $s \in S$, then $s \notin pA$. For if $s = pa$, where $a = \Sigma m_i s_i$ (the m_i are integers), then $s = \Sigma p m_i s_i$. This is impossible, since the elements of S are linearly independent. Hence $s + pA$, $s \in S$, are a basis of the vector space A/pA. Now $s \to s + pA$ is a bijection of S and this basis of A/pA. Now all bases of the same vector space have the same number of elements. Hence S, and therefore any basis of A, contains the same number of elements as a basis of A/pA.

Definition. The number of elements in a basis of a free abelian group is called the **rank** of the group. If the basis is an infinite set, then the rank is said to be infinite.

Since the basis of a free abelian group A is the index set of the family of infinite cyclic groups whose direct sum is isomorphic to A, we have

THEOREM 26. A free abelian group A is of rank n if and only if A is isomorphic to a direct sum of n infinite cyclic groups.

THEOREM 27. Every abelian group A is isomorphic to a quotient group of a free abelian group.

Proof: Let S be a subset of A that generates A. Construct the free abelian group (F,f) on S. The U.F.P. of (F,f), with $X = A$ and g taken to be the inclusion of $S \rightarrow A$, yields a unique homomorphism h of $F \rightarrow A$ such that $g = fh$. For $s \in S$, $sg = s$ and these elements generate A. We have $s = sg = (sf)h$. Since every element of A can be expressed in terms of the elements of S, it follows that every element of A is the image under h of an element of F. Hence h is an epimorphism $F \rightarrow A$ of the two groups. Therefore A is isomorphic to the quotient group $F/\ker h$.

4-8 FINITELY GENERATED ABELIAN GROUPS

An abelian group that can be generated by a finite number of elements is quite naturally called a **finitely generated abelian group.**

THEOREM 28. A subgroup G of a free abelian group F of rank n is a free abelian group of rank $m \leq n$. There exists a basis $u_1, \ldots, u_n$ of F and integers $t_1, t_2, \ldots, t_m$ such that (1) $t_1 u_1, t_2 u_2, \ldots, t_m u_m$ is a basis of G, and (2) $t_i \mid t_{i+1}$ for each i in $1 \leq i \leq m - 1$.

Proof: The theorem is true for $n = 1$, since F is an infinite cyclic group. We use induction on n, assuming the theorem to be true for all free abelian groups of rank $n - 1$.

For every basis of F, each element of G is a finite sum with integral coefficients of the basis elements. For all possible bases of F, let t_1 be the least positive integer occurring as a coefficient in all the expressions for the elements of G as such finite sums. Then for some $v_1 \in G$ and for some basis $x_1, x_2, \ldots, x_n$ of F,

$$(1) \qquad v_1 = t_1 x_1 + k_2 x_2 + \cdots + k_n x_n.$$

Now t_1 divides all k_i, $2 \leq i \leq n$. For if

$$k_i = q_i t_1 + r_i, \; 0 \leq r_i < t_1,$$

then

$$v_1 = t_1 (x_1 + q_2 x_2 + \cdots + q_n x_n) + r_2 x_2 + \cdots + r_n x_n.$$

But $u_1 = x_1 + \sum_{i=2}^{n} q_i x_i, x_2, \ldots, x_n$ is a basis of F (it is easy to see these elements form a linearly independent set of generators of F), and $r_i < t_1$. This constitutes a contradiction of the choice of t_1. Hence all $r_i = 0$ and $k_i = q_i t_1$, for all $i = 2, \ldots, n$, and therefore $v_1 = t_1 u_1$. Moreover, for any $y \in G$,

$$(2) \qquad\qquad y = h_1 x_1 + \cdots + h_n x_n.$$

We now prove that $t_1 \mid h_1$. For if $h_1 = q t_1 + r$, $0 < r < t_1$, the from (1) we have

$$y - q v_1 = r x_1 + (h_2 - q k_2) x_2 + \cdots + (h_n - q k_n) x_n$$

Since $y - q v_1 \in G$ and $r < t_1$, this is a contradiction of the choice of t_1, unless $r = 0$. Hence every $y \in G$ can be expressed in the form

$$(3) \qquad\qquad y = s_1 v_1 + s_2 x_2 + \cdots + s_n x_n$$

where of course the s_i are integers. Now let H be the free abelian group generated by $x_2, \ldots, x_n$, then (3) shows that $y - s_1 v_1 \in G \cap H$. Now $G \cap H$ is a subgroup of H, and by the induction hypothesis its rank $m - 1 \leq n - 1$, and there exists a basis $u_2, \ldots, u_m$ of H and integers $t_2, \ldots, t_m$ such that $v_i = t_i u_i$, $i = 2, \ldots, m$, is a basis of $G \cap H$, where $t_i \mid t_{i+1}$, $i = 2, \ldots, m$. Hence $y - s_1 v_1 = a_2 v_2 + \cdots + a_m v_m$; that is, y (and therefore any element of G) is a linear sum of the v_i, and therefore $v_1, v_2, \ldots, v_m$ generate G. But they are linearly independent, for $b_1 v_1 + \cdots + b_m v_m = 0$ implies $b_1 t_1 u_1 + b_m t_m u_m = 0$, and this implies $b_i = 0$ for $1 \leq i \leq m$, since the u_i form a basis of F. Hence G has the basis $v_1, v_2, \ldots, v_m$, $m \leq n$, and is therefore a free abelian group. Now $u_1, u_2, \ldots, u_n$ is a basis of F and $v_i = t_i u_i$, $i = 1, \ldots, m$. We know $t_i \mid t_{i+1}$ for all $i = 2, \ldots, m$, and so all there remains to do is to prove $t_1 \mid t_2$. Let $t_2 = q t_1 + r$, $0 \leq r < t_1$. Now $u_1 - q u_2, u_2, \ldots, u_n$ is easily seen to be a basis of F and $v_2 - v_1 = -t_1(u - q u_2) + r u_2$. Since $v_2 - v_1 \in G$, we have a contradiction of the choice of t_1, unless $r = 0$. Hence $t_1 \mid t_2$. The proof is now complete.

THEOREM 29. An additive abelian group A with n generators is isomorphic to a direct sum of n cyclic groups of orders $t_1, t_2, \ldots, t_n$ such that

$$1 \leq t_1 \leq t_2 \leq \cdots \leq t_n \leq \infty$$

where $t_i \mid t_{i+1}$, $i = 1, 2, \cdots, n - 1$, if t_{i+1} is finite. (Some of the t_i may be 1, in which case the corresponding cyclic groups are trivial, that is consist of the zero element only. This will occur if and only if there are superfluous generators.)

Proof: Let A be generated by $x_1, x_2, \ldots, x_n$. Then $A \approx F/G$, where F is the free abelian group generated by $x_1, x_2, \ldots, x_n$. The rank of F is n and let the rank of G be $m \leq n$. There exists a basis $u_1, u_2, \ldots, u_n$ of F and a basis $v_1, \ldots, v_m$ of G such that $v_i = t_i u_i$, $i = 1, 2, \cdots, m$, where the t_i are positive integers and $t_i \,|\, t_{i+1}$. Construct m cyclic groups $[y_i]$ of orders t_i for $i \leq m$, and $n - m$ infinite cyclic groups $[y_i]$ for $m < i \leq n$. (The y_i, $i = 1, 2, \cdots, n$ are merely symbols for the generators of the cyclic groups.) We shall now prove that F/G is isomorphic to the direct sum C of these n cyclic groups.

Apply the U.F.P. of F using the basis $u_1, u_2, \ldots, u_n$ for F and with $X = C$.

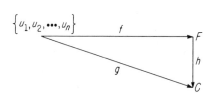

The mapping f is the inclusion $u_i f = u_i$, $i = 1, 2, \cdots, n$ and choose the mapping g to be $u_i g = y_i$, $i = 1, 2, \cdots, n$. We get a unique homomorphism h of $F \to C$ such that $fh = g$, and hence $u_i h = y_i$, $i = 1, 2, \ldots, n$. Clearly h is an epimorphism, for every element $y \in C$ has the form

$$y = \sum_1^n r_i y_i,$$ the r_i are integers, and hence $y = \sum_1^n r_i(u_i h) = (\Sigma r_i u_i)h$,

since h is a homomorphism. The kernel of h is the set of all $\sum_1^n k_i u_i$, the k_i are integers, such that $\sum_1^n k_i y_i = 0$. For $\Sigma k_i y_i = 0$ we must have all $k_i = 0$ for $m < i \leq n$, since these y_i generate infinite cyclic groups, while for $i \leq m$ the k_i must be multiples of the orders t_i of the finite cyclic groups. Thus $k_i = 0$, $m < i \leq n$ and $k_i = t_i q_i$ for $i \leq m$. Hence $\Sigma k_i u_i \in \ker h$ if and only if $\sum_1^m k_i u_i = \sum_1^m q_i t_i u_i = \sum_1^m q_i v_i \in G$. Hence $G = \ker h$. Hence $F/G \approx C$ and therefore $A \approx C$. Thus

(4) $$A = \sum_{i=1}^n \oplus \, [y_i].$$

If the abelian group A is written as a multiplicative group, then the result of the theorem would be expressed in the form

$$A = \prod_{i=1}^{n} [y_i] = [y_1] \times [y_2] \times \cdots \times [y_n].$$

Since infinite cyclic groups are indecomposable, while the finite cyclic groups decompose into indecomposable cyclic groups (cyclic groups whose orders are powers of primes) we have at once the

THEOREM 30 (Decomposition Theorem). Every finitely generated abelian group A is decomposable into a direct sum of indecomposable cyclic groups. The direct summands which are infinite cyclic groups form a direct sum which is a free abelian group F and the rest form a direct sum which is a finite group H, called the **torsion group of A.**

$$A = F \oplus H.$$

Thus a finite abelian group A is indecomposable (that is admits no nontrivial decomposition $A = B \oplus C$) if and only if it is cyclic of prime power order.

Example. Let A be the finitely generated abelian group generated by the elements x, y, z where $3x = 0$, $4y = 0$ and z is of infinite order.

The free abelian group F generated by x, y, z is the direct sum of infinite cyclic groups

$$F = [x] \oplus [y] \oplus [z].$$

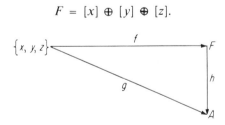

The U.F.P. for F, with f the inclusion mapping into F, and g the inclusion mapping into A, yields a unique homomorphism h and $xh = x$, $yh = y$, $zh = z$. The elements of F are $\alpha x + \beta y + \gamma z$, where α, β, γ are integers. The subgroup ker $h = G$ of F is the set of all $\alpha x + \beta y + \gamma z$ such that $(\alpha x + \beta y + \gamma z)h = \alpha x + \beta y + \gamma z = 0$. Hence $\alpha = ek$, $\beta = 4m$ and $\gamma = 0$, where k and m are positive integers. Thus ker h is the free group generated by $3x$ and $4y$ as a basis.

Choose $u_1 = x + 4y$, $u_2 = x + 3y$, and $u_3 = z$ as a new basis for F, and $v_1 = x + 4y$, $v_2 = 2x + 6y$ as a new basis for G. Then $v_1 = u_1$ and $v_2 = 2u_2$, so that $t_1 = 1$, $t_2 = 2$. These are the bases selected in Theorem 29.

The decomposition of a finitely generated abelian group into indecomposable cyclic groups is not unique.

Example. Let A be the abelian group of order 9 generated by elements x and y where $3x = 0$ and $3y = 0$. $B = \{0, x, 2x\}$, $C = \{0, y, 2y\}$ are cyclic subgroups that generate A and $B \cap C = 0$. Hence $A = B \oplus C$. Also $B' = \{0,\ x + y,\ 2x + 2y\}$, $C' = \{0,\ x + 2y,\ 2x + y\}$ are cyclic subgroups of A that generate A, and $B' \cap C' = 0$. Hence

$$A = B' \oplus C'.$$

However, it is not difficult to prove that this decomposition is unique in this sense: any two decompositions of a finitely generated abelian group into indecomposable subgroups have the same number of terms and their terms can be put into 1–1 correspondence such that corresponding terms are isomorphic groups.

We conclude this section with a brief mention of the elementary divisor and invariant factor theorems for finitely generated abelian groups.

Definition. The *rank* of a finitely generated abelian group is the number of infinite cyclic groups in its decomposition—that is, the rank of its free direct summand.

An additive abelian group A with n generators is isomorphic to the direct sum of n cyclic groups. In general m, say, of the cyclic groups are finite and $n - m$ of them are infinite. If $m = 0$, A is a free abelian group and if $m = n$, A is a finite abelian group. The orders t_i of the finite cyclic groups have the property $t_i \mid t_{i+1}$, $i = 1, 2, \ldots, m - 1$. Those $t_i \neq 1$ are called the **torsion coefficients** or **invariant factors of A.** It can be shown that the invariant factors of A together with the rank $n - m$ of A constitute a complete system of invariants of A, in the sense that two finitely generated abelian groups of the same rank and with the same invariant factors are isomorphic. This is the statement of the so-called Invariant Factor Theorem.

A finite additive abelian group $A \neq 0$ is a direct sum of indecomposable cyclic groups—that is, cyclic groups of prime power orders $p_1^{\alpha_1}$, $p_2^{\alpha_2}, \ldots, p_s^{\alpha_s}$, called the **elementary divisors** of the decomposition of A. It can be proved that any two such decompositions of a finite abelian group have the same set of elementary divisors. This is called the Elementary Divisor Theorem. Moreover, two finite abelian groups can be shown to be isomorphic if and only if they have the same set of elementary divisors.

The Invariant Factor Theorem for finite abelian groups states that the torsion coefficients and the number of (nontrivial) cyclic groups corresponding to these torsion coefficients in the direct sum of a finite abelian group are invariant for the group. It can also be proved that for finite abelian groups the Invariant Factor Theorem implies the Elementary Divisor Theorem and conversely.

The set of generators y_i of A in (4) are often referred to as a "basis" of the finitely generated group A, in the following sense: if

$$k_1 y_1 + k_2 y_2 + \cdots + k_n y_n = 0,$$

the k_i are integers, then $k_i y_i = 0$ for each i. For those generators y_i of infinite cyclic groups this means $k_i = 0$, while for the generators y_i of the finite cyclic groups this means $t_i \mid k_i$. This is then a weaker requirement than that for a basis of a free abelian group. It is easy to prove that the elements $y_1, y_2, \ldots, y_n$ of a finitely generated abelian group A form a "basis" (in this weaker sense) for A if and only if A is the direct sum of the cyclic groups generated by the y_i.

EXERCISES

1. Construct and describe the free abelian group F (a) on the set $S = \{a\}$ consisting of the one symbol a, (b) on the set $S = \{a, b\}$, and (c) on a finite set of n elements. In each case describe the mapping f of $S \to F$ and characterize the group.

2. Prove that the uniqueness of the homomorphism h of $F \to X$ in the U.F.P. for the free abelian group is equivalent to the condition that im f generates F.

3. If B is a subgroup of an abelian group A, prove that if A/B is a free abelian group then B is a direct summand of A.

4. Define a free group and state and prove the analogous theorems (up to the construction of the free group) to those for a free abelian group.

5. Give the details of the following outline of a proof of the existence of a free group on an arbitrary set S and validate it.

If $a, b, c, \ldots$ are elements of S, form a new set in 1–1 correspondence with S and designate these new elements $a^{-1}, b^{-1}, c^{-1}, \ldots$. Form "words," $a_1^{\epsilon_1} a_2^{\epsilon_2} \ldots a_r^{\epsilon_r}$, $a_i \in S$, $\epsilon_i = \pm 1$, and call the length of this word r. Call a word reduced if no a stands next to an a^{-1}. Introduce a word e, called the empty word, which has no letters and has the length 0. Let F be the set of all reduced words together with the empty word e.

Define a binary operation in F as follows: if u and $v \in F$, form their product by juxtaposition and then cancellation (if necessary) of any adjacent a and a^{-1}, this product is either e or a reduced word. Agree that $uv = v$ if $u = e$ and $uv = u$ if $v = e$.

In order to show that this operation is associative, let u, v, w be reduced words. If one of them is e, it is easy to prove that $(uv)w = u(vw)$. If neither of them is e, use induction on the length of v to prove associativity. Check first that $(uv)w = u(vw)$ if the length of $v = 1$. Thus F is a group with e as the neutral element.

Next, define the mapping f of $S \to F$ by $af = a$, $a \in S$. In order to prove (F, f) is a free group on S, let g be any mapping of S into an arbitrary group X.

Define a mapping h of $F \to X$ by $(a_1^{\epsilon_1} \ldots a_r^{\epsilon_r})f = (a_1 g)^{\epsilon_1} \ldots (a_r g)^{\epsilon_r}$. Now show that (a) h is a group homomorphism, (b) $fh = g$, and (c) h is unique.

Hence (F, f) is the free group generated by S.

6. Is an infinite cyclic group a free abelian group?

4-9 SUBNORMAL AND COMPOSITION SERIES

Subnormal series and composition series yield important information about the structure of groups. They also lead to a type of group called a solvable group, which is of very great importance in Galois theory (the application of group theory to the solution of equations). The central series yields another type of group called the **nilpotent group,** a generalization of the abelian group. In fact it turns out that abelian groups form a subclass of the class of nilpotent groups and nilpotent groups a subclass of the class of solvable groups. These are important classifications of groups.

Definition. A **subnormal series** of a group G is a finite descending chain of subgroups of G.

$$(1) \qquad G = G_0 \supset G_1 \supset G_2 \supset \cdots = G_r \supset 1,$$

beginning with G and ending with the unit subgroup 1, such that each G_{i+1} is a normal subgroup of G_i, $0 \leq i \leq r - 1$.

Note well that, except for G_1, the definition does not imply the G_i are normal subgroups of G itself. If all the subgroups in a subnormal series are normal subgroups of G, it is called a **normal series.** Every group has a subnormal series, for example $G \supset 1$.

The factor groups G_i/G_{i+1} are called the **factors** of the subnormal series and the number of factors $\neq 1$ is called the **length** of the series.

A group may have many different subnormal series. Any subnormal series that contains all the subgroups occurring in (1) is called a **refinement** of (1). Every subnormal series is a refinement of itself, but a proper refinement of (1) would contain new subgroups interposed between those of (1), if this is possible.

Definition. Two subnormal series of the same group are called **equivalent** if they have the same length and if there is a one-to-one correspondence between their factors, in which corresponding factors are isomorphic groups.

Our objective in this section is the proof of the celebrated Jordan-Hölder theorem. We reach it in two stages. First we prove a third theorem on group isomorphisms, which will, in turn, enable us to prove Schreier's theorem. The Jordan-Hölder theorem is then an immediate consequence of the Schreier theorem.

Lemma 7. Let D be a subgroup of a group G. Let A, B, C be subgroups of D and assume $A \triangle B$ and $C \triangle D$. Then $AC \triangle BC$.

Since $C \triangle D$ and $A \subseteq D$, we know that AC is therefore a subgroup of D and also that $AC = CA$. For the same reasons, BC is a subgroup of D and, again, $BC = CB$. Hence we have $BC(AC) = (AC)BC$. Since AC is a subgroup of BC, this last equation proves that $AC \triangle BC$.

We shall use this result in proving the following theorem, often called the Zassenhaus lemma.

THEOREM 31 (Third Isomorphism Theorem for Groups). Let G_1 and G_2 be subgroups of G. Let G_i' be a normal subgroup of G_i, $i = 1, 2$. then the following factor groups are isomorphic:

$$(G_1 \cap G_2)G_1'/(G_1 \cap G_2')G_1' \approx (G_1 \cap G_2)G_2'/(G_1' \cap G_2)G_2'$$

Proof: First we note that $G_1' \cap G_2 \triangle G_1 \cap G_2$. Since $G_2' \triangle G_2$ and $G_1 \cap G_2$ is a subgroup of G_2, the lemma above shows that $(G_1' \cap G_2)G_2' \triangle (G_1 \cap G_2)G_2'$. In the same way it follows that $(G_1 \cap G_2')G_1' \triangle (G_1 \cap G_2)G_1'$.

Since $G_1' \cap G_2$ and $G_1 \cap G_2'$ are normal subgroups of $G_1 \cap G_2$, we see next that $(G_1' \cap G_2)(G_1 \cap G_2') \triangle G_1 \cap G_2$. Write

$$T = (G_1 \cap G_2)/(G_1 \cap G_2')(G_1' \cap G_2)$$

and consider the mapping ϕ of $(G_1 \cap G_2)G_1' \to T$, defined by

$$(xg_1')\phi = x(G_1 \cap G_2')(G_1' \cap G_2),$$

where

$$x \in G_1 \cap G_2 \text{ and } g_1' \in G_1'.$$

The mapping ϕ is well-defined. For if $xg_1' = yh_1'$, $x, y \in G_1 \cap G_2$, $g_1', h_1' \in G_1'$, then $y^{-1}x = h_1'g_1'^{-1} \in G_1'$. Since $y^{-1}x \in G_1 \cap G_2$, we have $y^{-1}x \in G_1 \cap G_2 \cap G_1' = G_1' \cap G_2$. And now, because $G_1' \cap G_2 \subset (G_1 \cap G_2')(G_1' \cap G_2)$, we see that xg_1' and yh_1' are mapped into the same coset in T.

The mapping ϕ is a homomorphism, by virtue of

$$(xg_1')(yh_1') = xy(y^{-1}g_1'y)h_1', x, y \in G_1 \cap G_2, g_1', h_1' \in G_1',$$

for $y^{-1}g_1'y \in G_1'$, since $G_1' \triangle G_1$. Clearly ϕ is surjective. Hence ϕ is an epimorphism.

For $x \in G_1 \cap G_2$, $g_1' \in G_1'$, xg_1' belongs to the kernel of ϕ if and only if $x \in (G_1 \cap G_2')(G_1' \cap G_2)$; that is, if and only if $xg_1' \in (G_1 \cap G_2')G_1'$. This follows, since clearly $(G_1' \cap G_2)G_1' = G_1'$. Hence we have $\ker \phi = (G_1 \cap G_2')G_1'$.

Now applying the second isomorphism theorem for groups, we

obtain immediately the result

$$(G_1 \cap G_2)G_1'/(G_1 \cap G_2')G_1' \approx T.$$

Similarly,

$$(G_1 \cap G_2)G_2'/(G_1' \cap G_2)G_2' \approx T.$$

Since isomorphism of groups is transitive, the statement of the theorem follows at once.

THEOREM 32 (Schreier's Theorem). Any two subnormal series of the same group have isomorphic refinements.

Proof: Let

$$(1) \quad G = G_0 \supset G_1 \supset G_2 \supset \cdots \supset G_k = 1$$

$$(2) \quad G = H_0 \supset H_1 \supset H_2 \supset \cdots \supset H_m = 1$$

be two subnormal series of the group G. Set $G_{ij} = G_i(G_{i-1} \cap H_j)$, $H_{ij} = H_j(H_{j-1} \cap G_i)$, $i = 1, 2, \ldots, k$, $j = 1, 2, \ldots, m$. Note that

$$G_{im} = G_i, \qquad H_{kj} = H_j$$
$$G_{io} = G_{i-1}, \qquad H_{oj} = H_{j-1}.$$

Now $G_i \vartriangle G_{i-1}$ and $G_{i-1} \cap H_j$ is a subgroup of G_{i-1}, hence G_{ij} is a subgroup of G_{i-1}. Similarly, H_{ij} is a subgroup of H_{i-1}. Moreover, $G_{ij} \vartriangle G_{i,j-1}$ and $H_{ij} \vartriangle H_{i,j-1}$. This follows at once from Lemma 7. Hence we can write

$$G_{i-1} \supset G_{io} \supset G_{i1} \supset \cdots \supset G_{im} = G_i$$
$$H_{j-1} \supset H_{oj} \supset H_{ij} \supset \cdots \supset H_{kj} = H_j.$$

Furthermore, by the third isomorphism theorem, we have

$$(3) \qquad\qquad G_{i,j-1}/G_{ij} \approx H_{i-1,j}/H_{ij}.$$

Hence, if in (1) we insert all the subgroups G_{ij}, $j = 1, 2, \ldots, i - 1$ between G_{i-1} and G_i for each $i = 1, 2, \ldots, k$, the result is a refinement of (1) which, in general, has repetitions, since some subgroups $G_{i,j-1}$ and G_{ij} may be equal. Similarly we refine (2) by inserting the subgroups H_{ij}. By reason of (3), it is seen that these refinements are equivalent. Repetitions can be eliminated as follows. If $G_{i,j-1} = G_{ij}$, then it follows from (3) that $H_{i,j-1} = H_{ij}$, and therefore the repetitions can be simultaneously removed from the refinements without any injury being done to their equivalence. This proves the Schreier theorem.

Note that there are $km + 1$ subgroups in each subnormal series of Theorem 32 after the subgroups G_{ij} and H_{ij} are inserted and before any possible repetitions are removed.

Definition. A proper subgroup H of a group G is called a **maximum normal subgroup of** G if (1) $H \triangle G$ (2) there exists no proper normal subgroup K of G such that $H \subset K \subset G$.

If $H \triangle G$, then the normal subgroups of G/H have the form K/H, where $H \subset K$ and $K \triangle G$. If H is a maximum normal subgroup of G, then G/H contains no proper normal subgroups. A group that contains no proper normal subgroups is called a **simple group.** Conversely, if G/H is simple, then H is a maximum normal subgroup of G. Thus H is a maximum normal subgroup of G if and only if G/H is a simple group.

Definition. A **composition series** of a group G is a subnormal series whose factors are simple groups $\neq 1$, that is $G = G_0 \supset G_1 \supset G_2 \supset \cdots \supset G_r = 1$ is a composition series if each G_{i+1} is a maximum normal subgroup of $G_i, 0 \leqq i \leqq r - 1$.

Thus a composition series admits of no refinements. Clearly, a finite group always has a composition series.

Example. If A_3 is the alternating subgroup of the symmetric group S_3 of degree 3, then $S_3 \supset A_3 \supset 1$ is a composition series of S_3.

Example. $S_4 \supset A_4 \supset V \supset T \supset 1$ is a composition series of the symmetric group of degree 4. Again, A_4 is the alternating group, V is the group e, (12)(34), (13)(24), (14)(23), and T is the group e, (12)(34).

Not every group has a composition series. An infinite cyclic group has no composition series, for the next-to-last term in a subnormal series for such a group is an infinite cyclic subgroup, and it therefore contains another infinite cyclic group as a subgroup.

Lemma 8. An abelian group A is simple if and only if it is cyclic of prime order.

Proof: Let $a \neq e$ be an element of A. If A is simple then $[a] = A$ and hence A is cyclic. Moreover A cannot be infinite, for it would then contain infinite cyclic subgroups. Let $\#A = m$. If $m = pq, p > 1$, $q > 1$, then a^p is of order q and $[a^p]$ would be a subgroup of A of order q. Hence m must be prime. The converse is obviously true.

An infinite abelian group G has no composition series. For if $G = G_0 \supset G_1 \supset \cdots G_r = 1$ is a composition series, then the G_i/G_{i+1} are simple and hence, by the lemma, are cyclic groups of prime order p_i. Therefore $\#G_{r-1} = p_{r-1}$, $\#G_{r-2}/G_{r-1} = p_{r-2}$ and hence $\#G_{r-2} = p_{r-1}p_{r-2}$. It is not hard to see that this leads inevitably to $\#G = p_0 p_1 p_2 \cdots p_{r-1}$, which contradicts G being an infinite group.

THEOREM 33 (Jordan-Hölder Theorem). Any two composition series of the same group are equivalent.

Proof: By the Schreier theorem, two such series have equivalent refinements. Since neither can be refined further, they must therefore be equivalent.

Thus the Jordan-Hölder theorem associates with any finite group a sequence of simple groups which yield some insight into the structure of G.

Definition. A group G is called **solvable** if G has a subnormal series whose factors are abelian groups.

Since A_3 is a subgroup of index 2 in S_3, the factor group S_3/A_3 is abelian. Moreover, A_3 is itself an abelian group, and so the series $S_3 \supset A_3 \supset 1$ has abelian factors and therefore S_3 is a solvable group. Of course every abelian group is a solvable group.

THEOREM 34. A finite group G is solvable if and only if the composition factors of G are cyclic of prime order.

Proof: Assume G is solvable. Then it has a subnormal series

$$G = G_0 \supset G_1 \supset \cdots \supset G_r = 1$$

in which all the G_i/G_{i+1} are abelian. Since G is a finite group, this refines to a composition series for G whose factors are simple. We show the composition factors are abelian. Suppose that in the refinement a subgroup G' is interposed between G_i and G_{i+1}, $G_i \supset G' \supset G_{i+1}$, where $G' \vartriangle G_i$ and $G_{i+1} \vartriangle G'$. G'/G_{i+1} is a subgroup of the abelian group G_i/G_{i+1} and hence is abelian. Next we need to prove that G_i/G' is abelian. It is easy to see that the mapping $xG_{i+1} \rightarrow xG'$ is an epimorphism of $G_i/G_{i+1} \rightarrow G_i/G'$. Hence as a homomorphic image of an abelian group, G_i/G' is abelian. Thus the refinement of a series whose factors are abelian has abelian factors. Thus the composition series of G has simple abelian factors and hence they must be cyclic of prime orders.

The converse is obviously true.

We leave it as exercises for the reader to prove that every subgroup of a solvable group is solvable and every factor group of a solvable group is solvable.

Let $f(x) = a_0 + a_1 x + \cdots + a_n x^n$ be a polynomial whose coefficients a_i are elements of a field F. An **extension field** F' of F is a field containing F as a subfield. The least extension field E of F in which the polynomial factors into linear factors is called the **splitting field** of the polynomial $f(x)$. By the least extension field is meant there exists no intermediate field K, $F \subset K \subset E$, in which $f(x)$ factors into linear factors.

The group of the equation $f(x) = 0$ is defined as the group of automorphisms of E over F, where E is the splitting field of $f(x)$. It can be shown (this is called Galois theory, after the mathematician Evariste Galois) that $f(x) = 0$ is "solvable by radicals" if and only if its group is solvable, in the sense of the above definition of solvability. The phrase "solvable by radicals" means the splitting field E of $f(x)$ is obtained by an extension by radicals—that is, by successive adjunctions to the base field F of roots of equations of the form $x^n - b = 0$, where at each stage b is an element of the extension field obtained by the previous adjunctions. Some elementary results of Galois theory are that the general equation of the 5th degree is not solvable by radicals and that an arbitrary angle cannot be trisected by compass and straight edge alone.

4-10 NILPOTENT GROUPS

Let $Z_0, Z_1, Z_2, \ldots$ be a sequence of normal subgroups of a group G such that $Z_0 = 1$; $Z_1 = C(G)$, the center of G; $Z_2/Z_1 = C(G/Z_1)$, the center of G/Z_1; $\ldots, Z_{i+1}/Z_i = C(G/Z_i)$; and so on. Clearly

$$Z_0 \subset Z_1 \subset Z_2 \subset Z_3 \subset \cdots \subset Z_i \subset \cdots$$

This called an **ascending central series** of G.

If for some n, $C(G/Z_n) = 1$, then the series terminates in Z_n; that is, $Z_n = Z_{n+1} = \cdots$, and if Z_n is a proper subgroup of G then the series never reaches G. If $C(G/Z_n) = G/Z_n$ then the series ends in G; that is,

$$Z_0 \subset Z_1 \subset \cdots \subset Z_{n+1} = G.$$

In this latter case (that is, if for some n, G/Z_n is an abelian group) then $Z_{n+1} = G$, and G is called a **nilpotent group.** Thus a nilpotent group is one whose ascending central series ends in the group.

Clearly an abelian group is always nilpotent. The other extreme to an abelian group would be a group G for which $C(G) = 1$ and G would not be nilpotent.

Even for a finite group G the ascending central series need not terminate in G. For example, $C(S_3) = 1$ and therefore S_3 is not nilpotent. We have seen however that S_3 is solvable.

The nilpotent groups form a class of groups containing the abelian groups and contained in the class of solvable groups. The next theorem proves this.

THEOREM 35. A nilpotent group G is solvable.

Proof: Let $Z_0 \subset Z_1 \subset Z_2 \subset \cdots \subset G$ be the ascending central series of G. Then $G \supset \cdots \supset Z_2 \supset Z_1 \supset Z_0$ is a subnormal series (in fact even a normal series) for G. Its factors are obviously abelian and so G is a solvable group. We have then

abelian groups $\subset$ nilpotent groups $\subset$ solvable groups.

A finite p-group G is nilpotent. First note that we know the order of G is a power of p. Now form the ascending central series of G. Then $Z_0 = 1, Z_1 = C(G) \neq 1$. (It was proved earlier that the center of a p-group always contains an element distinct from the neutral element of the group.) Hence G/Z_1 is again a finite p-group and its center $Z_2/Z_1 \neq 1$; that is, $Z_2 \supset Z_1$. Since $C(G/Z_1) \neq 1$, the ascending central series continues to G, for eventually $C(G/Z_n) = G/Z_n$.

If A and B are subgroups of a group G, let $[A, B]$ stand for the group generated by all commutators of the form $aba^{-1}b^{-1}$, $a \in A, b \in B$.

THEOREM 36. If G is a nilpotent group, then $[Z_{i+1}, G] \subset Z_i$ and $Z_i \, \Delta \, G$.

Proof: Since Z_{i+1}/Z_i is the center of G/Z_i, $z_{i+1}Z_igZ_i = gZ_iz_{i+1}Z_i$ for all $z_{i+1} \in Z_{i+1}$ and all $g \in G$. Hence $z_{i+1}^{-1}g^{-1}z_{i+1}g \in Z_i$ which proves $[Z_{i+1}, G] \subset Z_i$.

Next we have $[Z_i, G] \overset{\subseteq}{=} [Z_{i+1}, G] \subset Z_i$. Hence $z_igz_i^{-1}g^{-1} \in Z_i$ and $gz_i^{-1}g^{-1} \in Z_i$ for all $g \in G$ and $z_i \in Z_i$. Hence $Z_i \, \Delta \, G$.

THEOREM 37. Every subgroup H of a nilpotent group G is nilpotent.

Proof: Let $1 = Z_0 \subset Z_1 \subset \cdots \subset Z_n = G$ be an ascending central series for G.

Then $[Z_{i+1} \cap H, H] \overset{\subseteq}{=} [Z_{i+1}, G] \cap H \overset{\subseteq}{=} Z_i \cap H$ (by the previous theorem). If repetitions are omitted, the subgroups $Z_i \cap H$ form an ascending central series for H and thus H is nilpotent.

THEOREM 38. The direct product of a finite number of nilpotent groups is nilpotent.

Proof: Use induction on the number of direct factors. Assume the theorem true for $n - 1$ factors,

$$G' = G_1 \times \cdots \times G_{n-1}.$$

We show that $G' \times G_n$ is nilpotent if G' and G_n are nilpotent. Let

$$1 = Z_0' \subset Z_1' \subset \cdots \subset Z_s' = G'$$

and

$$1 = Z_0 \subset Z_1 \subset \cdots \subset Z_t = G.$$

Then $Z_0' \times Z_0 \subset Z_1' \times Z_1 \subset \cdots \subset Z_s' \times Z_s = G' \times G_n$ (where we suppose $s > t$ and so we set $G_n = Z_{t+1} = Z_{t+2} = \cdots = Z_s$) is an ascending central series for $G = G' \times G_n$. For $Z_{i+1}' \times Z_{i+1}/Z_i' \times Z_i = $ center of $G' \times G_n/Z_i' \times Z_i$ (since Z_{i+1}'/Z_i' and Z_{i+1}/Z_i are the centers of G'/Z_i' and G_n/Z_i respectively). Hence G is nilpotent.

We have seen earlier that a finite abelian group is the direct product of its Sylow p-subgroups for the different primes p that divide the order of the group. This is not true of arbitrary finite groups, but it is a defining condition of finite nilpotent groups.

Lemma 9. A proper subgroup H of a nilpotent group G is properly contained in its normalizer $N(H)$. (That is, $H \subset N(H)$ and $H \neq N(H)$.)

Proof: In the ascending central series for G, $1 = Z_0 \subset H$ and $H \subset Z_n = G$. Hence there exists i such that $Z_i \subset H$ and $Z_{i+1} \not\subset H$. For $z_{i+1} \in Z_{i+1}$ and $h \in H$, we have $z_{i+1}h^{-1}z_{i+1}^{-1}h \in Z_i \subset H$. Hence $Z_{i+1} \subset N(H)$ and therefore $H \neq N(H)$.

THEOREM 39. Each Sylow p-subgroup of a finite nilpotent group G is normal in G and G is the direct product of its Sylow subgroups.

Proof: If P is a Sylow p-subgroup of G then $N(P)$ is its own normalizer and hence, by the previous lemma, $N(P) = G$. If $P, Q, \ldots$ are Sylow subgroups for the distinct primes $p, q, \ldots$ dividing the order of G, then P, Q are normal subgroups and we have

$$PQ \cdots \approx P \times Qx \ldots.$$

But the product of the orders of $P, Q, \ldots$ is the order of G. Hence $G \approx P \times Q \times \ldots.$

The converse of this theorem is also true, for every p-group is nilpotent and a direct product of nilpotent groups is nilpotent. Hence we have

THEOREM 40. A finite group G is nilpotent if and only if it is a direct product of its Sylow groups. (For each distinct prime p dividing the order of G exactly one Sylow p-subgroup occurs in the direct product.)

EXERCISE

1. If $H \triangle G$, prove that K/H is the center of the G/H if and only if the subgroup of commutators $[K, G] \subset H$.

4-11 GROUPS WITH OPERATORS

Definition. A group G, a set Σ, a mapping f of $G \times \Sigma \rightarrow G$ is called a **group with operators** if

(1) $(xy, \sigma)f = (x, \sigma)f \cdot (y, \sigma)f, x, y \in G, \sigma \in \Sigma.$

Observe that for every fixed $\sigma \in \Sigma$, f is an endomorphism of G. Σ is called the **set of operators**.

If we denote $(x, \sigma)f$ by $x\sigma$ then (1) becomes

$$(2) \qquad\qquad (xy)\sigma = (x\sigma)(y\sigma).$$

We call $x\sigma$ the **operator product**. It follows easily from (2) that $e\sigma = e$, $x^{-1}\sigma = (x\sigma)^{-1}$, and $x^n\sigma = (x\sigma)^n$ where n is an integer.

Example. A vector space V over a field F is an additive abelian group with F as the operator set. The operator product $x\sigma$ is simply scalar multiplication.

Let G be an operator group with the operator set Σ. Since each $\sigma \in \Sigma$ determines a unique endomorphism $\bar{\sigma}$ of G, then $\sigma \to \bar{\sigma}$ is a mapping of Σ into the set $E(G)$ of all endomorphisms of G. Conversely, let G be a group, Σ a set, and ϕ a mapping of $\Sigma \to E(G)$; that is for $\sigma \in \Sigma$, $\sigma\phi$ an endomorphism of G. Then an operator product $x\sigma$, $x \in G$, $\sigma \in \Sigma$, is defined in G by $x\sigma = x(\sigma\phi)$ and it is easily verified that (2) above is satisfied. Hence G becomes a group with operators. Thus a group G with operator set Σ can be defined as a group for which there exists a mapping of Σ into the set $E(G)$ of endomorphisms of G.

This latter definition enables us to very easily construct important groups with operators. For instance, if G is a group, select some subset S of $E(G)$ as the operator set Σ and take the mapping of $\Sigma = S \to E(G)$ to be simply the inclusion mapping. Then G becomes an operator group with the subset S as its set of operators.

An ordinary group (as distinguished from an operator group) can be regarded however as an operator group whose operator set is empty. A ring (not necessarily commutative) can be regarded as an operator group, for a ring is an additive abelian group and right multiplication (or left), the ring product, taken as the operator product. This interpretation makes the ring its own operator set. A module over a ring is an operator group whose set of operators is a ring. The operator product again is simply scalar multiplication.

Thus many different algebraic systems can be regarded as operator groups and their theory subsumed within the theory of operator groups. This is done by appropriate definitions for operator subgroup and operator homomorphism.

A subgroup H of a group G with a set Σ of operators is called an **operator subgroup** of G if $h\sigma \in H$ for all $h \in H$ and all $\sigma \in \Sigma$.

If G and G' are two groups with the same set Σ of operators, then a homomorphism f of $G \to G'$ is called an **operator homomorphism** if $(x\sigma)f = (xf)\sigma$, $x \in G$, $\sigma \in \Sigma$.

With these definitions, for instance, a vector space V over a field F becomes an operator group whose operator subgroups are the subspaces of V, while the linear transformations are the operator homomorphisms.

Example. Let G be a group. If we select the inner automorphisms of G as the set of operators, then, under the second definition of an operator group, G becomes an operator group whose operator subgroups are the normal subgroups.

EXERCISES

1. An additive abelian group G can be regarded as an operator group with the set of integers as the operator set. What is the endomorphism of G that corresponds to the integer n of the operator set? (See the second definition of an operator group.) Prove your answer. What subgroups of G are the operator subgroups?

2. If T is an endomorphism of a vector space V and if α is a fixed vector of V, the mapping H of $V \rightarrow V$ defined by $\gamma H = \gamma T + \alpha$, $\gamma \in V$, is called an **affine transformation** of V. H is said to be nonsingular if T is an automorphism of V.

Prove the set $A(V)$ of all nonsingular affine transformations of V is a group under multiplication. That is if H, $K \in A(V)$ and $\gamma K = \gamma T' + \alpha'$, where α' is a fixed vector, then $HK \in A(V)$ and under this binary operation $A(V)$ is a group. $A(V)$ is called the **affine group of V.** Those nonsingular affine transformations for which the endomorphism T is always 1_V, the identity mapping on V, are called **translations of V.** Prove the set of all translations of V is a normal subgroup of the affine group $A(V)$.

3. Show that the group S_3 has 6 inner automorphisms and no outer automorphisms. What check can you use to prove that these are all of the inner automorphisms of S_3?

4. Find all the inner and outer automorphisms of the cyclic group of order 8.

5. The centralizer $C(x)$ of an element x of a group G is the set of all elements of G that commute with x. Prove $C(x)$ is a subgroup of G and show $C(x)$ is the normalizer of x. Prove that the number of distinct conjugates of the element x is equal to the index of $C(x)$ in G.

6. Find the numbers of distinct conjugates of the elements (1234) and (12)(34) of S_4. Find the centralizers of these elements.

7. Write out the class equation for each of the following groups (a) S_3 (b) S_4 (c) the group of symmetries of the square (d) A_4.

8. If E is a family of endomorphisms of a group G then a subgroup H of G is said to be *invariant under E* if $H\phi \subseteq H$ for all $\phi \in E$. For example the normal subgroups are those which are invariant under all the inner automorphisms of G.

A subgroup that is invariant under all the automorphisms (inner and outer) of G is called a **characteristic subgroup of G.** Naturally a characteristic subgroup is a normal subgroup.

(a) Let H be a characteristic subgroup of G and let $H \subset N$ where N is a normal subgroup of G. Prove H is a normal subgroup of N.

(b) Prove that the center of a group is a characteristic subgroup.

9. A subgroup H of G is called **fully invariant** if it is invariant under every endomorphism of G.

(a) Prove that the commutator subgroup of G is fully invariant.

(b) Prove that any subgroup of a cyclic group is fully invariant.

(c) Prove that the higher commutator subgroups (that is commutator subgroups of commutator subgroups) are fully invariant.

10. Prove that an abelian group is simple if and only if it is of prime order.

11. If G is a group of transformations on a set E, the *stabilizer* G_x of an element $x \in E$ is defined as the set of all $\phi \in G$ such that $x\phi = x$.

(a) Prove G_x is a subgroup of G.

(b) If two elements of E are in the same G-orbit prove that their stabilizers are conjugate subgroups.

(c) Prove that the number of elements in the G-orbit containing $x \in E$ is the index of G_x in G.

(d) If X is a G-orbit in E, call two elements of X *equivalent* if and only if they have the same stabilizer. Prove this is an equivalence relation.

12. Prove that the direct product of abelian groups is an abelian group.

13. Prove that every finite cyclic group has a composition series whose factors have prime orders.

14. Find the numbers of different Sylow groups in a group of order 405.

15. Use Cauchy's theorem to aid in determining all groups of order p^2, where p is a prime.

16. If $\cdots \supset A \supset B \supset C \supset \cdots$ is a refinement of a subnormal series $\cdots \supset A \supset C \supset \cdots$ in which A/C is abelian, prove B/C and A/B are abelian. Thus any refinement of a subnormal series that has abelian factors will also have abelian factors.

17. Use Schreier's theorem to prove that every factor group of a solvable group is solvable.

18. A matrix group is a group of matrices under the binary operation of matrix multiplication. Prove that every finite group is isomorphic to a matrix group. This is called a matrix representation of the finite group.

19. G is a group generated by two elements x and y which satisfy the relations $x^6 = 1$, $y^2 = 1$, and $x^k y = yx^{6-k}$, $1 \leq k \leq 5$. Find all the Sylow subgroups of G.

20. G is a group of order $2p$, where p is a prime and $p > 2$. Find all the subgroups of G and prove that G is either a cyclic group or a dihedral group.

21. Let A be the free abelian group generated by the set S and let B be the free abelian group generated by the set T, where $S \cap T$ is empty. Prove that $A \oplus B$ is the free abelian group generated by $S \cup T$.

22. If G is the abelian group generated by the elements x, y, z where $x^{10} = 1$, $y^{12} = 1$ and $z^{18} = 1$, express G as a direct product (a) of cyclic groups (b) of indecomposable cyclic groups.

23. If $G = H_1 \times H_2$ (direct product) and $K_1 \, \Delta \, H_1$, $K_2 \, \Delta \, H_2$, prove $K_1 \times K_2 \, \Delta \, G$ and that $G/(K_1 \times K_2) \approx H_1/K_1 \times H_2/K_2$.

24. An abelian group A is called **divisible** if for every $a \in A$ and every $n \in Z$, there exists $b \in A$ such that $a = nb$.

(a) Prove the additive group of rational numbers is divisible and that a cyclic group is not divisible.

(b) Prove that a direct sum of abelian groups is divisible if and only if each direct summand is divisible.

(c) Prove that a quotient group of a divisible abelian group is divisible.

25. If B is a subgroup of a finitely generated abelian group prove that the rank of A/B is equal to the difference of the rank of A and the rank of B.

26. G is a group. If all the automorphisms of G are taken as a set of operators, show G becomes an operator group and find all its operator subgroups.

27. Prove that the homomorphic image of a nilpotent group is nilpotent.

28. If H is a normal subgroup of a nilpotent group G, prove that G/H is a nilpotent group.

29. Define a free semigroup (F, f) on a set S where F is the semigroup and f maps $S \rightarrow F$. Prove f is injective and Sf generates F. Show F is unique up to isomorphism.

30. Let F be the set of all finite sequences $(a_1, \ldots, a_n)$ of elements $a_i \in S$. Define $(a_1, \ldots, a_n) \circ (b_1, \ldots, b_m) = (a_1, \ldots, a_n, b_1, \ldots, b_m)$. Show F is a semigroup. If f is defined by $af = (a)$, $a \in S$, prove (F, f) is the free semigroup on S.

31. Show the set of positive integers is the free semigroup (under addition) on a singleton set (that is, a set of one element). Describe the free semigroup on a set of two elements.

chapter 5

Rings

We come now to a different type of algebraic system, called a ring. The most striking feature of a ring in contrast to a group is that a ring has two binary operations. Actually a ring is, in a natural and orderly way, the next step in the development of varieties of algebraic systems. It originates as an additive abelian group, to which we add a second binary operation, "multiplication." In the most general type of ring this multiplication is not assumed either commutative or associative, nor is it required to have a neutral element. Two further conditions, called the **distributive laws,** are imposed, and we have a ring. These laws serve to relate the two otherwise independent binary operations. All this yields what is known as a nonassociative ring; however throughout this book we shall deal only with rings, whose multiplication is associative, associative rings as they are called.

We now proceed to the precise definition of a ring.

5-1 DEFINITIONS

Definition. A *ring R* is an additive abelian group and a mapping of $R \times R \to R$ that defines a binary operation in R called a **product,** and this product is required to be doubly distributive with respect to the addition.

Designate the product of two elements x and y of R by xy. The definition states that this product is to have the distributive properties.

$$x(y + z) = xy + xz \quad \text{and} \quad (x + y)z = xz + yz$$

for all elements x, y, and z of R.

As we have observed then, a ring, in contrast to a group, is an algebraic system with two binary operations. The ring is an abelian group with respect to one of them (so that this binary operation is both commutative and associative) and the second binary operation must be doubly distributive with respect to the first one. In the above definition addition has been used as the group operation and multiplication as the second operation. The two distributive properties relate the two binary operations of the ring.

We continue to employ addition and multiplication for our two ring operations.

Definitions. If the multiplication in a ring is associative, the ring is called an **associative ring,** and if it is commutative the ring is called a **commutative ring.** The ring R is said to be a ring with identity (unit element) e if $e \in R$ and $ex = xe = x$, for all $x \in R$. *We emphasize that henceforth all rings under consideration are assumed to be associative.*

Definition. If $x \neq 0$ and $y \neq 0$ are elements of R such that $xy = 0$, then x and y are called **zero-divisors.** A ring may or may not have zero-divisors.

The Arithmetic of a Ring

Denote by 0 the unique element of R that is the neutral element of R when regarded as a group, and denote by $-x$ the (additive) inverse of the element x of R. This means $x + (-x) = (-x) + x = 0$. The notation x^n, where n is a positive integer, means a product of n factors each equal to x, while nx means the sum of n terms each equal to x. If R has an identity 1, we define $x^0 = 1$. Note that $(xy)^2 = xyxy$ and $(xy)^2 \neq x^2y^2$ unless $xy = yx$. Bearing these facts in mind and using the two distributive laws it is easy to prove the following very important results. We write $x - y$ for $x + (-y)$.

1. (a) $x0 = 0x = 0$ for all $x \in R$ [HINT: $x(y + 0) = xy$].
(b) $x(-y) = -(xy) = (-x)y$ for all $x, y \in R$. [HINT: $x(y - y) = 0$.]
(c) $(-x)(-y) = xy$ for all $x, y \in R$ [HINT: $(x - x)(-y) = 0$.]

2. Use induction on n with m held fixed to prove
(a) $x^n x^m = x^{n+m}$.
(b) $(x^m)^n = x^{mn}$. [Now turn' m loose and it is seen that (a) and (b) hold for all nonnegative integers.

3. If a ring R has an identity e, show by assuming the existence of a second identity e' that e is unique.

Examples of rings are the following:
1. The integers $0, \pm 1, \pm 2, \dots$ form a commutative ring with identity under the ordinary operations of addition and multiplication. This ring has no zero-divisors.

2. The rational numbers and the real numbers form commutative rings with identities and no zero-divisors, again under ordinary addition and multiplication.

3. The set of all 2×2 matrices $\begin{pmatrix} a & b \\ c & d \end{pmatrix}$ whose elements are integers, say, form a ring under matrix addition and matrix multiplication. This is

a noncommutative ring with the identity $\begin{pmatrix} 1 & 0 \\ 0 & 1 \end{pmatrix}$. Its zero element is the

matrix $\begin{bmatrix} 0 & 0 \\ 0 & 0 \end{bmatrix}$. Note that $\begin{bmatrix} 2 & 1 \\ 4 & 2 \end{bmatrix} \begin{bmatrix} -1 & 3 \\ 2 & -6 \end{bmatrix} = \begin{bmatrix} 0 & 0 \\ 0 & 0 \end{bmatrix}$, so that this ring

has zero-divisors.

4. Let n be a positive integer greater than 1. Define an equivalence relation on the set of all integers by $x \equiv y \pmod{n}$ if and only if $x - y$ is divisible by n. This is read as x is congruent to y modulo n. Let $\bar{x}$ denote the equivalence class containing the integer x. An equivalence class for this case is called a residue class modulo n. Define an addition and multiplication on the set of residue classes as follows: $\bar{x} + \bar{y} = \overline{x + y}$ and $\bar{x} \cdot \bar{y} = \overline{xy}$. It is easy then to show that the set of all residue classes forms a commutative ring with the identity $\bar{1}$. If n is not a prime number, then $n = ab$, where $a > 1$ and $b > 1$, and $\bar{a}\bar{b} = \bar{n} = \bar{0}$. Thus this ring has zero-divisors if n is not a prime. It is called the **ring of residue classes** modulo n.

5. The set of all $n \times n$ matrices, $n > 1$, over the integers, or rational numbers or real numbers, forms a noncommutative ring with identity and possessing divisors of zero.

6. The real functions (that is, functions whose values are real numbers) defined on some interval (a, b) will form a ring under the following conventions:

For all $x \in (a, b)$ and any such functions f and g, define (a) addition by $x(f + g) = xf + xg$; (b) multiplication by $x(fg) = (xf)(xg)$; (c) the zero function θ by $x\theta = 0$; (d) the function $-f$ by $x(-f) = -(xf)$.

We see that $-f$ is the additive inverse of f, that is $f + (-f) = \theta$. Clearly these functions form an additive abelian group. The multiplication of the functions is seen to be associative by reason of the fact that multiplication of real numbers is associative. For if f, g, h are such functions, then

$$\begin{aligned} x[f(gh)] &= (xf)[x(gh)] = (xf)(xg)(xh) \\ &= [(xf)(xg)](xh) = [x(fg)](xh) \\ &= x[(fg)h]. \end{aligned}$$

Since x is an arbitrary element of (a, b), it follows that $f(gh) = (fg)h$.

Let us next verify the distributive laws. Since the multiplication of the functions is seen to be commutative, it is enough to verify one of these laws. Consider $f(g + h)$. Let $x \in (a, b)$. Now

$$\begin{aligned} x[f(g + h)] &= [xf][x(g + h)] = (xf)(xg + xh) \\ &= (xf)(xg) + (xf)(xh), \end{aligned}$$

since xf, xg, and xh are real numbers. Hence

$$x[f(g + h)] = x(fg) + x(fh) = x(fg + fh),$$

and therefore $f(g + h) = fg + fh$.

Thus these functions form a commutative ring. The ring has an identity, namely the function e which maps every $x \in (a, b)$ into the real number 1.

Some special and important kinds of rings are the following:

Definition. An **integral domain** is a commutative ring with identity and having no zero-divisors.

Example. The integers, the rational numbers, the real numbers.

Definition. A **division ring** is a ring with identity such that every nonzero element x has a multiplicative inverse x^{-1}, that is $xx^{-1} = x^{-1}x = e$. This means then that in a division ring the nonzero elements form a multiplicative group. Note that the multiplication is not assumed commutative.

Example 1. Let G be a simple additive abelian group (that is, G has no proper subgroups). If f is an endomorphism of G, then ker $f = 0$ or ker $f = G$. If ker $f = G$, then f is the zero endomorphism, the one that maps every element of G into the neutral element. If ker $f = 0$, then f is injective. Moreover, im $f = Gf = 0$ or im $f = G$. If im$f = 0$, then f is again the zero endomorphism. Otherwise im $f = G$ and f is surjective. Thus any nonzero endomorphism of G is bijective and hence is an automorphism of G. Let E be the set of all endomorphisms of G. For $f, g \in E$ define addition and multiplication in E, by $x(f + g) = xf + xg$ and $x(fg) = (xf)g$, for all $x \in G$. With the understanding that for $f \in E$, $-f$ is defined by $x(-f) = -(xf)$, $x \in G$, we can very easily show E is a noncommutative ring. Since every nonzero endomorphism f is an automorphism of G, and therefore has an inverse f^{-1},

$$ff^{-1} = f^{-1}f = 1_G$$

where 1_G is the identity map on $G \rightarrow G$, we see that E is a division ring.

Example 2. The quaternions at the end of this section.

Definition. A **field** is a commutative ring with identity such that every nonzero element has a multiplicative inverse.

Examples. The rational numbers, the real numbers, the complex numbers.

The set of all 2×2 matrices of the form $\begin{bmatrix} x & 0 \\ 0 & x \end{bmatrix}$, where x is any rational number is easily seen to be a field.

The ring of residue classes modulo 5 is a field. The inverse of $\overline{2}$ is $\overline{3}$ and the inverse of $\overline{4}$ is $\overline{4}$.

Thus a field is a division ring in which the multiplication is commutative. Since in both a field and a division ring, nonzero elements have inverses, it follows that there are no zero-divisors in a field or a division ring. For if $xy = 0$ and $x \neq 0$ then, since x has an inverse, $x^{-1}xy = x^{-1}0 = 0$; that is, $ey = y = 0$.

If a ring has no zero-divisors, then if $ab = ac$ and $a \neq 0$, it follows that $b = c$. (This is called the **cancellation law of multiplication**.) For then $a(b - c) = 0$, and hence $b - c = 0$. Conversely if the cancellation law is valid in a ring, then the ring has no zero-divisors. For if $ab = 0$ and $a \neq 0$, then $ab = a0$ and hence $b = 0$, canceling a. Thus in an integral domain we can perform this type of cancellation.

We next prove a very simple theorem about a *finite* ring—that is, a ring with a finite number of elements.

THEOREM 1. A finite ring R with an identity e and having no zero-divisors is a division ring.

> **Proof:** Let x be a nonzero element of R. Form $x^2, x^3, x^4, \ldots$. Since R is finite some $x^k = x^j, j < k$. Thus $x^{k-j} = e$. If $k - j = 1$, then $x = e$ and $x^{-1} = e$. If $k - j > 1$, then $x^{k-j-1} = x^{-1}$. Thus each nonzero element of R has an inverse and so R is a division ring.

> **Corollary.** A finite integral domain is a field.

EXERCISE

1. Prove that the ring of residue classes modulo a prime number p is a field. (This makes use of the fundamental property of a prime p—that is, if p divides a product of integers then p divides at least one of the factors of this product.)

Example. Historically one of the most interesting examples of a division ring is the so-called **quaternions.**

Let i, j, k be symbols which satisfy the following multiplication table

	i	j	k
i	-1	k	$-j$
j	$-k$	-1	i
k	j	$-i$	-1

That is, for example, $j \cdot j = j^2 = -1$, $k \cdot j = -i$.

We also assume that for any real number x, $xi = ix$, $xj = jx$, $xk = kx$.

Now form the set Q of all symbols q, where $q = x + yi + zj + wk$ and x, y, z, w are real numbers.

The element $\bar{q} \in Q$ defined by $\bar{q} = x - yi - zj - wk$ is called the **conjugate** of q.

Define equality in Q by

$$x + yi + zj + wk = x' + y'i + z'j + w'k$$

if and only if $x = x'$, $y = y'$, $z = z'$, $w = w'$. Thus $x + yi + zj + wk = 0$ if and only if $x = y = z = w = 0$. Let addition in Q be defined by

$$(x + yi + zj + wk) + (x' + y'i + z'j + w'k)$$
$$= x + x' + (y + y')i + (z + z')j + (w + w')k$$

and let multiplication be defined by following the rules of the multiplication table, bearing in mind that a real number commutes with each symbol i, j, k. It is now a straightforward task to verify that Q is a ring with the identity 1. Since $ij = -ji$, we see that Q is a noncommutative ring. The elements of Q are called **quaternions**.

Define the *norm* of $q \in Q$ as $q\bar{q}$. Then if $q = x + yi + zj + wk$, $q\bar{q} = x^2 + y^2 + z^2 + w^2$. Thus if $q \neq 0$, then $q\bar{q} \neq 0$. Hence if $q \neq 0$, the quaternion

$$q' = \frac{\bar{q}}{x^2 + y^2 + z^2 + w^2}$$

is the multiplication inverse of q; that is, $qq' = q'q = 1$. This shows that Q is a division ring.

There exists an interesting and elegant matrix form for a quaternion, which we can discover as follows. Since $ij = k$,

$$x + yi + zj + wk = (x + yi) + (z + wi)j.$$

Regarding $x + yi$ and $z + wi$ as complex numbers, write down the matrix

$$\begin{bmatrix} x + yi & z + wi \\ -(z - wi) & x - yi \end{bmatrix}$$

and note that its determinant equals $x^2 + y^2 + z^2 + w^2$, the norm of the quaternion $q = x + yi + zj + wk$. This is a matrix over the field of complex numbers. If we associate this matrix with the quaternion

$$q = x + yi + zj + wk = x + yi + (z + wi)j = \alpha + \beta j,$$

where $\alpha = x + yi$, $\beta = z + wi$, we get a 1–1 correspondence

$$q \leftrightarrow \begin{bmatrix} \alpha & \beta \\ -\bar{\beta} & \bar{\alpha} \end{bmatrix}$$

of Q with a subset of the 2×2 matrices over the complex field. Here $\bar{\alpha} = x - yi$, $\bar{\beta} = z - wi$ are the conjugates of the complex numbers α and β.

If $q' = \alpha' + \beta'j$ is another quaternion, it is easily verified that for $q + q' = \alpha + \alpha' + (\beta + \beta')j$,

$$q + q' \leftrightarrow \begin{bmatrix} \alpha & \beta \\ -\bar{\beta} & \bar{\alpha} \end{bmatrix} + \begin{bmatrix} \alpha' & \beta' \\ -\bar{\beta}' & \bar{\alpha}' \end{bmatrix},$$

and that for $qq' = \alpha\alpha' - \beta\bar{\beta}' + (\alpha\beta' + \beta\bar{\alpha}')j$

$$qq' \leftrightarrow \begin{bmatrix} \alpha & \beta \\ -\bar{\beta} & \bar{\alpha} \end{bmatrix} \begin{bmatrix} \alpha' & \beta' \\ -\bar{\beta}' & \bar{\alpha}' \end{bmatrix}.$$

This proves Q is isomorphic to the ring of matrices of the form $\begin{bmatrix} \alpha & \beta \\ -\bar{\beta} & \bar{\alpha} \end{bmatrix}$, where α, β are complex numbers.

We can therefore regard such a matrix as a quaternion. In fact the isomorphism can be considered as a matrix representation of the division ring of quaternions.

EXERCISES

1. Let $R \times R'$ be the Cartesian product of two rings R and R'. Define an addition and multiplication in the set $R \times R'$ by

$$(x, x') + (y, y') = (x + y, x' + y')$$
$$(x, x') \cdot (y, y') = (xy, x'y'),$$

where $x, y \in R$ and $x', y' \in R'$. Prove that with these binary operations $R \times R'$ is a ring. It is called the direct sum of the rings R and R' and written $R \oplus R'$.

2. Generalize the idea of a direct sum to an arbitrary finite number of rings $R_1, R_2, \ldots, R_n$.

5-2 SUBRINGS AND IDEALS

Definition. A **subring** S of a ring R is a nonempty subset of R that itself forms a ring under the binary operations defined in R.

EXERCISE

1. Prove that a subgroup S of a ring R which has the property that if x, y are in S then xy is in S is a subring of R.

THEOREM 2. A subset S of a ring R is a subring of R if and only if $x, y \in S$ implies $x - y$ and xy are in S.

The proof is left as an exercise with this observation: if $x \in S$ then $x - x = 0 \in S$ and so $0 - x = -x \in S$, thus if $x, y \in S$ then $x - (-y) = x + y \in S$.

Subrings Generated by a Subset of a Ring

It is readily proved that the intersection of a family of subrings of a ring R is a subring of R. Now suppose T is a subset of R. We define the **subring of R generated by** T as the smallest subring containing T. Let $[T]$ denote the subring generated by T. Clearly it is the intersection of all subrings that contain T. What do the elements of $[T]$ look like? They must be finite sums of finite products of elements of T and the negatives of such products, that is they must have the form $\Sigma(\pm \, t_1 t_2 \ldots t_n)$, where each $t_i \in T$. For the set of all such elements forms a subring by Theorem 2, and any subring containing T would have to contain all such elements.

One cannot emphasize too strongly that the reader must beware of making unwarranted assumptions in his reasoning. For instance a ring R may not have an identity and yet one of its subrings S may have an identity. This merely signifies that there is an element of R that serves as an identity for the elements of S but not for all elements of R. An example of this is the following.

Example. Consider the set R of all matrices of the form $\begin{bmatrix} x & 0 \\ y & 0 \end{bmatrix}$ whose elements are integers. It is easy to show that R is a ring and that it does not have an identity. A subring of R is the set S of all matrices of the form $\begin{bmatrix} x & 0 \\ 0 & 0 \end{bmatrix}$ and this subring does have the identity $\begin{bmatrix} 1 & 0 \\ 0 & 0 \end{bmatrix}$. At this point the reader may need to refer to the definition of an identity. Note that while $\begin{bmatrix} x & 0 \\ y & 0 \end{bmatrix} \begin{bmatrix} 1 & 0 \\ 0 & 0 \end{bmatrix} = \begin{bmatrix} x & 0 \\ y & 0 \end{bmatrix}$, we have $\begin{bmatrix} 1 & 0 \\ 0 & 0 \end{bmatrix} \begin{bmatrix} x & 0 \\ y & 0 \end{bmatrix} = \begin{bmatrix} x & 0 \\ 0 & 0 \end{bmatrix}$.

Definition. An **ideal** A of a ring R is a subring that is both a left and $RA \subseteq A$, and a **right ideal** A of R is a subring of R such that $AR \subseteq A$. Here RA stands for the set of all elements of R of the form ra where $r \in R$ and $a \in A$.

Definition. An **ideal** A of a ring R is a subring that is both a left and a right ideal.

Thus left ideals, right ideals, and ideals are special kinds of subrings. Not all subrings are ideals. For example, in the ring of rational numbers the subring of integers is not an ideal.

Clearly, if a ring R is commutative, then all left and right ideals of R are ideals of R.

Example. In the ring of integers, all the multiples of some fixed integer *m* form an ideal.

Example. In the ring of real functions [see Sec. 5-1, Examples of rings, (6)] the subset of all real functions that vanish at the same point *c* of (a, b) is an ideal. This follows at once from the definition of an ideal.

On the other hand the constant functions (a constant function is one that maps every $x \in (a,b)$ into the same real number) form a subring, but not an ideal.

EXERCISE

1. Prove that a subset *A* of a ring *R* is a left ideal in *R* if and only if (i) $x, y \in A$ implies $x - y \in A$ (ii) $r \in R$, $x \in A$ imply $rx \in A$. Since this is an "if and only if" condition, it can be used as a definition of a left ideal. A similar property holds for right ideals.

Ideals

Ideals—that is, two-sided ones (left and right), are by far the most important kind of subring. They play a role in ring theory analogous to that of the normal subgroups in group theory. By use of ideals new rings can be constructed out of a ring *R*, called **quotient** or **factor rings**. This is the purpose of our next discussion.

Quotient Rings

Let *A* be an ideal in a ring *R*. An equivalence relation on *R* is defined as follows: Two elements *x* and *y* of *R* are said to be **congruent modulo** *A*, written $x \equiv y \pmod{A}$ if and only if $x - y \in A$. The equivalence class containing $x \in R$ is written $x + A$ and consists of all elements of *R* of the form $x + a$, $a \in A$. It is called a **coset** of *A* in *R*. Note that $y \in x + A$ if and only if $y \equiv x \pmod{A}$. Designate by R/A the set of all cosets of *A* in *R*. Since the ideal *A* is a normal subgroup of the abelian group *R*, it follows that R/A is an additive abelian group. We plan to define on R/A the structure of a ring. To do this we next need to define a multiplication of its elements—that is, of the cosets of *A* in *R*.

If $x \equiv y \pmod{A}$ and $u \equiv v \pmod{A}$ then $x - y \in A$ and $u - v \in A$. Since *A* is an ideal, $x(u - v) = xu - xv \in A$ and $(x - y)v = xv - yv \in A$, and therefore $yv \in xu + A$. Thus if any element $y \in x + A$ is multiplied by any element $v \in u + A$, the product $yv \in xu + A$. Hence we define the product of two cosets by

$$(x + A)(u + A) = xu + A$$

This is to be understood as a definition of the product of two cosets and not as a statement about the ring product of the two sets $x + A$ and

$u + A$. It can next be easily verified that this product is doubly distributive with respect to addition, and hence R/A now has the structure of a ring. R/A is called the **quotient ring of R with respect to A.** Its zero element is the element A and the additive inverse of $x + A$ is $-x + A$. It is important to bear in mind that A has to be an ideal in order for R/A to form the quotient ring. If A were merely a left or a right ideal or merely a subring, R/A would form a quotient group but not a quotient ring.

Note that it follows at once that if R is a commutative ring then so is a quotient ring R/A, and if R has an identity e, then R/A has the identity $e + A$.

The zero element alone forms an ideal N in a ring R and R itself is an ideal in R. These two ideals are referred to as **trivial** or **improper ideals.** For these trivial ideals we see that $R/N = R$ and $R/R = N$.

To avoid errors in reasoning it is important to stress that an ideal A in a subring S of a ring R is not necessarily an ideal in R. For this means $AS = SA \subset A$ but not that $AR = RA \subset A$.

The zero element 0 alone forms a subring of the ring R, and in this trivial ring the zero element has the property of an identity element as well. To avoid such a trivial type of ring one can assume that the ring contains a nonzero element, or if a nontrivial ring with identity is desired, it is customary to write: Let R be a ring with $1 \neq 0$.

EXERCISES

1. Prove that the set A of all matrices of the form $\begin{bmatrix} a & b \\ 0 & c \end{bmatrix}$ over the ring of integers is a ring and show that the set B of matrices of the form $\begin{bmatrix} d & e \\ 0 & 0 \end{bmatrix}$ is an ideal in A. What about the matrices of the form $\begin{bmatrix} a & 0 \\ 0 & c \end{bmatrix}$?

2. Show that all 2×2 matrices over the rational field form a noncommutative ring with an identity and having proper zero-divisors. Show that all matrices of the form $\begin{bmatrix} a & 0 \\ 0 & a \end{bmatrix}$ form a subring of this ring. Is this subring an ideal? is it a field?

3. If an element of a noncommutative ring with an identity has both a left and a right multiplication inverse, prove they are equal.

4. If a ring R has a unique left identity element e ($ex = x$ for all $x \in R$), prove that e is an identity for the ring. [Consider $(e + y - ye)x$ for all x and y.]

5. Let A be an ideal in the ring Z of integers. Clearly there are positive integers in A, and by the well-ordering theorem there is a least positive integer m in A. If n is any integer in A, then $n = mq + r$, where $0 \leq r < m$. Prove that unless $r = 0$, there is a contradiction of the choice of m. Hence $A = (m)$, where (m) is the set of all multiples of m.

6. Prove that all subrings of Z are ideals.

7. Let A and B be ideals in a ring R. Define $A + B$ as the set of all elements $a + b$, $a \in A$ and $b \in B$. Define AB as the set of all finite sums of the form $\Sigma a_i b_i$, $a_i \in A$, $b_i \in B$. Prove $A + B$ and AB are ideals in R.

8. If (m) and (n) are ideals in the ring Z of integers prove $(m) + (n) = (d)$ where d is the G.C.D. of m and n, and prove $(m)(n) = (h)$ where h is the L.C.M. of m and n.

9. Denote by Z_m, m a positive integer, the quotient ring $Z_m = Z/(m)$, where Z is the ring of integers. Prove that all ideals of Z_m have the form $(k)/(m)$ where $k \mid m$.

10. Prove that all subrings of Z_m are ideals.

11. Find all the ideals of the ring Z_{12}.

12. If M is the ring of all 2×2 matrices over a ring R prove that the set of all 2×2 matrices over an ideal A in R is an ideal in M.

13. Prove that a ring with identity element $1 \neq 0$ is a division ring if and only if it has no proper left (right) ideals.

14. A function of period k is one for which $f(x + k) = f(x)$ for all x of its domain. Prove the set S of all real functions of period 2π forms a ring. Does it have an identity? Show the set T of all real functions of period π is a subring of S. Is T an ideal in S?

15. Is it possible for a commutative ring with identity 1, that has proper zero divisors, to have a subring that is an integral domain? Prove your answer.

16. Describe the ideals of the ring $Z \times Z$. (See Exercise 1, Sec. 5-1.)

17. Describe the ideals of the ring $R \times R'$, where R and R' are rings.

18. If R and R' are integral domains, is $R \times R'$ an integral domain?

5-3 THE CHARACTERISTIC OF A RING

As we have seen earlier, important information about the structure of a ring is the knowledge of whether it has zero-divisors or not. Equally important information is a knowledge of what is called the characteristic of a ring. The elements of the ring of residue classes modulo 6 have the property that $6x = 0$ for every element x of this ring. On the other hand in the ring of integers there exists no positive integer n such that $nx = 0$, for x a nonzero integer. In order to perform computations in a ring we therefore must know this information.

Let R be a ring. Then R is an additive abelian group. Designate by R_+ the ring R regarded as an additive abelian group. Every $x \in R$ generates a cyclic subgroup $[x]$ of R_+. Let $\alpha(x)$ stand for the order of the subgroup $[x]$. If $n > 0$ is an integer, then $nx = 0$ if and only if (i) $\alpha(x) \mid n$, when $\alpha(x)$ is finite, (ii) $n = 0$, when $\alpha(x)$ is infinite.

Two mutually exclusive cases can occur.

(1) There exists at least one integer $N > 0$, such that $Nx = 0$, for all $x \in R$. Hence no $\alpha(x)$ is infinite and each $\alpha(x) \mid N$, and so $\alpha(x)$ takes on only a finite number of values. If m is the least common multiple of these values, then $mx = 0$ for all $x \in R$ and $m \mid N$ if $Nx = 0$ for all $x \in R$. In this case the ring R is said to be of *characteristic m*, written char $R = m$.

(2) There exists no N such that $Nx = 0$, for all $x \in R$. This occurs if the $\alpha(x)$ are not bounded above (although they may all be finite). Then $nx = 0$, for all $x \in R$, implies $n = 0$. In this case R is said to be of *characteristic 0*, char $R = 0$.

There are some special but important circumstances which can occur.

A. If R has an identity e and if e has finite order m, $\alpha(e) = m$, then char $R = m$. For if x is any element of R, then $mx = (me)x = 0$.

B. If $d \in R$ is not a left zero-divisor of R and if $\alpha(d) = m$, then char $R = m$. For $0 = (md)x = d(mx)$, for all $x \in R$. Hence $mx = 0$. The same is true of right zero-divisors.

C. If R has no zero-divisors, then char $R = 0$ or char $R = p$ where p is a prime.

To see this first observe that in such a ring every element has the same order. For if $\alpha(x) = m$ and $\alpha(y) = n$, then by (B) we have $m \mid n$ and $n \mid m$. Hence $m = n$. Hence char $R = 0$ or char $R = m > 0$. If

$$\text{char } R = m > 0,$$

suppose $m = hk$, where $h > 1$, $k > 1$. If $x \neq 0$ is an element of R, then $0 = mx^2 = (hx)(kx)$ and this implies either $hx = 0$ or $kx = 0$. Either of these is a contradiction, since each $x \in A$ has the same order m. Hence m must be a prime.

Thus the characteristic of an integral domain, a field, or a division ring must be either zero or a prime p. An ordered integral domain has characteristic zero, for $ne = e + e + \cdots + e > 0$, since $e > 0$.

If the characteristic of a ring is positive, then the characteristic of a subring can never be greater than that of the ring; in fact it is a divisor of the characteristic of the ring. For if S is a subring of R and if

$$\text{char } R = n > 0,$$

and char $S = m$, then $mx = 0$, for all $x \in S$, where m is the smallest such integer for S. Since $nx = 0$, we must have $m \mid n$.

If the characteristic of a ring is 0, then a subring may have any characteristic. On the other hand if the characteristic of a subring is 0, then the characteristic of the ring has to be 0.

EXERCISES

1. A Boolean ring B is a ring such that $x^2 = x$ for every $x \in B$. Prove that a Boolean ring is commutative and that it has characteristic 2. Prove that if $x, y \in B$ and $x + y = 0$ then $x = y$.

2. Prove that the set B of all subsets of an infinite set S (inclusive of S itself and the empty set) is a Boolean ring under the binary operations:

$$\alpha + \beta = \alpha \cup \beta - \alpha \cap \beta, \; \alpha \cdot \beta = \alpha \cap \beta$$

for all subsets α and β of S.

3. Can any commutative ring with an identity be ordered? (See Sec. 1-6.)

4. If a ring R has at least two elements and if R has no proper zero divisors, prove that the characteristic of R is either 0 or a prime.

5. Prove that any subdomain of an integral domain D has the same characteristic as D.

5-4 RING HOMOMORPHISMS

Mappings of one ring into another are most important, useful, and interesting when they preserve the two ring operations. Such mappings are called **ring homomorphisms,** and it will be noted that in their many properties there is a great similarity to group homomorphisms and their properties.

Definition. A *homomorphism* f of a ring R to a ring R' is a mapping of $R \rightarrow R'$ such that for all elements x and y of R,

$$(x + y)f = xf + yf \quad \text{and} \quad (xy)f = (xf)(yf).$$

We denote by ker f the **kernel of the homomorphism** f, that is the subset of elements of R for which $xf = 0'$. Here $0'$ is the zero element of R'. We denote by im $f = Rf$ the *image of* f, that is the subset of elements x' of R' for which there exists at least one element x of R such that $x' = xf$. Thus im f is simply the image under f of the ring R. In general im $f \subset R'$ and if im $f = R'$ then f is said to be an **epimorphism** (that is, a surjective homomorphism). Note that we might have im $f = 0'$, in which case f is the zero homomorphism, that is the homomorphism which maps every element of R into the zero element of R'. The reader can easily check that this is actually a homomorphism.

THEOREM 3. If f is a homomorphism $R \rightarrow R'$ of two rings then

$$0f = 0' \text{ and } (-x)f = -(xf).$$

Proof: For any $x \in R$, $xf = (x + 0)f = xf + 0f$. Hence $0f = 0'$. Also $0' = 0f = (x - x)f = xf + (-x)f$, and hence $(-x)f = -(xf)$.

THEOREM 4. A homomorphism f of two rings $R \to R'$ is injective if and only if ker $f = 0$.

Proof: If ker $f = 0$, then if $xf = yf$, that is $(x - y)f = 0'$, then $x - y \in$ ker f and so $x = y$. Hence f is injective. Conversely if f is injective and if $x \in$ ker f, then $xf = 0'$ and since $0f = 0'$, it follows that $x = 0$. Hence ker $f = 0$.

Definition. A homomorphism f is an **isomorphism** if f is bijective— that is, if f is both injective and surjective.

If two rings are isomorphic, they are algebraically indistinguishable, so far as their ring structure goes, and one can be regarded as a copy of the other. It is easy to see that if one ring of an isomorphism is commutative or has an identity or has zero-divisors, then so will the other ring have these properties. Thus if one ring is an integral domain or a division ring, it follows that its isomorphic copy will be an integral domain or a division ring.

If a homomorphism f of two rings $R \to R'$ is injective, then clearly R and im $f = Rf$ are isomorphic rings. In such a case it is quite common therefore to identify im f with R and to speak of R being embedded in R', or even to go so far as to speak of R as a subring of R'.

THEOREM 5. If f is a homomorphism of two rings $R \to R'$, then ker f is an ideal in the ring R and im f is a subring of R'.

Proof: If $x, y \in$ ker f, that is $xf = 0'$, and $yf = 0'$, then clearly $(x - y)f = 0'$ and $(xy)f = 0'$. Thus $x - y$ and xy are in ker f which makes ker f a subring of R. If $r \in R$ and $x \in$ ker f then $(rx)f = (rf)(xf) = 0'$ and hence $rx \in$ ker f. Thus $R(\text{ker } f) \subseteq$ ker f. Similarly, $(\text{ker } f)R \subseteq$ ker f and hence ker f is an ideal in R.

Next let $x', y' \in$ im $f = Rf$. Then $x' = xf$ and $y' = yf$ for elements $x, y \in R$. Clearly $x' - y' = (x - y)f$ and $x'y' = (xy)f$ and so $x' - y'$ and $x'y'$ are in im f. Hence im f is a subring of R'.

THEOREM 6. If A is an ideal in a ring R then the mapping f of $R \to R/A$ defined by $xf = x + A, x \in R$, is an epimorphism. (It is known as the **natural** or **canonical homomorphism** of R on to R/A.)

Proof: $(x + y)f = x + y + A = x + A + y + A = xf + yf$.

$$(xy)f = xy + A = (x + A)(y + A) = (xf)(yf).$$

Thus f is a homomorphism. Clearly it is surjective, since for every coset $x + A$ we have $xf = x + A$.

Now A is the zero element of R/A and so the kernel of f consists of all elements x of R such that $x \in A$. We speak then of A being the kernel of f.

In particular, if f is a homomorphism of $R \to R'$ then $R/\ker f$ is a quotient ring of R and, of course, $x \to x + \ker f$ is an epimorphism of $R \to R/\ker f$.

THEOREM 7. If A is a subring of R and f is a homomorphism of the two rings $R \to R'$ then Af is a subring of R'.

Proof: For $x', y' \in Af$ imply $x' = xf$ and $y' = yf$, where $x, y \in A$. Hence $x' - y' = (x - y)f$ and $x'y' = (xy)f$ and therefore are in Af. Thus Af is a subring of R'.

Definition. A ring R' is called a **homomorphic image** or **homomorph** of a ring R if there exists an epimorphism f of $R \to R'$.

THEOREM 8. A ring R' is a homomorph of a ring R if and only if R' is isomorphic to a quotient ring of R.

Proof: Let f be an epimorphism of $R \to R'$. Consider the mapping g of $R/\ker f \to R'$ defined by $(x + \ker f)g = xf$, $x \in R$. It is well-defined, that is single-valued, since $x + \ker f = y + \ker f$ means $x - y$ is in $\ker f$ and hence $xf = yf$.

Under the rules for adding and multiplying cosets, it is easy to check that g is a homomorphism. Moreover g is injective, since $\ker g$ is the set of all $x + \ker f$ such that $xf = 0'$. Thus $x \in \ker f$, and so $\ker g$ consists of the single element $\ker f$, the zero element of $R/\ker f$. That g is surjective follows at once from the fact that f is. Hence g is an isomorphism. Thus if R' is a homomorphic image of R, then R' is isomorphic to a quotient ring of R.

Conversely, let the ring R' be isomorphic to a quotient ring R/A of the ring R. We want to show that R' is a homomorphic image of R. To do this we need to determine an epimorphism f of $R \to R'$. Let g be the given isomorphism of $R/A \to R'$, and let h be the natural epimorphism of $R \to R/A$. It is easy to verify that the composite mapping hg is a homomorphism of $R \to R'$ and that it is surjective. If we set $f = hg$, we have the epimorphism f of $R \to R'$ which makes R' a homomorphic image of R.

This last theorem essentially determines all the possible homomorphic images of a given ring R, for it shows that every homomorphic image is isomorphic to some quotient ring of R, and that for any ideal A in R we get a ring R/A that is isomorphic to a homomorphic image of R. This means that all homomorphic images of a ring R are isomorphic copies of the quotient rings of R. Homomorphic images of a ring inherit many of the properties of the ring itself. This is not surprising, since a homomorphic image of a ring is a picture (usually in miniature) of the ring.

Let f be a homomorphism $R \to R'$ of two rings. If S' is a subset of R', define $f^{-1}S'$ as the subset of R such that $x \in f^{-1}S'$ if and only if $xf \in S'$. Observe that we are not defining a mapping f^{-1} but a set $f^{-1}S'$. (f^{-1} is the usual notation for the inverse of f, and we are not assuming an inverse exists.) If S is a subset of R and if $S' = Sf$, then $S \subset f^{-1}S'$ and it does not follow that $S = f^{-1}S'$. This is because elements not in S may be mapped under f into S'. In other words, $f^{-1}S'$ catches all elements of R that map into S', whether they are in S or not. Of course if $f^{-1}S' = S$ then $Sf = S'$.

Lemma 1. Let f be a homomorphism $R \to R'$ of two rings. Let S be a subring of R that contains the kernel of f. Then if $S' = Sf$ we have $S = f^{-1}S'$.

Proof: Let $x \in f^{-1}S'$. Then $xf = x' \in S'$. Hence there exists $y \in S$ such that $yf = x'$. Hence $xf = yf$ and $(x - y)f = 0'$. Thus $x - y \in \ker f \subset S$. Since $y \in S$ and S is a subring, it follows that $x \in S$. Thus $f^{-1}S' \subset S$. Since $S \subset f^{-1}S'$, we have $S = f^{-1}S'$.

The following theorem gives us fundamental and valuable information about the relation between a ring and its homomorphic image.

THEOREM 9. Let f be an epimorphism $R \to R'$ of two rings, then

1. If A is an ideal in R, Af is an ideal in R'.
2. If A' is an ideal in R', then $A = f^{-1}A'$ is an ideal in R.
3. $A \to Af$ is a bijection (1-1 correspondence) of those ideals A in R that contain the kernel of f on to the set of all the ideals $A' = Af$ or R'.
4. If A and A' are corresponding ideals under this bijection then R/A and R'/A' are isomorphic rings.

Proof: 1. We know already that Af is a subring of R'. Moreover $R'(Af) = (Rf)(Af) = (RA)f \subseteq Af$ and similarly $(Af)R' \subseteq Af$. Hence Af is an ideal.

2. Let $x, y \in A$. Since $Af = A'$, xf and yf are in A'. Hence $(x - y)f$ and $(xy)f$ are in A' and hence $x - y$ and xy are in A. Hence A is a subring. Now $(RA)f = (Rf)(Af) = R'A' \subseteq A'$. Hence $RA \subseteq A$. Similarly, $AR \subseteq A$. Hence A is an ideal in R.

3. Consider the mapping $A \to Af$. Since A is an ideal in R then Af is an ideal in R'. Since A contains $\ker f$ we know that $f^{-1}(Af) = A$. [Here we can write this as $(Af)f^{-1} = A$, since f is a mapping with an inverse f^{-1}, when it is regarded as a mapping of these two sets of ideals.] Thus the mapping $A \to Af$ is injective. But it is also sur-

jective, for if A' is any ideal in R' then $A = f^{-1}A'$ is an ideal in R containing ker f (for $0' \in A'$). Hence $A' = Af$. Thus the mapping $A \rightarrow Af$ is a bijection.

4. Define a mapping g of $R/A \rightarrow R'/A'$ by $(x + A)g = xf + A'$. Again using the usual rules for adding and multiplying cosets it is easy to see that g is a homomorphism. Now g is surjective since f is. To prove g is an isomorphism we now need only to show that it is injective. Let $x + A \in $ ker g. Then $(x + A)g = A'$ (the zero element of R'/A') and this means $xf \in A'$. Hence $x \in A$ and therefore $x + A = A$ (the zero element of R/A). Hence g is injective and hence an isomorphism.

If we apply the foregoing theorem to the natural epimorphism $x \rightarrow x + A$ of a ring R on its quotient ring R/A, we see that every ideal B' in R/A is the image of an ideal B in R that contains A (the kernel of the natural epimorphism). Thus the ideal B' consists of elements $b + A$, $b \in B$; that is, $B' = B/A$. This informs us that every ideal in the quotient ring R/A has the form B/A, where B is an ideal in R that contains A. Moreover part (4) of the theorem tells us that R/B and $R/A/B/A$ are isomorphic rings. Here $R/A\big/B/A$ is the quotient ring of the ring R/A with respect to the ideal B/A.

The foregoing homomorphism theorems for rings are fundamental since they are valid for arbitrary rings.

We conclude the section with the so-called **second isomorphism theorem** for rings.

Second Isomorphism Theorem for Rings

If A and B are ideals in a ring R, then

$$A + B/B \approx A/A \cap B.$$

Proof: $A + B$ is the set of all $a + b$, $a \in A$, $b \in B$. It is easily verified that $A + B$ and $A \cap B$ are ideals in R. The mapping $a \rightarrow a + B$ is an epimorphism of $A \rightarrow A + B/B$ and its kernel is seen to be $A \cap B$. Hence by Theorem 8,

$$A + B/B \approx A/A \cap B$$

the isomorphism being explicitly given by $a + B \rightarrow a + A \cap B$.

5-5 THE RING OF ENDOMORPHISMS OF AN ABELIAN GROUP

Let E be the set of all endomorphisms of an abelian additive group G. This means that if $\alpha, \beta, \gamma, \ldots \in G$, then they are mappings of $G \rightarrow G$ such that $(x + y)\alpha = x\alpha + y\alpha$ for all x and y of G. An addition is defined in E as follows: For $\alpha, \beta \in E$ define the sum $\alpha + \beta$ as the mapping of $G \to G$ given by $x(\alpha + \beta) = x\alpha + x\beta$. It is easy to see that $\beta + \alpha = \alpha + \beta$ and that $\alpha + \beta$ is an endomorphism of G. Let σ stand for the zero endomorphism, that is the element of E that maps every element of G into the zero element of G. We can then define an additive inverse $-\alpha$ of α by $x(-\alpha) = -x\alpha$, for all $x \in G$. (The reader can easily check that $-\alpha \in E$.) Clearly $\alpha + (-\alpha) = \sigma$. E has now the structure of an abelian group, for the addition of endomorphisms is clearly associative. Next we can define a multiplication in E by use of map composition. Define $\alpha\beta$ by $x(\alpha\beta) = (x\alpha)\beta$. Clearly $\alpha\beta \in E$. Of course in general $\alpha\beta \neq \beta\alpha$. It is now quite easy to verify that the two distributive laws hold, that is $\alpha(\beta + \gamma) = \alpha\beta + \alpha\gamma$ and $(\alpha + \beta)\gamma = \alpha\gamma + \beta\gamma$. Hence E is a noncommutative ring. E has an identity ι. This is the endomorphism that maps every x of G into itself, $x\iota = x$ for all $x \in G$. E is called the **ring of endomorphisms of the abelian group G.**

Does E have zero-divisors? If $\alpha \neq \sigma$ and $\beta \neq \sigma$, can $\alpha\beta = \sigma$? This would mean $x(\alpha\beta) = (x\alpha)\beta = 0$ for every $x \in G$ and would occur if $G\alpha = \text{im } \alpha \subseteq \ker \beta$, that is if the image of α is a subgroup of the kernel of β. Hence in general E is a noncommutative ring with identity and having zero-divisors.

We next show that if R is any ring we can determine a homomorphism of R into a ring of endomorphisms of an abelian group.

Let $a \in R$ and define a group endomorphism a_R of R (that is a mapping of $R \to R$, since R forms an additive abelian group, which preserves addition) by $xa_R = xa$, $x \in R$. Clearly a_R is actually a group endomorphism of R. For each $a \in R$ there corresponds an a_R. Furthermore it is easy to verify that

$$(a + b)_R = a_R + b_R \quad \text{and} \quad (ab)_R = a_R b_R,$$

which means that A_R is a subring of endomorphisms of the ring $E(R)$ of all group endomorphisms of R. Moreover this also means that the mapping f of $R \to A_R$ defined by $af = a_R$, $a \in R$, is a homomorphism. Thus f is a homomorphism of the ring R into the ring $E(R)$ of group endomorphisms of R. The kernel of f is the ideal in R of all elements a of R such that $Ra = 0$ (since $a \in \ker f$ if and only if $xa_R = 0$ for all x of R, that is $xa = 0$ for all x of R). Clearly the zero element 0 of R is such an element, but $\ker f$ may contain nonzero elements of R. If R has zero-divisors, this is possible. (The set of all elements a of R for which $Ra = 0$ can easily be seen to be an ideal. It is called the **right annihilator** of R.) However, if R has an identity e, then $Ra = 0$ implies $ea = a = 0$ and in this case the kernel of f contains only the zero element so that f becomes an isomorphism. Thus if R is a ring with identity, then R is isomorphic to a ring of endomorphisms of an abelian group.

A homomorphism of a ring R on to a ring of endomorphisms of an abelian group is called a **representation** of R. In the theory of rings representations play a most important role, for many properties of a ring are studied and determined by its representations. Furthermore, the abstract theory is often enriched by the study of such concrete models.

EXERCISES

1. Show that the ring of group endomorphisms of the additive group of integers Z is isomorphic to Z.

2. If F is a field show that the ring of group endomorphisms of the additive group F is isomorphic to F.

3. Prove that the ring of group endomorphisms of the ring $Z \times Z$ (see Exercise 1, Sec. 5-1) is not commutative and that it has proper divisors of zero.

5-6 EXTENSIONS OF RINGS

We consider next two embedding theorems for rings which produce extensions of the rings. Their proofs make instructive use of the homomorphism properties of rings, derived in Sec. 5-4.

Definition. A ring R is said to be *embedded* in a ring S if the ring S contains a subring R' that is isomorphic to R. The ring S is called an **extension** of the ring R.

THEOREM 10. Every ring R can be embedded in a ring with identity.

> **Proof:** Let Z be the ring of integers. Form the set $Z \times R$ of all ordered pairs (n, r), $n \in Z$ and $r \in R$. We wish to define an equality relation and two binary operations in $Z \times R$ to convert it into a ring B. Here are the definitions:
>
> *Equality:* $(n, r) = (m, s)$ if and only if $n = m$ and $r = s$.
> *Addition:* $(n, r) + (m, s) = (n + m, r + s)$.
> *Multiplication:* $(n, r)(m, s) = (nm, rs + mr + ns)$.

It is easy to verify that each binary operation is associative and that the addition is commutative, that the two distributive laws are true, that $(0, 0)$ is the zero element of B, that $(-n, -r)$ is the additive inverse of (n, r), and that, especially, B has the identity $(1, 0)$. Thus B is a ring with identity. Moreover, it is easy to verify that the set R' of all elements of B of the form $(0, r)$ form a subring of B and that the mapping $r \to (0, r)$ of $R \to R'$ is an isomorphism of the two rings. Thus R is embedded in B and, as we have pointed out earlier, R and R' can be identified, and then we speak of R being a subring of B.

Observe that B is commutative if R is commutative. Thus any ring can be regarded as a subring of a ring that has an identity.

THEOREM 11. An integral domain D can be embedded in a field.

Proof: Form the set F of all pairs (x, y) where $x, y \in D$ and $y \neq 0$. Proceeding as in the previous theorem, we seek to turn F into a field by suitable definitions of equality and addition and multiplication. Define

Equality: $(x, y) = (u, v)$ if and only if $xv = uy$.

Addition: $(x, y) + (u, v) = (xv + uy, vy)$.

Multiplication: $(x, y)(u, v) = (xu, yv)$.

[Note that the definition of equality does not require $x = u$ and $y = v$, so that this equality is an equivalence relation whose equivalence classes contain more than one element of F. We could at this point use (x, y) for the equivalence class containing the element (x, y) of F, and construct our field with these equivalence classes as its elements. This, however, we shall not do, since there is no algebraic disadvantage in using the elements (x, y) themselves of the equivalence classes, and the notation is less cumbersome.]

Again it is easy to show that with these definitions above, F is a field. Addition and multiplication are both commutative and associative. A zero element is $(0, 1)$ and an additive inverse of (x, y) is $(-x, y)$, since $(x, y) + (-x, y) = (0, y^2) = (0, 1)$. An identity is $(1, 1)$, and a nonzero element (x, y), $x \neq 0$, has the inverse (y, x). For $(x, y)(y, x) = (xy, xy) = (1, 1)$. Thus F is actually a field. F is called the **quotient field** of D. The subset D' of F comprising all elements of the form $(x, 1)$ is an integral domain (a field has no zero-divisors) and $x \longrightarrow (x, 1)$ is an isomorphism of $D \longrightarrow D'$. Thus D can be regarded as a subdomain of F. Exactly this method is used to embed the integers (they form an integral domain) in a field, familiarly known as the **field of rational numbers**. It will be observed that, in the definitions of addition and multiplication for the quotient field F of D, the models used are merely the ordinary addition and multiplication of fractions as practised in any respectable school.

Moreover F is the smallest field containing D; that is, F is the intersection of all fields containing D. Any field containing D would have to contain all elements that are in F.

Since $(1, y)$ is the inverse of $(y, 1)$, and $(y, 1)$ of D' corresponds to y of D in their isomorphism, and since we are identifying D' and D, it is then a natural notation to put $1/y$ or y^{-1} for $(1, y)$. This then leads at once to putting x/y or xy^{-1} for (x, y). Hence it is quite natural to write the elements of the quotient field F of the integral domain D in the form x/y, where $x, y \in D$ and, of course, $y \neq 0$.

We have seen that an arbitrary ring R always can be embedded in a ring S with identity, that is R can be regarded as a subring of S. We have also shown that a ring S with identity is isomorphic to a ring E of endomorphisms of an abelian group, and hence this same isomorphism makes R (as a subring of S) isomorphic to a subring of endomorphisms, that is to a subring of E. Hence we see that every ring (whether it has an identity or not) is isomorphic to a ring of endomorphisms of an abelian group.

A field is said to be *ordered* if it is an ordered integral domain. If Q is the quotient field of an ordered integral domain D, then for $(a, b) \in Q$, where $a, b \in D$, define (a, b) as positive if and only if $ab > 0$. Since $(b^2, 1) = (b, 1)^2$, it follows that $(b^2, 1)$ is positive. Since $(b^2, 1) = (b, 1)^2$, it follows that $(b^2, 1)$ is positive. Hence $(a, b) > 0$ if and only if $(a, b)(b^2, 1) = (ab^2, b) = (ab, 1) > 0$. Since $(ab, 1)$ corresponds to ab, our definition of positive is the natural one. The reader will have no difficulty in proving that this is the only possible ordering of Q and furthermore that it is an ordering compatible with that of D.

The Real and Complex Fields

We conclude this section with a very brief commentary on the real and the complex numbers.

We define a **real number system** to be an ordered field, such that every nonempty subset, that is bounded above, has a supremum (least upper bound). The supremum of such a set can be proved to be unique for this set.

An ordered field F is called **archimedean,** if given any pair of positive elements a and b, where $a < b$, there exists an integral multiple na of a such that $na > b$. A real number system can be proved to be archimedean.

Every element of a real number system can be proved to have a decimal expansion, that is, every element can be represented by means of a decimal fraction. Finally it can be shown that two real number systems are order-isomorphic as fields, and hence a real number system is unique up to isomorphism. We are then considered to be entitled to write of **the** real number system. Since a real number system can be actually constructed, its existence is not in doubt.

The **complex number system** can be introduced in a way that is very similar to that of the rational number system. In this case, however, we consider the set C of all ordered pairs (a, b) of real numbers a and b. If we define equality, addition, and multiplication by

$$(a, b) = (c, d) \quad \text{if, and only if, } a = c \text{ and } b = d,$$
$$(a, b) + (c, d) = (a + c, b + d)$$
$$(a, b) \cdot (c, d) = (ac - bd, ad + bc),$$

it is quite easy to show that C is a field and, moreover, that C contains a subfield (all elements of the form $(a, 0)$) which is isomorphic to the real field. C is called the **complex field** or the **field of complex numbers.**

The notation (a, b) for a complex number can be changed to the more familiar form $a + bi$, where $i = (0, 1)$. We see then that $i^2 = (-1, 0)$.

The complex field C differs in two very important respects from the real field. First of all C cannot be ordered. To see this we note that the real number $(1, 0) > 0$ and that $i^2 = (-1, 0) = -(1, 0) < 0$. But if C were an ordered field the square of any nonzero number would have to be positive. Secondly C is an algebraically closed field which the real field is not.

A field F is said to be **algebraically closed** or **algebraically complete** if the roots of every polynomial equation, that is equation of the form

$$a_n x^n + a_{n-1} x^{n-1} + \cdots + a_0 = 0$$

(where the $a_i \in F$ and not all $a_i = 0$) has all its roots in F. If it has one root in F then it is easy to show that it will have all its roots in F.

The real field is not algebraically closed, as the equation $x^2 + 1 = 0$ shows. The so-called fundamental theorem of algebra proves that if F is the complex field C then every polynomial equation with coefficients in C has a root in C. This implies that all its roots are in C, that is, the complex field is algebraically closed. This makes the complex field somewhat of a "culmination" field in algebra.

EXERCISES

1. Find all the nontrivial (nonzero) homomorphisms of (1) $Z_3 \rightarrow Z_4$, (2) $Z_4 \rightarrow Z_6$, (3) $Z_{10} \rightarrow Z_8$.

2. Under what restraint on m and n does there exist a nontrivial homomorphism of $Z_m \rightarrow Z_n$?

3. Let $B \subset C$ be two ideals in a ring R. Prove that the quotient ring C/B is an ideal in the quotient ring R/B and that

$$R/C \approx R/B/C/B.$$

4. If A is the set of all matrices of the form $\begin{bmatrix} a & b \\ 0 & c \end{bmatrix}$ over Z and B is the ideal in A of all matrices of the form $\begin{bmatrix} d & e \\ 0 & 0 \end{bmatrix}$, prove $A/B \approx Z$.

5. Find an epimorphism of the ring $Z_{12} \rightarrow Z_3$.

6. Find all the homomorphic images of the rings (a) Z_{10}, (b) $Z_3 \times Z_4$, (c) Z.

7. Is the set of all real numbers $x + y \sqrt[3]{2}$, where x and y are rational, an integral domain under ordinary addition and multiplication?

8. If a Boolean ring R contains no proper zero divisors, show that $R = 0$ or $R \approx Z_2$.

9. Let m and n be relatively prime positive integers. Prove that the rings Z_{mn} and $Z_m \times Z_n$ are isomorphic.

If m and n are not relatively prime prove that the two rings cannot be isomorphic. (Find a nonzero element of Z_{mn} that has to map into the zero element of $Z_m \times Z_n$ under any isomorphism.)

Thus the two rings are isomorphic if and only if m and n are coprime.

10. Prove that the intersection of a family R_i, $i \in I$, of subrings of a ring R is a subring of R.

11. Prove that the intersection of a family D_i, $i \in I$, of subdomains of an integral domain is an integral domain.

12. A prime subdomain of an integral domain D is one containing no proper subdomains.

(a) If e is the unit element of an integral domain D show that the set of all multiples $ne, n = 0, \pm 1, \pm 2, \ldots$ is a subdomain D' of D.

(b) Is D' a prime subdomain?

(c) Show that every integral domain in D contains a unique prime subdomain that is either isomorphic to Z_p, p a prime, or isomorphic to Z.

13. If R and R' are rings show that the ring $R \times R'$ contains subrings isomorphic to R and R'.

5-7 POLYNOMIAL RINGS

Let S be any ring with an identity. If $a_0, a_1, \ldots, a_n, x$ belong to S, then an expression of the form $a_0 + a_1 x + \cdots + a_n x^n$ is called a **polynomial in x over** S [*over* S means with coefficients in S]. Let R be a subring of S and assume R also contains the identity 1 of S. Let $x \in S$ and assume that x commutes with every element of R, that is $xr = rx$, for every $r \in R$. Let $R[x]$ denote the set of all polynomials in x over R. Then $R[x]$ is a ring. For the equality, addition and multiplication in $R[x]$ are all inherited from the ring S and it is merely necessary to verify that the sum and product of two polynomials are also in $R[x]$; that is,

$$(a_0 + a_1 x + \cdots + a_m x^m) + (b_0 + b_1 x + \cdots + b_n x^n)$$
$$= (a_0 + b_0) + (a_1 + b_1)x + (a_2 + b_2)x^2 + \cdots$$

and

$$(a_0 + a_1 x + \cdots + a_m x^m)(b_0 + b_1 x + \cdots + b_n x^n)$$
$$= a_0 b_0 + (a_0 b_1 + a_1 b_0)x + (a_0 b_2 + a_1 b_1 + a_2 b_0)x^2 + \cdots$$

In performing these operations, bear in mind that x commutes with all the elements of R, that is $xa = ax$ for all $a \in R$ and hence $x^2 a = x(xa) = x(ax) = (xa)x = (ax)x = ax^2$, since the multiplication is associative. And now by induction, it follows that $(ax^m)(bx^n) = abx^{m+n}$.

Thus $R[x]$ is a ring and is a subring of S. $R[x]$ is the subring of S *generated* by R and the element x. This means $R[x]$ is the smallest subring of S that contains R and x—that is, $R[x]$ is the intersection of all subrings of S containing R and x. This is easy to see, since any subring of S that contains R and x would obviously have to contain all the polynomials in x over R, for such a polynomial is manufactured by the addition and multiplication of elements of R with x.

Definition. The element x of S is called **transcendental** over R or **variable** over R if no polynomial in x over R can vanish unless all its coefficients $a_i = 0$. If x is not transcendental it is called **algebraic over** R. Thus if x is algebraic over R, then for some polynomial we have

$$a_0 + a_1 x + \cdots + a_n x^n = 0$$

where $n > 0$ and $a_n \neq 0$. Thus if S is the real field and R the subfield of rational numbers, the element $\sqrt{2} + \sqrt{3}$ of S is algebraic over R, since

$$(\sqrt{2} + \sqrt{3})^4 - 10(\sqrt{3})^2 + 1 = 0.$$

We are not going to prove it here, but it has been proved that the real numbers π and e (the base of the natural logarithms) are transcendental over the rational field, and so is the real number $2^{\sqrt{2}}$, known as **Hilbert's number**.

Definition. If x is a variable over the ring R (that is x is transcendental) then $R[x]$ is called the **ring of polynomials in x over R.**

In the ring of polynomials $R[x]$, note that two polynomials are equal if and only if the corresponding coefficients of the two polynomials are equal. For if $a_0 + a_1 x + \cdots + a_n x^n = b_0 + b_1 x + \cdots + b_n x$, then $(a_0 - b_0) + (a_1 - b_1)x + \cdots + (a_n - b_n)x^n = 0$, and, since x is transcendental, we have $a_i - b_i$ for each i.

Definition. Let $f(x) = a_0 + a_1 x + \cdots + a_n x^n$ be a polynomial in $R[x]$. If $a_n \neq 0$, then a_n is called its **leading coefficient** and the polynomial is said to be of **degree** n, written $\deg f = n$.

THEOREM 12. If f and g are two polynomials in $R[x]$, then $\deg (f + g) \leq \max (\deg f, \deg g)$ and $\deg fg \leq \deg f + \deg g$.

Proof: For if $\deg f = n$ and $\deg g = m$, then if $n > m$ *clearly*

$$\deg (f + g) = n,$$

while if $m = n$, then $\deg (f + g) < n$ if the two leading coefficients happen to cancel one another out, otherwise

$$\deg (f + g) = n.$$

As for the product fg its degree is clearly $m + n$, unless the product of the two leading coefficients is zero (something which could occur if the ring R has zero-divisors) in which case

$$\deg fg < m + n$$

THEOREM 13. Let R be a subring of a ring S and let x be an element of S that is variable over R. Let t be any element of S that commutes with every element of R. Then the mapping γ of $R[x] \to R[t]$ defined by $(f(x))\gamma = f(t)$ is a ring homomorphism of the two rings $R[x]$ and $R[t]$. The mapping γ is an isomorphism if and only if t is also variable over R.

Proof: The verification that γ is a ring homomorphism is routine and left to the student as an exercise. Clearly γ is an isomorphism if and only if its kernel is 0, and this is true if and only if the only zero polynomial in $R[t]$ is the one whose coefficients are all zero.

THEOREM 14. For any ring R, there exists a ring S and an element $x \in S$, such that x is a transcendental element over a subring R' of S that is isomorphic to R. (Since R' is isomorphic to R, we can speak of S as containing R and the element x as transcendental over R.)

Proof: Form a sequence $(a_i) = (a_0, a_1, a_2, \ldots)$ of elements a_i of R in which almost all the a_i are 0. ("Almost all" means "all but a finite number.") By courtesy we shall call (a_i) a polynomial, althought it may not look like one at this stage. If in (a_i) we have $a_n \neq 0$ and $a_k = 0$ for all $k > n$, then (a_i) will be called a **polynomial of degree** n and a_n will be called its **leading coefficient.** Denote by S the set of all such polynomials (a_i) formed from the elements of the ring R. As the reader might suspect, we are going to define a ring structure on S by the following definitions:

Equality: $(a_i) = (b_i)$ if and only if $a_i = b_i$ for all i.
Addition: $(a_i) + (b_i) = (a_i + b_i)$.

Multiplication:

$$(a_i)(b_i) = (c_i), \text{ where } c_i = \sum_{j+k=i} a_j b_k \text{—that is,}$$

$$c_0 = a_0 b_0, c_1 = a_0 b_1 + a_1 b_0, c_2 = a_0 b_2 + a_1 b_1 + a_2 b_0, \cdots$$

It is easy to verify that addition and multiplication produce sequences (a_i) whose terms are almost all zero, so that they are in S. Moreover, addition can be checked to be commutative and associative and multiplication to be associative. The zero element of S is the sequence $(0, 0, 0, \ldots, 0, \ldots)$ all of whose terms are zero and the additive inverse of (a_i) is clearly $(-a_i)$. Thus S is an abelian group under addition. The two distributive laws can also be verified, although this is more laborious. Simply follow carefully and accurately the definitions of addition and multiplication above. We shall therefore consider S to be a ring. The same rules for the degrees of the sum and the product of two polynomials of S can be seen to be true as for $R[x]$ earlier. Put

$$x = (0, 1, 0, 0, \ldots, 0, \ldots);$$

that is, x is the polynomial (a_i) in which $a_i = 1$ and $a_i = 0, i \neq 1$. x is an element of S. Following the multiplication rule above for sequences we see that $x^2 = (0, 0, 1, 0, 0, \ldots, 0, \ldots)$ and by induction that $x^k = (0, 0, \ldots, 0, 1, 0, \ldots, 0, \ldots)$, where 1 is in the $(k + 1)$st spot of the sequence and all other terms are zero. Now put

$$a' = (a, 0, 0, \ldots, 0, \ldots).$$

This is the sequence whose first term is a and all of whose other terms are 0. Clearly $a' \in S$. If we denote by R' the set of all terms of S of the form a', we find quite easily that R' is a subring of S. By the rule for multiplication we find

$$a'x^k = (0, 0, \ldots, 0, a, 0, \ldots, 0, \ldots) = x^k a',$$

this being the sequence whose $(k + 1)$st term is a while all other terms are 0.

Furthermore, x commutes with every element of the subring R'. Since $a_0' + a_1'x + \cdots + a_n'x = (a_0, a_1, \ldots, a_n, 0, 0, 0, \ldots, 0, \ldots) = 0' = (0, 0, \ldots, 0, \ldots)$ if and only if each $a_i = 0$, we see that x is transcendental over R'. The mapping of $R \rightarrow R'$ defined by

$$a \rightarrow a' = (a, 0, 0, \ldots, 0, \ldots)$$

is clearly an isomorphism, and so we have proved that there exists a subring R' of the polynomial ring $R'[x]$ that is isomorphic to the given ring R. As is customary in the case of isomorphic rings, we identify the two rings R and R' and do not distinguish between them. We speak of x as a transcendental element over R and write the polynomials in x with their coefficients in R, that is we write $a_0 + a_1x + \ldots + a_nx^n$, $a_i \in R, i = 0, 1, \ldots, n$.

Note that $x^2, x^3, \ldots$ are also transcendental elements over R'; that is, over R. If the ring R is not commutative, then ax, $a \in R$, does not commute with the elements of R, that is for $b \in R$, $b(ax) \neq (ax)b$, and so ax is neither transcendental nor algebraic.

Earlier we gave an example of an algebraic element over a ring and proved it to be algebraic. Also examples of transcendental elements were given, but they were not proved transcendental (in fact it is not easy to do this). Hence doubts can very well arise concerning the actual existence of transcendental elements. The above theorem dispels these doubts. It demonstrates the existence of transcendental elements over an arbitrary ring.

THEOREM 15. Let x be a variable over a ring (we use interchangeably the terms transcendental, variable, indeterminate). Then $R[x]$ is an integral domain if and only if R is an integral domain.

Proof: First observe that since x is transcendental, the product $ax^k = 0, a \in R$, if and only if $a = 0$. (This would not be true if x were algebraic.) Thus possible zero-divisors in $R[x]$ cannot arise from such products. If R is an integral domain, then we already know that $R[x]$ is a commutative ring with identity. Moreover $R[x]$ has no zero-divisors, since

$$(a_n x^n + \cdots + a_0)(b_m x^m + \cdots + b_0) = a_n b_n x^{n+m} + \cdots$$

and if $a_n \neq 0$, $b_m \neq 0$, then $a_n b_m \neq 0$. Conversely if $R[x]$ is an integral domain, then R has to be, since R is a subring of $R[x]$.

Corollary. If R is an integral domain and x is transcendental over R, then for any two polynomials f and g of $R[x]$,

$$\deg (f + g) \leq \max (\deg f, \deg g)$$
$$\deg fg = \deg f + \deg g.$$

Proof: This follows from Theorem 12 and the fact that here $R[x]$ is an integral domain.

The **units** of any ring with an identity are those elements of the ring that have multiplicative inverses. If u and v are two units of a ring R then so is their product uv, for its inverse is $v^{-1}u^{-1}$. Thus the units of a ring form a multiplicative group, called the **group of units** of the ring. The units of a ring are very important, as we shall see later, in factorization theory. We make merely the following important remark here.

Let D be an integral domain and let x be a variable over D. Then the units of the ring $D[x]$ are precisely those of the ring D. That is, the only units in $D[x]$ are those it inherits from D itself. For if $f(x)$ is a unit of $D[x]$, then it has an inverse $g(x)$ and $f(x) \, g(x) = 1$. Hence, by the degree equation above, $\deg f + \deg g = 0$, and hence $\deg f = 0$ and $\deg g = 0$. Thus $f(x)$ belongs to D.

Polynomial Functions

We end this section with a brief discussion of polynomial functions, in contrast with the polynomial forms studied above.

Definition. Let D be an integral domain. A function (mapping) f of $D \to D$ is called a **polynomial function** if for all $t \in D$,

$$f(t) = a_0 + a_1 t + \ldots + a_n t^n,$$

where, the $a_i, i = 0, 1, 2, \ldots, n$, are fixed elements of D. [Here we write $f(t)$ in place of the usual notation tf.]

Thus if $p(x) = b_0 + b_1 x + \cdots + b_m x^m$ is a polynomial in $D[x]$, then the function f defined by

$$f(t) = b_0 + b_1 t + \cdots + b_m t^m \qquad \text{for all } t \in D$$

is a polynomial function on $D \to D$.

The polynomial functions on $D \to D$ will form a ring P in the following way. Since they are mappings, we know that for $f, g \in P$, $f = g$ if and only if $f(t) = g(t)$ for all $t \in D$. Define addition and multiplication by $(f + g)(t) = f(t) + g(t)$, $(fg)(t) = f(t) \cdot g(t)$, for all $t \in D$. The zero of P is the polynomial function 0, given by $0(t) = 0$ for all $t \in D$, and $-f$ is the polynomial function defined by $(-f)(t) = -f(t)$. It is now readily verified that with these conditions, P is a ring. Note that P has as identity element the polynomial function which maps every element of D into 1. In general P can have zero divisors.

If D is a finite integral domain then two polynomial functions can be equal without their coefficients being identical. For example, if D_3 is the integral domain of residue classes modulo 3 whose elements we write simply as $0, 1, 2$, then

$$f(t) = t + 1 \text{ and } g(t) = t^3 + 1$$

define equal polynomial functions f and g on $D_3 \to D_3$.

Moreover, observe that $f(t) = t^2 + t$, $g(t) = 2t^2 + t$ define two nonzero polynomial functions on $D_3 \to D_3$, whose product is the zero polynomial function.

However, if D is an infinite integral domain, then it is not hard to prove that two polynomial functions are equal if and only if they have identical coefficients. Furthermore when D is infinite, P is likewise an integral domain. In this case then, there is an isomorphism of polynomial form to polynomial function and they become equivalent concepts.

More generally we can form a ring of polynomial functions on $R \to R$, where R is any commutative ring with identity. We describe another interesting way of doing this. With the same definitions of addition and multiplication as before, it is easily proved that the set $F(R)$ of all functions on $R \to R$ forms a ring (it is called a **function ring**). Now denote by f_a, $a \in R$, the constant function on $R \to R$ defined by $f_a(r) = a$, for all $r \in R$. As usual let 1_R stand for the identity map on R. The constant functions f_a, $a \in R$, together with 1_R are now used to generate a subring P of $F(R)$.

Let us examine the form of the elements of P. First we note that the product of two constant functions f_a and f_b is the constant function f_{ab}. Define $1_R^0 = f_1$, where f_1 is the constant function mapping every $r \in R$ into 1. Then according to Sec. 5-2 we see that the elements of the generated subring P are finite sums of terms of the form $\pm f_a 1_R^k$, where k is a nonnegative integer. The definition of multiplication tells us that if $k > 0$, then 1_R^k is defined by $1_R^k(r) = r^k$, and therefore $f_a 1_R^k$ is defined by

$f_a 1_R^k(r) = ar^k$. Hence the elements of P are the polynomial functions on $R \to R$.

5-8 PROPERTIES OF POLYNOMIAL RINGS

We know by Theorem 15 that if F is a field and x is transcendental over F, then $F[x]$ is an integral domain. We are now going to establish properties about the polynomial domain $F[x]$ which are similar to those possessed by the integral domain Z of integers $0, \pm 1, \pm 2, \dots$. *In this section all polynomials will belong to $F[x]$. We stress this by stating that a polynomial must have its coefficients in F.*

Divisibility

A polynomial f of $F[x]$ is said to be a **divisor** of (or to divide) a polynomial g of $F[x]$, written $f \mid g$, if there exists a polynomial h of $F[x]$ such that $g = fh$. A polynomial f of degree > 0 is said to be **irreducible** if it cannot be factored as a product of two polynomials of positive degree. Thus a polynomial f would be reducible if it could be factored into the product of two polynomials, that is $f = gh$, in which the degree of g and the degree of h is each ≥ 1.

The Euclidean Algorithm

Let f and g be two polynomials and assume $g \neq 0$. Then there exist unique polynomials q and r such that $f = qg + r$ and either $r = 0$ or $\deg r < \deg g$.

First observe that in proving this theorem no loss of generality results in taking the leading coefficient of $g = 1$. We can dispose of two cases at the start. If $\deg f < \deg g$, then the theorem is seen to be true if we take $q = 0$ and $r = f$. If $\deg f = \deg g$, take $q = a_n$, where a_n is the leading coefficient of f, and $r = f - a_n g$, and then $r = 0$ or $\deg r < \deg g$.

Thus we shall now assume $\deg g < \deg f$. Let S be the set of all polynomials of the form $f - pg$, where p runs through all the polynomials of $F[x]$. Let k be the least nonnegative integer which is the degree of a nonzero polynomial of S and let $f - qg$ be a polynomial of degree k in S. Put $f - qg = r$. Now $k < \deg g$. For suppose $k \geq \deg g$. Let $g = ax^m + \cdots$ and let $r = bx^k + \cdots$. If $m = k$ then $\deg (r - ba^{-1}g) < k$. But $r - ba^{-1}g = f - (q + ba^{-1})g$ and so belongs to S, and this would contradict the choice of k. If $m < k$, then $\deg (r - ba^{-1}x^{k-m}g) < k$, and since $r - ba^{-1}x^{k-m}g = f - (q + ba^{-1}x^{k-m})g$ belongs to S, we again get a contradiction. Therefore, $k = \deg r < \deg g$. Hence $r = 0$ if there exists a polynomial p such that $f - pg = 0$, or we have $\deg r < \deg g$.

It is easy to show that r and q are unique, for suppose two other polynomials r' and q' exist with the same properties. That is, $f = q'g + r'$ and either $r' = 0$ or $0 < \deg r' < \deg g$. Then we have $r - r' = (q' -$

$q)g$ and $\deg(r - r') < \deg g$. This first one means g is a divisor of $r - r'$ and the second contradicts this unless $r - r' = 0$, in which case $q' - q = 0$, and we see that r and q are unique.

Greatest Common Divisor

The **greatest common divisor** (G.C.D.) of two polynomials f and g is defined to be a polynomial h with leading coefficient 1 which satisfies the following conditions: (1) $h \mid f$ and $h \mid g$, and (2) if $p \mid f$ and $p \mid g$, then $p \mid h$.

We now prove the existence in the domain $F[x]$ of the G.C.D. of any two elements f and g and a simple generalization of this leads to the existence of the G.C.D. of any finite number of elements of $F[x]$. Form the set S of all polynomials of the form $pf + qg$, where p and q run through all the polynomials in $F[x]$. Let k be the least nonnegative integer forming the degree of such a nonzero polynomial, and let h be that polynomial of degree k whose leading coefficient is 1. Clearly h is unique, since if h' is a second such polynomial then $h \mid h'$ and $h' \mid h$; and, since their leading coefficients are 1, we must have $h = h'$. Now let

$$h = pf + qg.$$

Then any divisor of both f and g has to divide h. We show that $h \mid f$ and $h \mid g$. By the Euclidean algorithm there exist polynomials s and t such that $f = sh + t$ and $t = 0$, or we have $0 < \deg t < \deg h$. But unless $t = 0$ we get a polynomial $t = f - sh = f - s(pf + qg) = (1 - sp)f + (-sq)g$ which belongs to S and yet whose degree is less than k. This is a contradiction, and so $t = 0$. Hence $f = sh$ and thus $h \mid f$. Similarly $h \mid g$. This completes the proof that h is the G.C.D. of f and g.

THEOREM 16. If t is an irreducible polynomial and if $t \mid fg$ but $t \nmid f$ then $t \mid g$. (This property is described by saying that t is a *prime* polynomial.)

Proof: If t is irreducible and $t \nmid f$ then the G.C.D. of t and f is 1. Hence there exist polynomials p and q such that $1 = pt + qf$. Hence $g = ptg + qfg$ and, since t divides each term on the right, $t \mid g$.

This property of a polynomial that if it divides a product it must divide at least one of the factors is described by saying the polynomial is **prime.** The last theorem thus proves that an irreducible polynomial is prime. This is not true in general of integral domains. There are domains [see $Z(\sqrt{-5})$, to be discussed later] where irreducible elements are not necessarily prime. Of course a prime polynomial f is always irreducible. For if it it were reducible, say $f = gh$, then g and h would be polynomials of degrees greater than 0, but each of degree less than that of f. Thus $f \mid gh$, and yet f can divide neither g nor h which would contra-

dict f being a prime. Hence f must be irreducible   In the particular
integral domain $F[x]$ we have thus shown that irreducible and prime are
equivalent attributes.

We can use induction to prove that every polynomial of $F[x]$ whose
degree is > 0 can be expressed as the product of a finite number of irreducible polynomials. Let f be a polynomial of degree n. If deg $f = 1$,
then f is already irreducible. We then suppose that $n > 1$. Assume the
statement is true for all polynomials of degree $< n$. If f is irreducible then
we are through. If not, let $f = gh$; then the degrees of g and h are both
$< n$, and hence the theorem is true for g and h and hence for $f = gh$.
Thus by induction the statement is true for all polynomials of positive
degree. Hence $f = p_1 p_2 \ldots p_r$, where the p_i are irreducible polynomials. This factorization is unique up to the order of the factors and up
to associates of the p_i. This is proved in the same way as for the integers.
Two polynomials g and h are called **associates** if $g \mid h$ and $h \mid g$. Thus in
the factorization of f, above, some of the p_i could be replaced by their
associates, just so long as the product remains equal to f.

EXERCISES

1. Prove that if F is a field and x is transcendental over F, then $F[x^2]$ is
a subring but not an ideal in $F[x]$.

2. Prove by induction that there exists a G.C.D. for any n polynomials
of $F[x], n \geq 2$.

Let x be a variable over an integral domain D and let $D[t]$ be a ring
of polynomials in an arbitrary element t. As shown earlier, the mapping
γ of $D[x] \rightarrow D[t]$ defined by $(f(x))\gamma = f(t), f(x) \in D[x]$, is a ring homomorphism, which is an isomorphism if t also is a variable over D. The
kernel of γ is the ideal of all polynomials $f(x) \in D[x]$ for which $f(t) = 0$.
Let $g(x)$ be the unique polynomial of least degree with leading coefficient
equal to 1 of the polynomials in this kernel. Clearly if $h(x) \in \ker \gamma$ then
$g \mid h$, and so the kernel of γ is the ideal composed of all the multiples
of $g(x)$, that is to say $g(x)$ generates the kernel of γ. An ideal that is
generated by a single element $g(x)$ is called a **principal ideal** and is denoted
by $(g(x))$. From the theory of rings we see then that

$$D[x]/(g(x)) \approx D[t].$$

If t also is variable over D then $g(x) = 0$ and $(g(x)) = (0)$, the zero ideal,
and $D[x]/(0) = D[x]$.

EXERCISES

1. Prove that any rational root of the equation

$$a_0 x^n + a_1 x^{n-1} + a_2 x^{n-2} + \cdots + a_n = 0,$$

where the a_i are integers, must be of the form r/s where $r \mid a_n$ and $s \mid a_0$. (Put $x = r/s$ where we can assume r and s are relatively prime. Then multiply through by s^n.)

  2. Prove that the equation $2x^3 + 4x^2 + x + 5 = 0$ has no rational roots. If its coefficients are regarded as being in the field Z_3, prove it has all its roots in Z_3.

  3. Let $f(x)$ and $g(x)$ be polynomials over a field F, and let deg $f(x) = m$, deg $g(x) = n$, $m \geqq n$. Divide $f(x)$ by $g(x)$ and obtain a remainder $h(x)$ for which $h(x) = 0$ or deg $h(x) = k < n$. If $h(x) \neq 0$, divide $g(x)$ by $h(x)$ and obtain a remainder of 0 or of degree less than that of $h(x)$. Continuing in this way prove that the G.C.D. of $f(x)$ and $g(x)$ can be determined over the field F.

  4. Find the G.C.D. of the polynomials $3x^5 - 8x^4 - x^2 + 6x - 2$ and $x^9 - 3x^8 + x^7 - 2x^2 + 6x - 2$ over the rational field.

  5. The polynomial $x^2 - 2x - 1$ is irreducible over the rational field. Is it reducible over the fields Z_3, Z_5 and Z_7? For what values of the prime p is it reducible over Z_p?

5-9 POLYNOMIAL RINGS IN TWO VARIABLES

Let S be a ring with identity and R a subring of S containing the identity. Let x and y be two elements of S that commute with each other and with every element of R. Then x and y are called **independent variables** over R (that is, independent transcendental elements over R) if a finite sum $\Sigma a_{ij} x^i y^j = 0$, $a_{ij} R$, implies each $a_{ij} = 0$. Note that $\Sigma a_{ij} x^i y^j$ is a symbol for a polynomial in x and y with coefficients in R. Such a polynomial has the form $a_{00} + a_{10}x + a_{01}y + a_{20}x^2 + a_{11}xy + a_{02}y^2 + a_{30}x^3 + \cdots$ to a finite number of terms.

THEOREM 17. The elements x and y are independent variables over the ring R if and only if x is variable over R and y is variable over the ring $R[x]$.

Proof: Let x and y be independent variables over R. Now every polynomial $f(x)$ in x can be regarded as a polynomial in x and y, whose y terms have zero coefficients. Thus $f(x) = 0$ implies each of its coefficients is 0 and so x is transcendental over R. Also any polynomial in y over $R[x]$—that is, with coefficients in $R[x]$, is a polynomial $g(x, y)$ in x and y. Since $g(x, y) = 0$ implies each of its coefficients is 0, it follows that y is variable over $R[x]$. Conversely, let x be variable over R, and let y be variable over $R[x]$. Let $\Sigma a_{ij} x^i y^j$ be any polynomial in x and y. Then this sum is a polynomial in y over $R[x]$. Hence if $\Sigma a_{ij} x^i y^j = 0$, then the coefficient of each power of y is 0. But the coefficient of each power of y is a polynomial in x over R and, since x is variable over R, such a polynomial is 0 if and only if each of its coefficients is 0. Thus every $a_{ij} = 0$ and hence x and y are independent variables over R.

Since x and y commute with each other, we see that if they are independent variables over R then y is variable over R and x is variable over $R[y]$. If $R[x][y]$ denotes the ring of polynomials in y over $R[x]$, then clearly $R[x][y] = R[y][x]$, where $R[y][x]$ is the ring of polynomials in x over $R[y]$. We write simply $R[x, y]$ for both of them.

Clearly if D is an integral domain, then so is $D[x, y]$. For $D[x]$ is an integral domain and, since y is variable over $D[x]$, then $D[x, y]$ is an integral domain. As one might conjecture this is true when we have n independent variables.

EXERCISES

1. A nilpotent element x of a ring R is an element for which $x^m = 0$ for some positive integer m. If R is a commutative ring with identity, prove that the nilpotent elements of R form an ideal in R. This ideal is called the **radical** of R.

2. Let A be an ideal in a commutative ring R with identity. Prove that the set of all $x \in R$ such that $x^m \in A$ for some positive integer m (depending on x) is an ideal in R. This ideal is called the **radical** $\sqrt{A}$ of the ideal A. Note $A \subseteq \sqrt{A}$. Prove the radical of R is $\cap \sqrt{A}$ where the intersection is taken over all the ideals A in R.

3. R is a commutative ring with 1. A is an ideal in R. A mapping f of $R/A \to R/\sqrt{A}$ is defined by $(r + A)f = r + \sqrt{A}$, $r \in R$. Show f is a ring homomorphism. Find its kernel and prove directly that this kernel is an ideal. In what ring is it an ideal?

4. Find the nilpotent elements of an integral domain and of the ring of residue classes modulo 4.

5. Prove that an ordered field is an infinite field of characteristic zero. Every ordered field contains a "rational" field, the quotient field of its "integers." Prove the rational subfields of all ordered fields are order-isomorphic, that is, order is preserved by the isomorphism. (Start the construction of the isomorphism by requiring the zero elements of the two fields to correspond and the unit elements to correspond.)

6. Give a counterexample to disprove the following statement: if f is an injective homomorphism of two rings $A \to B$ and if A has an identity then B has an identity.

7. Find the group of automorphisms and the ring of endomorphisms of the abelian group G generated by the two elements x and y, where $x^4 = 1$ and $y^5 = 1$.

8. If Z and Z' are two systems of integers then there is a unique ring isomorphism f of $Z \to Z'$ which preserves order—that is, Z and Z' are order-isomorphic.

We make a few observations intended to assist the reader in proving this theorem. Note that an isomorphism of two ordered integral domains does not necessarily preserve order. Hence the crucial axiom for this property is the finite induction axiom. The proof of the theorem leans heavily on this axiom.

If f is an isomorphism of $Z \to Z'$ and P is the set of positive elements of Z and P' the set of positive elements of Z', prove $Pf = P'$. Moreover by assuming a second order-preserving isomorphism f' of $Z \to Z'$, prove $f = f'$. This takes care of uniqueness for f.

To construct f, let f be a mapping of $Z \to Z'$ defined by (1) $0f = 0'$, and (2) $(x + 1)f = xf + 1'$, for $x \in P$; (3) $xf = -(-x)f$, $x \in -P$. Note (2) is a recursive definition, that is it defines $1f, 2f, 3f, \ldots$ successively.

Show that (3) and (2) hold for all $x \in Z$, and then prove $(x + y)f = xf + yf$ for all x, y where either $x > 0$ or $y > 0$ by using induction. Then extend it to $x < 0$, $y < 0$.

In the same way, prove $(xy)f = (xf)(yf)$. This proves f is a homomorphism. Finally, prove f is bijective.

9. If A is an ideal in a ring R and if B is a subring of R show that R/A contains a subring isomorphic to B if $B \cap A = 0$.

10. We have seen in Section 5-8 that the G.C.D. $h(x)$ of two polynomials $f(x)$ and $g(x)$ over a field F can be expressed in the form

$$h(x) = s(x)f(x) + t(x)g(x),$$

where $s(x)$ and $t(x)$ belong to $F[x]$. We can express this by saying the G.C.D. of f and g is a homogeneous linear combination of f and g. Prove this theorem in the following way. By the Euclidean algorithm, $f = q_1 g + r_1, g = q_2 r_1 + r_2, r_1 = q_3 r_2 + r_3, \ldots, r_i = q_{i+2} r_{i+1} r_{i+2}, \ldots$. Hence show r_1 and r_2 are homogeneous linear functions of f and g. Now use induction, assuming the theorem true for $r_1, r_2, \ldots, r_{i+1}$ and proving it true for r_{i+2}.

11. Find the G.C.D. $h(x)$ of $f(x) = x^4 - 12x^3 + 46x^2 - 60x + 16$ and $g(x) = x^3 - 4x^2 - 2x + 12$. Also determine the polynomials $s(x)$ and $t(x)$ such that $h = sf + tg$.

chapter 6

Factorization and Ideals

A few words about this chapter are appropriate here. It is occupied with a general treatment of factorization in algebraic domains and makes careful distinction between an irreducible element and a prime element. However after the definitions in the first four paragraphs of Sec. 6-1, one can proceed to Secs. 6-2, 6-3, and 6-5 (omitting Sec. 6-4) to obtain the result that a Euclidean domain is a unique factorization domain. It does not seem desirable to specialize the treatment of factorization to fit merely the case of the Euclidean domain, when, with such little extra effort, one can obtain the result for a principal ideal domain. The ascending-chain condition is a very interesting, and certainly not abstruse concept, that is most important in algebra.

6-1 PRIMES

Let D be an integral domain. An element u of D that has a multiplicative inverse u^{-1} is called a *unit* of D. Thus $u^{-1}u = uu^{-1} = 1$. The product uv of two units u and v of D is also a unit, for it has the inverse $v^{-1}u^{-1} = u^{-1}v^{-1}$. Thus the units of D form an abelian multiplicative group called the **group of units** of D.

Two elements x and y of D are called **associates** if $x = uy$, where u is a unit of D, in which case $y = u^{-1}x$, where u^{-1} is also a unit.

An element x of D is said to *divide* or to be a *divisor* of an element y of D, written $x \mid y$, if there exists an element z of D such that $y = zx$. It is easy to see that x and y are associates in an integral domain if and only if $x \mid y$ and $y \mid x$.

An element x of D, that is not a unit, is called **irreducible**, if $x = yz$ implies that either y or z (not both, since then $yz = x$ would be a unit) is a unit. To put it another way, x is irreducible if $x = yz$ implies that either y or z is an associate of x. Thus x is reducible if $x = yz$ and neither y nor z is a unit, in which case we say that y and z are *proper factors* (as distinguished from units) of x.

An element x of D is called a **prime** if $x \mid yz$ implies that either $x \mid y$ or $x \mid z$. A prime is irreducible. For if x is a prime and $x = yz$, then if $x \mid y$, x and y are associates and hence z is a unit. The same is true if $x \mid z$, for then y is a unit. However an irreducible element is not necessarily a prime.

It is very instructive at this point to study the following example. It has several features that are worthy of our attention.

Let Z be the integral domain of the integers $0, \pm1, \pm2, \ldots$ and denote by $Z(\sqrt{-5})$ the set of all numbers of the form $a + b\sqrt{-5}$, where a and b are integers. The study of the properties of $Z(\sqrt{-5})$ is greatly facilitated by introducing the notion of the norm. The *norm* $N(r)$ of $r = a + b\sqrt{-5}$ is defined by $N(r) = r\bar{r} = a^2 + 5b^2$. Here

$$\bar{r} = a - b\sqrt{-5}$$

and is called the **conjugate** of r. Note that $N(r) = 0$ if and only if $r = 0$, that is, if and only if $a = 0$ and $b = 0$. Note that the norm of any non-zero element is always a positive integer. Clearly $N(rs) = N(r)N(s)$ for any two elements r and s. Under the ordinary rules of addition and multiplication it is easy to check that $Z(\sqrt{-5})$ is a commutative ring with the identity 1. Moreover if $r, s \in Z(\sqrt{-5})$ and $rs = 0$, with $r \neq 0$, then $N(rs) = N(r)N(s) = 0$. $N(r) \neq 0$, and hence $N(s) = 0$, hence $s = 0$. Thus $Z(\sqrt{-5})$ is an integral domain. What are its units? Let $rs = 1$, then $N(r)N(s) = N(1) = 1$. Hence $N(r) = 1$. If $r = a + b\sqrt{-5}$, then $N(r) = a^2 + 5b^2 = 1$. The only integral solutions for a and b of this equation are $a = \pm1$ and $b = 0$. Thus the units of $Z(\sqrt{-5})$ are ±1. Hence the only associates of an element r are r itself and $-r$.

Consider now the factoring

$$(1) \qquad 9 = 3 \cdot 3 = (2 + \sqrt{-5})(2 - \sqrt{-5}).$$

The element 3 is irreducible. For if $3 = rs$, then $N(3) = 9 = N(r)N(s)$. Hence $N(r) = 3, 1$, or 9. If $r = a + b\sqrt{-5}$, then $N(r) = a^2 + 5b^2 = 3$ has no integral solutions. Hence $N(r) \neq 3$. On the other hand if $N(r) = 1$ then r is a unit, while if $N(r) = 9$, then $N(s) = 1$ and s is a unit. Thus $3 = rs$ implies either r or s is a unit. Similarly we see that $2 \pm \sqrt{-5}$ are irreducible, since their norms are also 9. Thus we observe that (1) is two distinct factorizations of 9 into irreducible elements.

Another illustration of this phenomenon is

$$21 = 3 \cdot 7 = (1 + 2\sqrt{-5})(1 - 2\sqrt{-5}) = (4 + \sqrt{-5})(4 - \sqrt{-5}),$$

which is an example of three distinct factorizations of 21 into irreducible elements.

Moreover, (1) demonstrates that $2 + \sqrt{-5}$ divides $9 = 3 \cdot 3$ but that $2 + \sqrt{-5}$ does not divide 3, since they are both irreducible and are not associates. Hence $2 + \sqrt{-5}$ is not a prime. The same is true of $2 - \sqrt{-5}$ and 3.

Does $Z(\sqrt{-5})$ have any prime elements? Consider $\sqrt{-5}$. Assume $\sqrt{-5}$ divides the product $(a + b\sqrt{-5})(c + d\sqrt{-5})$. Then $N(\sqrt{-5}) = 5$ must divide $N[(a + b\sqrt{-5})(c + d\sqrt{-5})] = (a^2 + 5b^2)(c^2 + 5d^2)$. This implies 5 divides a^2c^2, and hence $5 \mid a^2$ or $5 \mid c^2$. If $5 \mid a^2$ then $5 \mid a$

and hence $\sqrt{-5}$ must divide $a + b\sqrt{-5}$. Thus $\sqrt{-5}$ is a prime element in $Z(\sqrt{-5})$.

Let us prove that $1 + 6\sqrt{-5}$ is a prime in $Z(\sqrt{-5})$. We chose this element since its norm 181 is a prime integer. Let $r, s \in Z(\sqrt{-5})$. We can show $1 + 6\sqrt{-5}$ is a prime by proving that if $1 + 6\sqrt{-5}$ does not divide r nor s, then $1 + 6\sqrt{-5}$ does not divide their product rs. Now if $1 + 6\sqrt{-5}$ does not divide r, then 181 does not divide $N(r)$. Similarly, 181 does not divide $N(s)$. Hence 181 does not divide $N(r)N(s)$. This implies $1 + 6\sqrt{-5}$ does not divide rs. Hence it is a prime.

Thus $Z(\sqrt{-5})$ is an integral domain in which some irreducible elements are prime and some are not. Moreover, some elements have more than one factorization into irreducible elements, and these are distinct factorizations in that the factors are neither the same nor associates.

6-2 UNIQUE FACTORIZATION DOMAINS

Definition. A **unique factorization domain,** written U.F.D., is an integral domain such that (1) every element that is not a unit is the product of a finite number of irreducible elements, and (2) this factorization is unique up to the order of the factors and to associates of the irreducible elements.

A few words of explanation: this means that if D is a U.F.D. and a is a nonunit of D, then $a = p_1 p_2 \cdots p_r$, where the p_i are irreducible elements of D, not necessarily distinct but finite in number; moreover, this factorization is unique up to the order of the factors p_i and the possibility that associates of the p_i could occur in the factoring. This type of uniqueness is described by saying that the factoring is *essentially unique.*

An example of a U.F.D. is the integral domain of integers, for this is what the so-called "fundamental theorem of arithmetic" states. Moreover, a field is a U.F.D. Every nonzero element of a field is a unit, and so a field has no irreducible elements. Hence it is trivially a U.F.D., since there are no nonunit elements to violate condition (1) above.

THEOREM 1. If condition (1) in the definition above holds, then condition (2) is equivalent to the condition: (2′) every irreducible element is a prime.

> **Proof:** First we show that (2) implies (2′). Assume condition (2). Let p be irreducible and let $p \mid ab$ and $p \nmid a$. Then $ab = pc$ for some c in the domain. Now the element ab has unique factorization, and hence at least one of its factors must be p or an associate of p. The factorization of a cannot contain p or its associate, since $p \nmid a$. Therefore the factorization of b must obtain p or its associate. Hence $p \mid b$ and hence p must be a prime. We now show that (2′) implies (2).

Let a be a nonunit element. If a is irreducible, then we are through. Hence let a be reducible. We use induction and assume that the assertion is true for any element that can be factored into n irreducible elements.

Suppose $a = p_1 p_2 \cdots p_{n+1} = q_1 q_2 \cdots q_m$ are two factorizations of a, where the p_i and q_i are irreducible elements. Then we have

$$p_1 \mid q_1 q_2 \ldots q_m,$$

and hence p_1 must divide at least one of the q_i, which (by renumbering the q_i if necessary) we shall suppose is q_1. This is true since all the p_i and q_i are now primes. Since the elements are in an integral domain, we can cancel p_1 and q_1 and obtain

$$p_2 \cdots p_{n+1} = u q_2 q_3 \cdots q_m$$

where u is a unit and $p_1 = uq$.

Now the induction hypothesis takes over and assures us that the left-hand side (which is a product of n factors) is an element with an essentially unique factorization, and so by the induction this is true of all elements of the integral domain.

Three remarks are in order:

1. The combination of conditions (1) and (2′) can be used to define a U.F.D., for an integral domain is a U.F.D. if and only if these two conditions hold.

2. The fact that the integral domain contains no zero-divisors is the property that ensures the uniqueness of the factorization.

3. If a is an element of a U.F.D. and if

$$a = p_1 p_2 \ldots p_r = q_1 q_2 \ldots q_s,$$

where the p_i and the q_i are irreducible, then $r = s$, and, by renumbering the q_i, p_i and q_i are associates for $i = 1, 2, \ldots, s$.

6-3 PRINCIPAL IDEAL DOMAINS

Definition. Let D be a U.F.D. The **greatest common divisor,** written G.C.D., of two elements a and b of D is an element d of D such that (i) $d \mid a$ and $d \mid b$ and (ii) if $c \in D$ and $c \mid a$ and $c \mid b$ then $c \mid d$.

Clearly if any other element d' of D has these two properties then it would be an associate of d, for $d \mid d'$ and $d' \mid d$. Thus the G.C.D. is essentially unique, if it exists. (It does. See below.)

Definition. Two elements a and b of a U.F.D. D are called **relatively prime** if their only common divisors are units of D.

Definition. The **least common multiple,** written L.C.M., of two elements a and b of a U.F.D. D is an element m of D such that (i) $a \mid m$ and $b \mid m$ and (ii) if $a \mid c$ and $b \mid c$, then $m \mid c$.

It is easy to see that the L.C.M. is essentially unique, if it exists. We now prove the existence of the G.C.D. and L.C.M. of two elements of a U.F.D.

THEOREM 2. In a U.F.D. D any two elements a and b of D have a G.C.D. d and a L.C.M. m; that is, d and m *belong to* D.

Proof: Let

$$a = p_1^{r_1} \cdots p_k^{r_k}$$

$$b = p_1^{s_1} \cdots p_k^{s_k}$$

be factorizations of a and b into primes p. Here we are writing a and b as the products of powers of distinct primes. There is no loss of generality in writing them as products of the powers of the same p_i, since we can remove any factor or add any factor by taking its exponent to be zero. These are essentially unique factorizations. Then it is clear at once that

$$d = p_1^{t_1} \cdots p_k^{t_k} \qquad \text{where } t_i = \min (r_i, s_i)$$

and

$$m = p_1^{u_1} \cdots p_k^{u_k} \qquad \text{where } u_i = \max (r_i, s_i).$$

By induction we can now prove that the G.C.D. and L.C.M. exist for any finite number of elements of D. Consider the G.C.D. It exists for two elements. Assume it exists for n elements of D. Consider the $n + 1$ elements $a_1, a_2, \ldots, a_{n+1}$. Let d' be the G.C.D. of $a_1, \ldots, a_n$ and let d be the G.C.D. of d' and a_{n+1}. Then it is easy to prove that d is the G.C.D. of the $(n + 1)$ elements.

EXERCISE

1. Prove that the L.C.M. of any finite number of elements of D exists.

We are now going to show that in a rather large class of rings unique factorization into primes is a property of the ring.

Definition. A **principal ideal** in an integral domain is an ideal generated by a single element a of the domain. We write the principal ideal generated by a as (a). Its elements have the form xa, where x runs through all elements of the domain.

Definition. A **principal ideal domain,** abbreviated P.I.D., is an integral domain in which every ideal is a principal ideal.

Example. The integral domain Z of integers is a P.I.D. For let B be an ideal in this domain. Now B must contain positive integers. Let m be the least positive integer belonging to B. Let $n \in B$. By the Euclidean algorithm for integers, $n = qm + r$, where $0 \le r < m$. Since n and m

belong to B, r belongs to B_1. But $r < m$ and hence, unless $r = 0$, we have a contradiction of the choice of m. Hence $n = mq$ and B is a principal ideal. Therefore Z is a P.I.D.

THEOREM 3. If F is a field and x is a transcendental element with respect to F, then $F[x]$ is a P.I.D.

Proof: We know from our earlier work that $F[x]$ is an integral domain. Let B be an ideal in $F[x]$. Let $f[x]$ be a polynomial with leading coefficient 1 of the least positive degree m in B. By the Euclidean algorithm, we have for any polynomial $g(x)$ in B that

$$g(x) = q(x)f(x) + r(x)$$

where $r(x) = 0$ or $0 \leq \deg r < \deg f$. By exactly the same reasoning as for the integers we see that $r(x) = 0$.

THEOREM 4. Any two elements a and b of a P.I.D. D have a G.C.D.

Proof: Form the ideals (a) and (b). Then $(a) + (b)$ is an ideal, in fact the ideal generated by the elements a and b. It must be a principal ideal, and so let $(a) + (b) = (d)$. We now show that d is the G.C.D. of a and b. For elements x and y of D we have $xa + yb = d$, which shows that every common divisor of a and b is a divisor of d. Now $a \in (d)$, and hence $d \mid a$. Similarly, $d \mid b$. Hence d is the G.C.D. of a and b, and what is equally important, d can always be expressed in the form $d = ra + sb$, where $r, s \in D$.

We merely mention here (deferring proof for a few pages) that while every pair of elements a and b of a U.F.D. has a G.C.D. d, it does not follow that we can express d in the form $d = ra + sb$. We are next going to use the fact that d does have this form in a P.I.D., to prove that in a P.I.D., every irreducible element is a prime.

THEOREM 5. Every irreducible element a of a P.I.D. D is a prime.

Proof: Let $a \mid bc$ and suppose $a \nmid b$. Then a and b are relatively prime, and therefore $ar + bs = 1$, $r, s \in D$. Hence $acr + bcs = c$, and since a divides every term on the left, then $a \mid c$. Hence a is a prime. This shows that in a P.I.D. irreducible and prime are equivalent terms.

Our objective is to show that a P.I.D. is a U.F.D. However, we still need more equipment.

Definition. A ring R is said to have the **ascending chain condition** for ideals, written a.c.c., if for every increasing sequence

$$A_1 \subseteq A_2 \subseteq A_3 \subseteq \cdots \subseteq A_n \subseteq \cdots$$

of ideals A_i in R, there exists an integer N such that $A_N = A_{N+1} = A_{N+2} = \cdots$, that is if the chain is stationary, (breaks off or is finite).

Not all rings have this property. Commutative rings with identity that have this property are called **Noetherian rings** (after the mathematician Emmy Noether). We next show that a P.I.D. is a Noetherian ring.

THEOREM 6. A P.I.D. satisfies the a.c.c.

> **Proof:** Let $A_1 \subsetneqq A_2 \subsetneqq \cdots$ be an ascending chain of ideals in the P.I.D. D. These will of course all be principal ideals. Form the union $A = \bigcup A_i$. This union is an ideal because each $A_i \subsetneqq A_{i+1}$. (If this were not so, the union would not be an ideal. For example the union of two ideals is not an ideal, in general.) A is a principal ideal. Let $A = (a)$. Then a belongs to some A_k and hence to all A_i, $i > k$, and $A \subsetneqq A_k$. Thus for $i > k$, $A_k \subsetneqq A_i \subsetneqq A$ and $A \subsetneqq A_k \subsetneqq A_i$. Hence $A_i = A_k$ for all $i > k$.

We are going to use the a.c.c. for a P.I.D. to prove that every non-unit element of a P.I.D. is the product of a finite number of primes. At this point there is nothing to prohibit such a factorization to continue indefinitely.

We make this remark before proceeding to the theorem. Note that in an integral domain $a \mid b$ means that $b \in (a)$ and hence that $(b) \subseteq (a)$. If $(b) = (a)$, then $a \mid b$ and $b \mid a$ and a and b are associates. Conversely $(b) \subsetneqq (a)$ means $a \mid b$.

THEOREM 7. Let a be a nonunit of a P.I.D. D, then a is the product of a finite number of primes.

> **Proof:** If a is a prime, we are finished. So let us assume a is reducible, let $a = a_1 b_1$. If a_1 and b_1 are both irreducible then again we are finished. Let a_1 be reducible, let $a_1 = a_2 b_2$. Continuing in this way, we get an increasing sequence of ideals.
>
> $$(a) \subsetneqq (a_1) \subsetneqq (a_2) \subsetneqq \cdots$$
>
> which must terminate, say, at N, that is $(a_N) = (a_{N+1}) = \cdots$ since D satisfies the a.c.c. But the chain can only break off with (a_N) if a_N is irreducible; otherwise the chain would continue. Thus we have proved that a is the product of a finite number of primes.

THEOREM 8. A P.I.D. is a U.F.D.

> **Proof:** This follows at once from the last theorem and the fact that in a P.I.D. every irreducible element is a prime.

EXERCISES

1. Prove the set D of all numbers $m + n\sqrt{-7}$, where m and n are integers is an integral domain.

2. Find in D (see Exercise 1) two factorizations of 16 into primes.

3. Is D (see Exercise 1) a P.I.D.? Prove your answer.

4. If R is a P.I.D. and p is a prime in R, prove $D/(p)$ is a field.

5. If a, b, c are elements of a P.I.D. R, prove that their G.C.D. d exists and can be expressed in the form $d = ua + vb + wc, u, v, w \in R$.

6. For nonzero elements x and y of a P.I.D., show that $(x) \subset (y)$ if and only if $y \mid x$.

7. A is an ideal in a P.I.D. P. Describe the ideals of P/A and show P/A is a P.I.D. If A is an ideal of P/A find the element that generates it.

8. If d is the G.C.D. of the elements x and y of a P.I.D. P, and if A is an ideal of P, find the G.C.D. of the elements $x + A$ and $y + A$ of P/A.

9. Find all the ideals of Z_{10} (the ring of integers mod 10) and find the elements that generate these ideals.

10. If p is a prime element of the P.I.D. P, is $p + A$ a prime in P/A, where A is an ideal of P? Describe the prime elements of P/A in terms of those of P.

11. If an integral domain D is isomorphic to a P.I.D. P, is D a P.I.D.? Prove your answer.
If D is homomorphic to P, is D a P.I.D.? Prove your answer.

12. Prove that a U.F.D. in which every proper prime ideal is maximal is a P.I.D.

6-4 THE UNIQUE FACTORIZATION DOMAIN D[x]

Our next objective is to show that if D is a U.F.D. and if x is a transcendental element over D, then $D[x]$ is also a U.F.D. We first need some definitions and some more information.

Definition. Let D be a U.F.D. and let $f(x) \in D[x]$, where x is a variable or transcendental over D. Let

$$f(x) = a_n x^n + a_{n-1} x^{n-1} + \ldots + a_0.$$

The **content** $C(f)$ of $f(x)$ is defined to be the G.C.D. of the coefficients $a_n, a_{n-1}, \ldots, a_0$ of $f(x)$.

Definition. A polynomial $f(x)$ of $D[x]$ is called a **primitive polynomial** if $C(f)$ is a unit of D. Thus every polynomial in $D[x]$, where D is a U.F.D., is the product of a primitive polynomial and an element of D.

Before proceeding to some theorems we collect together some simple but important facts relating to polynomial rings in transcendental elements over rings that are unique factorization domains. Let D be a U.F.D., and let x be variable over D. Form the polynomial domain $D[x]$.

Fact I: This is a reminder from our earlier work that the only units of $D[x]$ are those of D itself. (This is because D has no zero-divisors.) See the last paragraph, Sec. 5-7.

Fact II: The irreducible elements of $D[x]$ are the primes of D and the irreducible primitive polynomials of $D[x]$. For if a polynomial $f(x)$ is not primitive, then $C(f) = a$, where a is not a unit and $f = af_1$, where f is primitive. This is already a factorization of f into nonunits of $D[x]$. This is all another way of saying that in $D[x]$ an irreducible polynomial has to be primitive.

THEOREM 9. Let D be a U.F.D. and let F be its quotient field. Let x be transcendental over F, and hence over D. Two primitive polynomials of $D[x]$ are associates in $D[x]$ if and only if they are associates in $F[x]$.

Proof: Let f and g be primitive polynomials that are associates in $D[x]$. Then $f = ug$, where u is a unit of D. But u is a unit of F, and hence a unit of $F[x]$. Hence f and g are associates in $F[x]$.

Conversely, let the primitive polynomials f and g be associates in $F[x]$. Then $f = ug$, where u is a unit of F. Let $u = b/c$, where $b, c \in D$. Then $cf = bg$. Since f and g are primitive polynomials of $D[x]$, it follows that c and b are associates in D and hence $u \in D$. Hence f and g are associates in $D[x]$.

THEOREM 10. The product fg of two primitive polynomials f and g is a primitive polynomial.

Proof: Let p be any prime of D. p does not divide all the coefficients of f nor all the coefficients of g, since f and g are primitive. If $f = \Sigma a_i x^i$, $g = \Sigma b_i x^i$, let a_r and b_s be the first coefficients of f and g respectively that are not divisible by p. Then in the coefficient $\cdots + a_{r-1}b_{s+1} + a_r b_s + a_{r+1}b_{s-1} + \cdots$ of x^{r+s} in the product fg, we see that p divides every term except the term $a_r b_s$. This term it cannot divide, since D is a U.F.D. and $p \nmid a_r$; therefore $p \nmid b_s$. Thus p does not divide the coefficient of x^{r+s} in the product fg. This is true of every prime p of D and hence of every nonunit of D. Thus the coefficients of fg have only units as common divisors and hence fg is primitive.

Corollary 1. If f and g are two polynomials of $D[x]$, then

$$C(fg) = C(f)C(g).$$

Proof: $f = cf_1$ and $g = dg_1$ where $c, d \in D$ and f_1 and g_1 are primitive polynomials. Hence $fg = cdf_1g_1$. Hence

$$C(fg) = cdC(f_1g_1) = cd,$$

since f_1g_1 is primitive. $C(f) = c$ and $C(g) = d$, and so the corollary is proved.

THEOREM 11. An irreducible primitive polynomial f of degree > 0 over a U.F.D. D is irreducible over the quotient field F of D.

Proof: Suppose $f = gh$, where $g, h \in F[x]$, and neither g nor h is a unit. $g = \dfrac{a_n}{b_n} x^n + \cdots + \dfrac{a_o}{b_o}$ where the $a_i, b_i \in D$. Let m be the L.C.M. of $b_o, b_1, \ldots, b_n$. Then $mg \in D[x]$. Hence $mg = cg_1$ where $m, c \in D$ and g_1 is a primitive polynomial of $D[x]$. Hence we have $g = ag_1$ where $a = m^{-1}c \in F$. Similarly, $h = bh_1$, where $b \in F$ and h_1 is a primitive polynomial of $D[x]$. Thus $f = abg_1h_1$, where $ab = r/s \in F$ and $r, s \in D$. Hence $sf = rg_1h_1$. Now $C(sf) = s$ and $C(rg_1h_1) = r$, hence r and s are associates in D and so $f + ug_1h_1$, where u is a unit of D. Since neither g nor h is a unit of $F[x]$ and, hence, not a unit of F, it follows that g and h are polynomials of positive degree belonging to $D[x]$. But this contradicts f being irreducible in $D[x]$. Hence f must be irreducible in $F[x]$.

THEOREM 12. Let D be a U.F.D., then every irreducible element of $D[x]$ is a prime.

Proof: The irreducible elements of $D[x]$ are the primes of the U.F.D. D and the irreducible primitive polynomials of $D[x]$. Let p be an irreducible primitive polynomial of $D[x]$ of degree > 0. Let $p \mid fg$ where f and g are polynomials of $D[x]$, and suppose $p \nmid f$. Now the polynomials $p, f, g \in F[x]$, where F is the quotient field of D. Since $F[x]$ is a P.I.D., then by Theorem 11, p is an irreducible element and therefore a prime element of $F[x]$. Hence $p \mid g$ over F; that is, $g = ph$, where $h \in F[x]$. Let $h = \dfrac{a_n}{b_n} x^n + \cdots + \dfrac{a_o}{b_o}$ and let m be the L.C.M. of $b_n, b_{n-1}, \ldots, b_o$. Thus $mh = k \in D[x]$. Now

$$k = ak_1 \quad \text{and} \quad g = bg_1,$$

where $a, b \in D$ and k_1 and g_1 are primitive polynomials of $D[x]$. Hence $mbg_1 = apk_1$; that is, $g_1 = m^{-1}b^{-1}apk_1$, where $m^{-1}b^{-1}a \in F$. Thus the two primitive polynomials g_1 and pk_1 are associates in $F[x]$, for $m^{-1}b^{-1}a$ is a unit of F. Hence by Theorem 9 g_1 and pk_1 are associates in $D[x]$. Therefore $g_1 = upk_1$, where u is a unit of D, and therefore $bg_1 = bupk_1$—that is $g = bupk$, where $bu \in D$.

Hence $p \mid g$ in D and therefore p is a prime element in $D[x]$. If p is a prime of D and $p \mid fg$ but $p \nmid f$, then $p \mid g$; for if not, then p would not divide fg. (See the proof of Theorem 10). Hence all irreducible elements of $D[x]$ are primes.

THEOREM 13. If D is a U.F.D. then every nonunit element of $D[x]$ is a finite product of irreducible elements of $D[x]$.

Proof: If a nonunit element is an element of D then we know this is true, since D is a U.F.D. Let $f(x)$ be a polynomial of degree > 0 of $D[x]$. Then $f = cf_1$, where $c \in D$ and f_1 is a primitive polynomial. If c is not a unit of D then c is the product of a finite number of primes of D, and these primes are irreducible elements of $D[x]$. Let deg $f_1 = n$. We proceed by induction and assume the theorem true for all polynomials of degree $< n$. If f_1 is irreducible, then being primitive it is an irreducible element of $D[x]$ and we are through. If f_1 is not irreducible, then $f_1 = gh$, where deg $g < n$ and deg $h < n$. By the induction hypothesis the theorem is true for g and h and hence for $gh = f$. Hence the theorem is true for every polynomial of positive degree.

And now we have the main theorem:

THEOREM 14. If D is a U.F.D. and x is transcendental over D, then $D[x]$ is a U.F.D.

Proof: This follows at once from the results of the last two theorems.

And now by induction on n we can easily prove that if D is a U.F.D. and $x_1, x_2, \ldots, x_n$ are independent variables (transcendentals) over D, then $D[x_1, x_2, \ldots, x_n]$ is a U.F.D.

While every pair of elements a and b of a U.F.D. has a G.C.D. d, it does not follow that we can express d in the form $d = ax + by$. Of course we know that we can do this in a P.I.D. But consider the integral domain $Z[x]$ of the polynomials in the variable x over the domain of integers Z. Now Z is a P.I.D. and hence a U.F.D. and so, by the last theorem, $Z[x]$ is a U.F.D. However $Z[x]$ is not a P.I.D. For consider the ideal in $Z[x]$ generated by the elements x and 3, that is the ideal $(x, 3)$.

Now $(x, 3) \neq Z[x]$, and we shall show that it is not a principal ideal. If $(x, 3)$ is a principal ideal, then it is generated by some polynomial $h(x)$ belonging to $Z[x]$; that is, $(x, 3) = (h(x))$. Hence there exist polynomials $r(x)$ and $s(x)$ such that $x = r(x)h(x)$ and $3 = s(x)h(x)$. Now the second equation forces $h(x) = 3$. (For if $h(x) = 1$, then $(x, 3) = (1) = Z[x]$). But if $h(x) = 3$, then there can exist no polynomial $r(x)$ satisfying the first equation. Thus $(x, 3)$ cannot be a principal ideal.

In $Z[x]$ consider the elements $2x + 1$ and 3. Their G.C.D. is 1. However there are no polynomials $f(x)$ and $g(x)$ such that

$$(2x + 1)f(x) + 3g(x) = 1.$$

For this equation would be an identity and so for $x = 1$ we would have $3f(1) + 3g(1) = 1$, where $f(1)$ and $g(1)$ are integers, and clearly this is impossible. Thus $Z[x]$ is an example of a U.F.D. that is not a P.I.D. It is also an example of a ring where the G.C.D. d of two elements a and b cannot be expressed in the form $d = ra + sb$.

In some rings it can happen that two elements do not have a G.C.D. while another pair of elements in the same ring do have. Consider the behavior of E the ring of even integers. $6 \in E$ and has no divisors, that is 6 is a prime in E. Thus 4 and 6 have no G.C.D., since $1 \notin E$. However while 8 and 12 have the G.C.D. 2, there are no even integers x and y such that $8x + 12y = 2$. The irreducible elements in E have the form

$$4k + 2 = 2(2k + 1).$$

Hence E is not a principal ideal ring.

Algebraic Domains

A root of a polynomial equation over the field of rational numbers is called an **algebraic number.** An algebraic number, then, is a complex number that is a root of an equation of the form

$$a_0 + a_1 x + \cdots + a_n x^n = 0$$

where the a_i are rational numbers, not all zero.

Let Q be the rational field and let $\alpha_1, \alpha_2, \ldots, \alpha_n$ be algebraic numbers. We shall denote by $Q[\alpha_1, \ldots, \alpha_n]$ the smallest subfield of the complex field that contains Q and these algebraic numbers, that is, it is the intersection of all subfields of the complex field containing Q and these numbers. An extension field E of a field F is a field that contains F. E can be regarded as a vector space over F, and is called a finite extension of F if it is a finite dimensional vector space over F. Now $Q[\alpha_1, \ldots, \alpha_n]$ is an extension of Q and in Chapter 9 we shall prove that it is a finite extension. Finite extensions of the rational field Q are called **algebraic number fields.**

A root of a monic irreducible polynomial over Q; that is, a root of an equation of the form

$$x^n + a_{n-1} x^{n-1} + \cdots + a_0 = 0$$

where the a_i are (rational) integers, is called an **algebraic integer.** In any algebraic number field, the algebraic integers form an integral domain. For example in the algebraic number field $Q[\sqrt{5}]$, it can be shown that the algebraic integers have the form $m + \dfrac{n}{2}(1 + \sqrt{5})$ where m, n are (rational) integers. These form an integral domain within this algebraic number field. Such integral domains are called **algebraic domains.** In general, since factorization is not unique in algebraic domains, they are not principal ideal domains.

Example. $\alpha = \frac{1}{2} + \sqrt{-1}$ is an algebraic number, since it satisfies the equation $4\alpha^2 - 4\alpha + 5 = 0$, while $\beta = \sqrt{2} + \sqrt{-1}$ is an algebraic integer, since $\beta^4 - 2\beta^2 + 9 = 0$.

6-5 EUCLIDEAN DOMAINS

A special kind of principal ideal ring is the Euclidean domain. The unique factorization theorem is proved for the integers and for the polynomial rings by use of the long division property, that is the Euclidean algorithm. This involves the notion of magnitude, (absolute value for integers, degree for polynomials). In the general principal ideal ring the unique factorization theorem is proved without introducing the notion of magnitude, the fact that every ideal is a principal is strong enough to achieve it on its own. The notion of magnitude reappears with Euclidean domains. However since they will be shown to be principal ideal domains we shall not need to use magnitude to establish uniqueness of factorization in a Euclidean domain.

Definition. A **Euclidean domain** D is an integral domain such that to every element $a \in D$ there corresponds a unique nonnegative integer $\beta(a)$ with the following properties: (i) $\beta(a) = 0$ if and only if $a = 0$, (ii) $\beta(ab) = \beta(a)\beta(b)$, for all $a, b \in D$, and (iii) for any a and $b \neq 0$ of D, there exists $q, r \in D$ such that $a = bq + r$, where $\beta(r) < \beta(b)$.

This last property endows D with a Euclidean algorithm.

THEOREM 15. A Euclidean domain D is a principal ideal domain.

Proof: Let B be an ideal in D. If $B = 0$, then $B = (0)$. Let then $B \neq (0)$, in which case there exists $b \in B$ such that $\beta(b) > 0$. Among all such nonzero elements of B choose an element b such that $\beta(b)$ is the least positive integer for all the elements of B. Hence for every $c \neq 0 \in B$ we have $\beta(b) < \beta(c)$ and we have $c = bq + r$, where $\beta(r) < \beta(b)$. Now $r = c - bq$ is in B and hence, unless $r = 0$, we have a contradiction of the choice of b. Hence $r = 0$ and so B is the principal ideal (b). Hence D is a P.I.D.

In view of the remarks above about magnitude we would expect the domain Z of integers to be a Euclidean domain. It is, as can be easily verified, with $\beta(n) = |n|$, the absolute value of the integer n. Also if F is a field and x is a variable over F, then $F[x]$ is a Euclidean domain, again as can be quite easily verified, with $\beta(f(x)) = 2^{\deg f(x)}$ where we make the convention that if $f(x) = 0$ then degree of $f(x)$ is defined as $-\infty$.

Example. Let D be the set of all complex numbers of the form $m + n\sqrt{-2}$ where m and n are integers. It is easy to verify that D is an integral domain. We prove now that it is Euclidean. Define

$$\beta(m + n\sqrt{-2}) = m^2 + 2n^2.$$

Then axioms (1) and (2) for a Euclidean domain are easily verified. Let us show (3) is satisfied. Let $a, b \neq 0 \in D$ and put $\rho = \sqrt{-2}$. Then $\dfrac{a}{b} = s + t\rho$, where s and t are rational numbers. Integers u and v can be determined such that $|s - u| \leq \frac{1}{2}$ and $|t - v| \leq \frac{1}{2}$. Then $a = b[u + (s - u) + v\rho + (t - v)\rho] = bq + r$, where $q = u + v\rho$ and $r = b[s - u + (t - v)\rho]$. Observe that q and r belong to D. Now $\beta(r) = \beta(b)[(s - u)^2 + 2(t - v)^2] \leq \dfrac{3\beta(b)}{4}$. Hence $\beta(r) < \beta(b)$. This proves that axiom (3) holds and hence D is Euclidean.

A Euclidean domain can be proved directly to be a U.F.D. without requiring the more general theorem that a P.I.D. is a U.F.D.

We do need the fact that a Euclidean domain D is a P.I.D. but only to prove that every ideal in the domain is principal. Then by induction on $\beta(b)$ it can be easily proved (by the method already used) that if $b \in D$ and is not a unit, then it is a finite product of primes. This leaves only uniqueness to prove. Since every ideal in D is principal it follows easily that if $a \in D$ is a prime then if $a \mid bc$, either $a \mid b$ or $a \mid c$. This fundamental property is then all that is needed to prove the uniqueness, which is again demonstrated by assuming the existence of two distinct prime decompositions for the same element.

The reader is asked to work through all this carefully and obtain a rigorous proof.

EXERCISES

1. Prove that the set D of complex numbers of the form $m + ni$, $i = \sqrt{-1}$, m and n are integers, is a Euclidean domain.

2. Prove $3 + 2i$ is a G.C.D. of $-23 + 2i$ and $13i$ in D. Find a G.C.D. of $29 + 33i$ and $22 + 4i$.

3. Prove the set of all real numbers of the form $m + n\sqrt{2}$, where m and n are integers, is a Euclidean domain.

6-6 PRIME IDEALS AND MAXIMAL IDEALS

We conclude this chapter with a brief introduction to prime and maximal ideals in commutative rings. This leads to two simple but very important theorems and to some further observations about unique factorization.

Throughout this discussion, it is assumed that the ring R *is commutative, but not necessarily having an identity.* Thus left and right ideals are simply ideals.

Definition. An ideal P in a commutative ring R is called a **prime ideal** if $ab \in P$ implies either $a \in P$ or $b \in P$.

If two elements are not in a prime ideal, then their product is not in the ideal. The ideal (0) is prime if and only if R contains no zero-divisors.

Let $a_1, a_2, \ldots, a_n$ be elements of a commutative ring R. The ideal generated by these n elements is the smallest ideal in R containing these elements. We denote this ideal by $(a_1, a_2, \ldots, a_n)$. It is thus the intersection of all ideals of R that contain $a_1, a_2, \ldots, a_n$. It is easy to see that the elements of $(a_1, \ldots, a_n)$ have the form

$$(2) \qquad m_1 a_1 + m_2 a_2 + \cdots + m_n a_n + x_1 a_1 + \cdots + x_n a_n,$$

where the m_i are integers and the x_i belong to R. For the set of all elements of this form obviously forms an ideal containing $a_1, a_2, \ldots, a_n$ and moreover any ideal containing these elements would have to contain all elements of the form (2). If R has an identity e then the elements of $(a_1, \ldots, a_n)$ take on the simpler form

$$y_1 a_1 + y_2 a_2 + \cdots + y_n a_n, \quad y_i \in R.$$

For $m_i a = m_i(e a_i) = (m_i e) a_i$, and since $m_i e \in R$, we have

$$m_i a_i + x_i a_i = (m_i e + x_i) a_i = y_i a_i,$$

where we write $y_i = m_i e + x_i$. An ideal, such as $(a_1, a_2, \ldots, a_n)$, that is generated by a finite number of elements is called a **finitely generated ideal**.

As in the case of an integral domain (Sec. 6-3), an ideal generated by a single element a of a commutative ring R is called a **principal ideal** and, as before, we write it as (a). If R does not have an identity, the elements of (a) have the form $na + ra$, where n is an integer and $r \in R$. On the other hand if R has an identity these elements take the simpler form ra, $r \in R$, as explained above.

Definition. An ideal A in R is called a **maximal ideal** if $A \neq R$ and if A is not properly contained in any ideal except R itself. Thus A is a maximal ideal if and only if $A \subset B$, where B is an ideal, implies $B = R$.

One useful way of proving that an ideal A is maximal is to show that the ideal (A, r), generated by A and an arbitrary element $r \in R$ that is not contained in A, is the entire ring R. For this means that A is not properly contained in any ideal. The ideal (A, r) contains all elements of the form $na + xa + yr + mr$, where n and m are integers and $a \in A$, $x, y \in R$. Thus (A, r) is a maximal ideal if every element of R can be expressed in this form. If R has an identity e then it is sufficient to show that the element $e \in (A, r)$, for then $(A, r) = R$.

Example. Consider the ring Z of integers. We know that it is a principal ideal domain and so all its ideals have the form (n) where $n \in Z$. The elements of (n) are the multiples mn of n. The zero ideal is prime and the rest of the prime ideals of Z have the form (p), where p is a prime integer. For if $mn \in (p)$ and $m \not\in (p)$ then $n \in (p)$ if and only if p is a prime. In other words if $p \mid mn$ and $p \nmid m$ then $p \mid n$ if and only if p is prime.

Moreover those prime ideals of the form (p) are the maximal ideals of Z. For if $n \not\in (p)$, then n and p are relatively prime and hence there are integers r and s such that $rn + sp = 1$. Now $rn + sp$ belongs to the ideal generated by (p) and n and therefore this ideal is Z. Thus (p) is not properly contained in any ideal except Z itself and so is prime. Conversely if (p) is a maximal ideal and n is any integer not in (p) then the ideal generated by (p) and n is Z. Hence there are integers r and s such that $rn + sp = 1$ and this is true of every such n. Thus p and every such integer n are relatively prime. This forces p to be a prime.

Example. Consider the ring of polynomials $Z[x]$. Here Z is the ring of integers. We have seen earlier that it is not a principal ideal domain. In fact we showed that $(2, x)$ is not a principal ideal. It is of course a finitely generated ideal, being the ideal generated by the elements 2 and x. The principal ideal (x) in $Z[x]$ consists of all elements of the form $xf(x), f(x) \in Z[x]$, that is of all polynomials with a zero constant term. Clearly if the product of two polynomials belongs to (x) then at least one has to be in (x), and hence (x) is a prime ideal. However (x) is not a maximal ideal, for $(x) \subset (2, x) \subset Z$. Thus $Z[x]$ contains prime ideals that are not maximal.

The ideal $(2, x)$ is maximal as we shall now prove. The elements of $(2, x)$ are polynomials with constant term 0 or an even integer. Let $h(x)$ be any polynomial that is not contained in $(2, x)$. We can write $h(x)$ in the form $h(x) = xk(x) + 2n - 1$, where $2n - 1$ is any odd integer. Let B be the ideal generated by $(2, x)$ and $h(x)$. Then $h(x) - xk(x) \in B$, and so $2n - 1 \in B$. Since $2n \in B$, $2n - (2n - 1) = 1 \in B$ and we have $B = Z[x]$. This proves $(2, x)$ is a maximal ideal.

Example. In the ring E of even integers $0, \pm 2, \pm 4, \ldots$, the ideal (4) is maximal but is not prime. Any even integer that is not in (4) has the form $2(2n - 1)$. The ideal A generated by (4) and $2(2n - 1)$ contains the element $4n - 2(2n - 1) = 2$ and hence $A = E$. Thus (4) is a maximal ideal. However it is not prime, for $2 \cdot 2 = 4 \in (4)$ and $2 \notin (4)$. Thus there are rings in which maximal ideals need not be prime.

Before starting the next two theorems we make this reminder which is the clue to the method of proof of each of them: if R is a commutative ring with the identity 1 and A is an ideal in R, then the quotient ring R/A is a commutative ring whose zero element is the coset A and whose identity is the coset $1 + A$.

THEOREM 16. Let A be an ideal in a commutative ring R with identity. Then the quotient ring R/A is an integral domain if and only if A is a prime ideal.

Proof: In view of the reminder above, we need only show that R/A has no zero-divisors if and only if A is prime. Thus R/A is an integral domain if and only if, $(x + A)(y + A) = xy + A = A$ and $x + A \neq A$ imply $y + A = A$. In terms of the ideal A this simply translates into: R/A is an integral domain if and only if when $xy \in A$ and $x \notin A$ then $y \in A$, which means if and only if A is a prime ideal.

THEOREM 17. Let A be an ideal in a commutative ring R with identity 1. Then the quotient ring R/A is a field if and only if A is a maximal ideal.

Proof: Again here we need only show that R/A is a field if and only if a nonzero element $x + A$, $x \notin A$, of R/A has a multiplicative inverse $y + A$, $(x + A)(y + A) = xy + A = 1 + A$. Again in terms of the ideal A this means that R/A is a field if and only if for every $x \notin A$ there is a $y \in R$ such that $1 - xy = a \in A$. Since $xy + a$ belongs to the ideal (A, x) generated by A and the element x, we see that $1 \in (A, x)$ and hence $(A, x) = R$. Thus R/A is a field if and only if the ideal $(A, x) = R$ for every $x \in A$ and this means if and only if A is a maximal ideal.

These theorems are frequently used to prove that an ideal is prime or that an ideal is maximal. An important result that follows at once from these two theorems is that in a commutative ring with identity every maximal ideal is a prime ideal (for every field is an integral domain). However in rings without an identity, as we have shown, maximal ideals need not be prime. Note that in an integral domain the zero ideal (0) is a prime ideal, but it is not prime in a ring with zero-divisors. A ring is called a **prime ring** if (0) is a prime ideal in the ring.

Example. Another method of proving that the principal ideal (x) is a prime ideal in the domain $Z[\sqrt{-5}]$ is to use the theorems above. Consider the mapping of $Z[x] \rightarrow Z$ defined by $f(x) \rightarrow f(0)$. It is clear that this is a homomorphism whose kernel is (x). Hence by the fundamental homomorphism theorem for rings we have, $Z[x]/(x) \approx Z$. Since Z is an integral domain, this isomorphism proves that $Z[x]/(x)$ is an integral domain and hence that (x) is a prime ideal. Also since Z is not a field, then $Z[x]$ is not a field, and so (x) is not a maximal ideal.

EXERCISE

1. The mapping $f(x,y) \rightarrow f(0,y)$ is a homomorphism of the integral domains $F[x,y] \rightarrow F[x]$, where F is a field. Show that (x) is a prime, but not a maximal ideal in $F[x,y]$.

Example. The ideal (x,y) is a prime and a maximal ideal in the polynomial domain $F[x,y]$, where F is a field. This can be proved from the definitions of prime and maximal, since (x,y) is the ideal of all polynomials of $F[x,y]$ that have constant term 0. However it is easier to use the homomorphism $F[x,y] \rightarrow F$ defined by $f(x,y) \rightarrow f(0,0)$. For the kernel is clearly (x,y) and hence $F[x,y]/(x,y) \approx F$ and this isomorphism proves $F[x,y]/(x,y)$ is a field and hence that (x,y) is a maximal ideal. Since in a commutative ring every maximal ideal is a prime ideal, this also shows that (x,y) is a prime ideal.

THEOREM 18. A quotient ring R/A of a principal ideal ring R is also a principal ideal ring.

Proof: Let B be an ideal in R/A. Then if f is the natural (canonical) homomorphism of $R \rightarrow R/A$, we know that $f^{-1}B$ is a principal ideal of R. Hence $f^{-1}B = (r)$, where $r \in R$, and therefore $B = (rf)$, where $rf = r + A$. Hence B is a principal ideal.

6-7 SIMPLE OPERATIONS WITH IDEALS

If A and B are ideals in a commutative ring R, then the set $A + B$ consisting of all elements of the form $a + b$, $a \in A$ and $b \in B$, is evidently an ideal, called the *sum* of A and B. Clearly also the intersection $A \cap B$ of two ideals is an ideal.

Caution: If A and B are ideals in a commutative ring R, we define the set AB to be the set of all finite sums of the form

$$\Sigma a_i b_i, a_i \in A, b_i \in B.$$

It is seen at once that if AB is defined in this way, then AB is an ideal. This **product** AB is very important in the later factorization of ideals.

EXERCISES

1. Let $A = (a_1, a_2, \ldots, a_n)$ and $B = (b_1, b_2, \ldots, b_m)$ be finitely generated ideals in a commutative ring with identity. Prove that AB is the finitely generated ideal $AB = (a_1 b_1, \ldots, a_n b_1, \ a_1 b_2, \ldots, a_n b_2, \ldots, a_1 b_m, \ldots, a_n b_m)$.

2. Let A, B, and C be ideals in a commutative ring. Prove the following:
(i) $A(B + C) = AB + AC$
(ii) $AB \subset A \cap B$

If A is an ideal in a ring R with identity, then $AR = A = RA$. Note that $AB \subsetneq A$ and $AB \subsetneq B$ for all ideals. Moreover, if A, B, C are ideals then $A \subseteq BC$ implies $A \subseteq B$ and $A \subseteq C$. One strangeness in dealing with ideals should be noted. When we speak of A being a factor or a divisor of B, we naturally mean that $B = AC$ for some ideal C. Yet this means $B \subsetneq A$ and $B \subsetneq C$, that is B is contained in its factors. For example in Z, $(6) \subset (2)$ and yet $(6) = (2)(3)$.

THEOREM 19. In a principal ideal domain a nonzero ideal (p) is a prime ideal if and only if p is a prime (irreducible). (The ideal (0) is a prime ideal.) The proof is left as an exercise.

THEOREM 20. In a principal ideal domain D every nonzero prime ideal is maximal.

Proof: Let $P = (p)$ be a prime ideal. Then p is a prime element of D. Let $q \notin (p)$, then q and p are relatively prime elements and hence there exist $r, s \in D$ such that $rq + sp = 1$. Hence 1 is in the ideal (P, q) generated by P and the element q. Hence $(P, q) = D$ and therefore P is maximal.

THEOREM 21. A commutative ring R with identity is a field if and only if its only ideals are (0) and R.

Proof: Let R be a field and let $A \neq (0)$ be an ideal in R. Choose $a \in A$ such that $a \neq 0$. Then $a^{-1}a = 1 \in A$ and hence $A = R$.

Conversely, suppose R has no nontrivial ideals. Let $x \neq 0$ be an element of R. Clearly the set Rx is an ideal in R. Since $1 \in R, Rx \neq (0)$ and hence $Rx = R$. Therefore there exists $y \in R$ such that $yx = 1$, that is the nonzero element x has an inverse in R. Hence R is a field.

Corollary. A field has no nontrivial homomorphic images.

Proof: The only quotient rings of a field are the field itself and (0). Hence by the fundamental theorem of ring homomorphisms the only homomorphic images of a field are itself and (0).

Note that a *prime ideal* P can be defined as follows: if A and B are ideals and $AB \subseteq P$ then either $A \subseteq P$ or $B \subseteq P$. This is equivalent to the earlier definition as the reader can verify.

Lemma. If P is a prime ideal in an integral domain R and $P = AB$, then either $A = P$ and $B = R$ or $A = R$ and $B = P$.

Proof: We know that $P \subseteq A$ and $P \subseteq B$. Now if neither A nor B is in P, then there exists $a \in A$ and $b \in B$ such that $a, b \notin P$. Since $ab \in P$ and P is prime, this is impossible. Hence either $A \subseteq P$ and therefore $A = P$, and $B = R$ or $B \subseteq P$ and then $B = P$ and $A = R$.

In a commutative ring with no zero-divisors in which every nonzero prime ideal is a maximal ideal (for example, a principal ideal domain), we can show that if $P, Q,$ and T are all prime ideals and if $PQ \subseteq T$, then either $P = T$ or $Q = T$. For if neither P nor Q is in T, then there exists a product $pq \in T, p \in P, q \in Q$, for which neither p nor q is in T—something which is impossible, since T is a prime ideal. Hence either $P \subseteq T$ or $Q \subseteq T$, and since P and Q are also maximal ideals, this means either $P = T$ or $Q = T$.

Let R be a commutative ring with identity 1. Then by use of Zorn's axiom we can show that every ideal A in R is contained in some maximal ideal in R. For let S be the set of all ideals in R that contain A and that are not R itself. (If A is itself maximal, then we are through.) S is a poset under the relation inclusion. A totally ordered subset B_i, $i \in I$, of S has an upper bound $\cup B_i$. ($1 \notin B_i$ for any i and hence $1 \notin \cup B_i$.) Hence there exists a maximal ideal in S.

Any element of R that is not a unit clearly is contained in some maximal ideal in R.

EXERCISES

1. If Z is the ring of integers determine whether the ideals (x) and $(x, 2)$ are prime or maximal or both in the ring $Z[x]$.

2. Is the ideal (x, y) prime or maximal in the ring $Z[x, y]$?

3. If F is the rational field determine which of the following ideals are prime or maximal in the ring $F[x, y]$. (a) (x), (b) $(x - 2, y - 3)$, (c) $(x^2 + 1)$.

4. Prove that every prime ideal is maximal in a Boolean ring. (See Exercise 1, Sec. 5-3.)

6-8 QUADRATIC DOMAINS

Let Q denote the rational field. Let m be a rational integer, $m \neq 0$, $m \neq 1$, and denote by $Q(\sqrt{m})$ the set of all elements of the form

$a + b\sqrt{m}$, where $a, b \in Q$. Clearly no loss of generality results in assuming that m contains no factor (except 1) that is the square of an integer. If $\alpha = a + b\sqrt{m}$ then call $\bar{\alpha} = a - b\sqrt{m}$ the **conjugate** of α. $Q(\sqrt{m})$ is evidently an integral domain and that it is a field follows from the fact that a non-zero element $\alpha = a + b\sqrt{m}$, a and b not both 0, has the inverse

$$\alpha^{-1} = \frac{1}{\alpha} = \frac{\bar{\alpha}}{\alpha\bar{\alpha}} = \frac{a - b\sqrt{m}}{a^2 - mb^2}.$$

Note $a^2 - mb^2 \neq 0$ by the definition of m.

It can be proved that the set of all algebraic numbers forms a field. In fact, since α satisfies the equation $\alpha^2 - 2a\alpha + a^2 - mb^2 = 0$, it is an algebraic number and $Q(\sqrt{m})$ is a subfield of the field of all algebraic numbers. $Q(\sqrt{m})$ is called a **quadratic field.** The integers of $Q(\sqrt{m})$ are those elements α which satisfy an equation of the form $\alpha^2 + n_1\alpha + n_2 = 0$ where n_1 and n_2 are rational integers. Those elements that are integers of $Q(\sqrt{m})$ will depend on m, as the following example points out:

The integers of $Q(\sqrt{-5})$ can be proved to have the form $n_1 + n_2\sqrt{-5}$ where n_1 and n_2 are integers. However, the integers of $Q(\sqrt{5})$ turn out to have the form $(n_1 + n_2\sqrt{5})/2$, where $n_1 + n_2$ is an even integer. Thus $\alpha = (3 + \sqrt{5})/2$ is an integer of $Q(\sqrt{5})$, since $\alpha^2 - 3\alpha + 1 = 0$.

The integers of $Q(\sqrt{m})$ form an integral domain $D(\sqrt{m})$, called a **quadratic domain.** The integral domain $Z(\sqrt{-5})$, that we studied earlier, is clearly the integral domain of the quadratic field $Q(\sqrt{-5})$.

Now factoring of elements into prime elements is unique in some domains of quadratic integers. However we have seen that there is at least one where this is not true. The quadratic domain $Z(\sqrt{-5})$ does not have the property of unique factorization of its elements into primes. Of course we showed that $Z[x]$, where x is a transcendental element, is a U.F.D. The failure of unique factorization in quadratic domains of the elements led to the invention of ideals. We are going to show by an example how factorization of the elements into primes can be replaced by the factorization of the ideals of the domain uniquely into the product of prime ideals. That is the difficulty is overcome by using the ideals of the domain in place of the elements for factorization purposes. We illustrate this without establishing the uniqueness of the factorization.

Consider the quadratic domain $Z(\sqrt{-5})$. Here Z is, as usual, the domain of integers so that the elements of $Z(\sqrt{-5})$ have the form $a + b\gamma$ where we write $\gamma = \sqrt{-5}$, and $\gamma^2 = -5$. We found that for 9 we had the two factorizations into irreducible elements

$$9 = 3 \cdot 3 = (2 + \gamma)(2 - \gamma).$$

Consider the ideal $P = (3, 1 + \gamma)$. It is not hard to show that an element $a + b\gamma \in P$ if and only if $a \equiv b \pmod 3$ and hence that the quotient ring $Z(\sqrt{-5})/P$ is isomorphic to the field of residue classes modulo 3. Thus P is a prime ideal. In fact it is also a maximal ideal. Similarly it can be shown that $Q = (3, 2 + \gamma)$ is a prime ideal. Here $a + b\gamma \in Q$ if and only if $a + b \equiv 0 \pmod 3$. Moreover we see that the ideal (3) factors into $(3) = PQ$. Thus if we replace 9 by the principal ideal (9) generated by 9 we obtain the following factorization of the ideal (9) into prime ideals: $(9) = PQPQ = P^2Q^2$.

Another example from the same quadratic domain $Z(\sqrt{-5})$ is the factorization of 21 which we saw earlier had three distinct factorizations into irreducible elements. If we replace 21 by the principal ideal (21) then we find that $(21) = P_1 P_2 P_3 P_4$ where the P_i are the following prime ideals: $P_1 = (3, 1 + 2\gamma)$, $P_2 = (3, 1 - 2\gamma)$, $P_3 = (7, 1 + 2\gamma)$, $P_4 = (7, 1 - 2\gamma)$. Every element $\alpha \in Z(\sqrt{-5})$ determines a principal ideal (α) which is the unique product of prime ideals.

That the unique factorization of an ideal into the product of prime ideals is true in domains of algebraic integers follows from the special properties that prime ideals possess in such domains. For instance, it is a fact that in a domain of algebraic integers proper prime ideals are also maximal ideals. Moreover, in such domains if P is a prime ideal and A and B are ideals, then $PA = PB$ implies $A = B$.

EXERCISE

1. Prove in the domain $Z(\sqrt{-5})$ that $(6) = PQT^2$ where $P = (3, 1 + \gamma)$, $Q = (3, 2 + \gamma)$, $T = (2, 1 + \gamma)$ are prime ideals. Note that $T^2 = (2)$.

6-9 BASIS THEOREM

We conclude the chapter with a proof of the very interesting and important Basis Theorem of Hilbert. It states that if every ideal in a ring R is finitely generated, then every ideal in the polynomial ring $R[x]$ is finitely generated.

THEOREM 22 (Hilbert Basis Theorem). If R is a Noetherian ring with identity 1 and if x is transcendental over R, then $R[x]$ is a Noetherian ring.

Proof: We prove $R[x]$ is a Noetherian ring by showing that any ideal A in $R[x]$ is finitely generated.

For $i = 0, 1, 2, \ldots$, define a family of ideals B_i of R as follows:

B_i is the set of all elements $b \in R$ such that there exists a polynomial $g(x)$ of degree i,

$$g(x) = bx^i + b_{i-1}x^{i-1} + \cdots + b_0$$

in the ideal A of $R[x]$ with b as its leading coefficient. Since $b_1, b_2 \in B_i$ implies $b_1 - b_2 \in B_i$ and since $b \in B_i$, $r \in R$ implies $rb = br \in B_i$, it is easy to see that B_i is an ideal.

If $b \in B_i$, so that b is the leading coefficient of a polynomial such as $g(x)$ above, then clearly $xg(x) = bx^{i+1} + \cdots + b_0 x \in A$ and hence $b \in B_{i+1}$. Thus $B_i \subsetneqq B_{i+1}$ for all i.

Since R is Noetherian, the ascending chain

$$B_0 \subseteq B_1 \subseteq B_2 \subseteq \cdots \subseteq B_n \subseteq \cdots$$

breaks off, say at $n = N$, and

$$B_n = B_{N+1} = \cdots$$

Now each B_i is finitely generated, hence

$$B_i = (b_{i1}, \ldots, b_{ik_i}), b_{ij} \in R.$$

Let $f_{ij}(x)$ be a polynomial in A of degree i with the leading coefficient b_{ij}, $1 \leq j \leq k_i$. Since $b_{ij} \in B_i$ there must be at least one such polynomial in A.

We now claim the set of all these polynomials f_{ij} (there are a finite number $k_0 + k_1 + \cdots + k_N$ of them in all) generates the ideal A of $R[x]$. The method of proof is by induction on the degree r of a polynomial of A. Clearly all polynomials $g(x)$ of A of degree 0 are elements of R, so that B_0 comprises all elements of A that are in R and hence B_0 is an ideal in R and so is finitely generated.

We can assume that all polynomials of A of degrees $< r$ are generated by the f_{ij}. Let $g(x) \in A$ and be of degree r. Then

$$g(x) = c_r x^r + \cdots$$

If $r \leq N$, then $c_r \in B_r$ and hence $c_r = \sum_{j=1}^{k_r} c_{rj} b_{rj}, c_{rj} \in R$. Now

$$f_{rj}(x) = b_{rj} x^r + \cdots$$

and hence $g(x) - \sum_{j=1}^{k_r} c_{rj} f_{rj}(x) \in A$ and this polynomial has a degree $< r$. By the induction hypothesis it is generated by the f_{ij} and hence $g(x)$ is generated by the f_{ij}.

If $r > N$, $c_r \in B_r = B_N$ and therefore $c_r = \sum_{j=1}^{k_N} c_{Nj} b_{Nj}, c_{Nj} \in R$.

Hence $g(x) - \sum_{j=1}^{k_N} c_{Nj} x^{r-N} f_{Nj}(x)$ has a degree $< r$ and, as before, therefore $g(x)$ is generated by the f_{ij}.

Thus all polynomials of degree r are generated by the f_{ij} and hence the induction argument proves this is true of all polynomials of A. Hence A is generated by the finite number of polynomials f_{ij} and so is a finitely generated ideal. Since A is an arbitrary ideal in $R[x]$, this proves that every ideal in $R[x]$ is finitely generated. Therefore $R[x]$ is a Noetherian ring.

Corollary 1. If $x_1, \ldots, x_n$ are independent transcendental elements over the Noetherian ring R, then the ring $R[x_1, \ldots, x_n]$ is a Noetherian ring.

Proof: Use induction on the number n of transcendental elements.

Corollary 2. If R is a division ring or a P.I.D. then every ideal of $R[x_1, \ldots, x_n]$ is finitely generated; that is, $R[x_1, \ldots, x_n]$ is a Noetherian ring.

Proof: A division ring R has only two ideals, the zero ideal (0) and R itself. Both are finitely generated, R itself being generated by any nonzero element.

A P.I.D. is a Noetherian ring, for all its ideals are generated by a single element.

EXERCISES

1. Prove the existence of a G.C.D. d for any n elements $a_1, a_2, \ldots, a_n$ of a P.I.D. R.

2. Prove that d can be expressed in the form
$$d = b_1 a_1 + \cdots + b_n a_n, b_i \in R.$$

3. Prove that the ideal (x) is prime but not maximal in $F[x]$ and that the ideal $(x, 2)$ is prime and maximal in $F[x]$. F is the rational field.

4. Z is the ring of integers and F is the rational field. Show that $(x - 1)$ is a prime ideal in both $Z[x]$ and $F[x]$ and that it is a maximal ideal in $F[x]$ but not a maximal ideal in $Z[x]$. Prove $F[x]/(x - 1) \approx F$.

5. Let $f(x) = a_0 + a_1 x + \cdots + a_{n+m} x^{n+m}$ be a polynomial with integral coefficients. Prove that if there exists a prime p such that (1) $p \nmid a_{n+m}$, (2) $p \mid a_i$ for all $i < n+m$, and (3) $p^2 \nmid a_0$, then $f(x)$ is irreducible over the rational field. (Assume $f(x) = g(x)h(x)$ and reach an easy contradiction). This is known as **Eisenstein's irreducibility criterion.**

6. Apply the Eisenstein criterion to prove that the pth cyclotomic polynomial
$$f(x) = \frac{x^p - 1}{x - 1} = x^{p-1} + x^{p-2} + \cdots + 1$$
where p is a prime, is irreducible. This is not true if p is not a prime. (Set $x = y + 1$ and work with this polynomial.)

7. A *primary* ideal Q in a commutative ring R is an ideal such that if a, $b \in R$, and $ab \in Q$, $a \notin Q$, then $b^m \in Q$ for some positive integer m. If F is a field prove $B = (x^2, xy, y^2)$ is a primary ideal in $F[x,y]$. Since $x^2 - y^2 = (x + y)(x - y) \in B$, what does this prove about B? (Note that $f(x,y) \notin B$ if and only if $f(x,y)$ has a tail $ax + by + c$ where $a^2 + b^2 + c^2 \neq 0$.)

8. Prove that the radical of the ideal (x^2, y^2) in $F[x,y]$ is the ideal (x,y).

9. Let A be an additive abelian group. For $a, b \in A$ define a product in A by $ab = 0$. Show that this makes A a commutative ring with no unit element. What are the ideals of this ring?

Next prove that if A is of prime order that A contains no proper ideals and show that A contains an ideal that is maximal but not prime. This illustrates the fact that in rings without a unit element, a maximal ideal need not be prime.

10. Show in a principal ideal ring R that two elements $a, b \in R$ have a G.C.D. d that can be expressed in the form $d = au + bv, u, v \in R$. Then show that a prime decomposition exists for each element. What additional assumption must be made about R in order that this prime decomposition be unique? Prove your answer.

11. $F[x,y]$ is the polynomial ring in two variables over a field F. Show that the ideal (x,y) of $F[x,y]$ is not a principal ideal. This proves $F[x,y$ is not a P.I.D. (Assume $(x,y) = (f(x,y))$ for some $f(x,y) \in F[x,y]$. Then $x = g(x,y)f(x,y)$ and $y = h(x,y)f(x,y)$. Show this leads to a contradiction.) Is $F[x,y]$ a U.F.D.?

12. In a Euclidean domain prove
(0) if $a \neq 0$ then $\beta(0) < \beta(a)$.
(2) if $a \mid b$ and $\beta(a) = \beta(b)$ then a and b are associates.

13. Prove the direct sum of principal ideal rings is itself a principal ideal ring.

14. A commutative ring R is said to satisfy the descending chain condition for ideals if every descending chain $A_1 \supset A_2 \supset \cdots \supset A_n \cdots$ of ideals terminates. This means there exists a positive integer N such that $A_N = A_{N+1} = \cdots$.

If R is a principal ideal domain and if A is a proper ideal in R prove that R/A satisfies the descending chain condition for ideals.

15. If F is the rational field and p is a prime, prove that the ideal $(x^{p-1} + x^{p-2} + \cdots + x + 1)$ is a maximal ideal in $F[x]$.

16. Show that the set D of all complex numbers of the form $m + n\sqrt{-3}$, where m and n are integers, is an integral domain. If we define $\beta(m + n\sqrt{-3}) = m^2 + 3n^2$, show that the method used in Sec. 6-5 to prove D a Euclidean domain breaks down.

17. If D is the set of all complex numbers $m + n\sqrt{-3}$ where either m and n are integers or both of them are halves of odd integers, prove that D is a Euclidean domain.

18. If D is a Euclidean domain, prove that $a \in D$ is a unit of D if and only if $\beta(a) = 1$.

19. In any U.F.D. prove that the ideal generated by a prime element is a prime ideal and prove that every nonzero nonprime element generates a nonprime ideal. This proves that (a) is a prime ideal in a U.F.D. if and only if a is a prime of the domain.

20. Find all the units of the Euclidean domain of complex numbers $m + ni$, where m and n are integers. Prove that $2, 3$, and 5 factor into primes in this domain.

21. Let b be a nonzero element of a principal ideal ring. Prove that the residue classes modulo b form a group which consists of elements relatively prime to b.

22. If B is an ideal in a Noetherian ring A, prove that the quotient ring A/B is Noetherian.

23. Let B and C be ideals in a Noetherian ring. Prove that $B^n \subseteq C$ for some positive integer n if and only if $\sqrt{B} \subseteq \sqrt{C}$. (See Exercise 2 of Sec. 5-9.)

chapter 7

Modules

A module is a simple but important generalization of a vector space. In a vector space the set of scalars is a field or a division ring. In a module it is an arbitrary ring. That is, like a vector space, a module is an additive commutative group with scalar multiplication, only the scalars in a module are elements of a ring. Modules are important algebraic systems in themselves but are also closely related to their rings in the development of ring theory. Later we shall study algebras, which are modules over commutative rings with some additional requirements. The structure of a module is common to many algebraic systems.

7-1 DEFINITIONS

Definition. Let R be any ring. An additive abelian group M is called a **right R-module** or **right module over R** if a mapping of the cartesian product $M \times R$ into M is defined, such that the "product" $mr = (m, r)f$, $m \in M, r \in R$ (corresponding in vector spaces to scalar multiplication) satisfies the three following requirements:

(i) $m(r_1 + r_2) = mr_1 + mr_2$,
(ii) $(m_1 + m_2)r = m_1 r + m_2 r$,
(iii) $m(r_1 r_2) = (mr_1)r_2$,

for all r, r_1, r_2 in R and all m, m_1, m_2 in M.

Similarly, one can define a left R-module in which now the product would take the form rm instead of mr, for $r \in R$ and $m \in M$.

Observe that with a module, as in a vector space, two zeros are actually involved, the zero 0_M of the module itself (this is the neutral element of the commutative group M) and the zero 0_R of the ring R. It is easy to see from the axioms that $m0_R = 0_M$ and $0_M r = 0_M$ for all $m \in M$ and all $r \in R$. We shall use the symbol 0 for both of these zeros indiscriminately, since it is usually evident from the context which one is intended.

EXERCISE

1. By induction on n, prove the following extensions of axioms (i) and (ii):

$$(m_1 + \cdots + m_n)r = m_1 r + \cdots + m_n r,$$

216

$$m(r_1 + \cdots + r_n) = mr_1 + \cdots + mr_n,$$

where the $r_i \in R$ and the $m_i \in M$.

Let us designate by MR the set of all finite sums of the form $\Sigma x_i r_i$ where $x_i \in M, r_i \in R$. From the definition of a right R-module we know that $MR \subseteq M$. However, if $x \in M$ and $r \in R$, while we know $xr \in M$, we cannot assert that $x \in MR$; that is, it is not true in general that $MR = M$.

Definition. A right R-module M is called **unitary** or **unital** if $MR = M$.

This means that M is a unital right R-module if and only if every element x of M can be expressed in the form of a finite sum

$$x = \Sigma x_i r_i,$$

where $x_i \in M$ and $r_i \in R$.

Lemma. If the ring R has an identity 1, then a right R-module M is unital if and only if $x \cdot 1 = x$ for every x of M.

Proof: Let M be unital and let x be any element of M. Then

$$x = \Sigma x_i r_i, \qquad x_i \in M, r_i \in R.$$

Hence $x \cdot 1 = (\Sigma x_i r_i) \cdot 1 = \Sigma(x_i r_i \cdot 1) = \Sigma x_i(r_i \cdot 1) = \Sigma x_i r_i = x.$ Thus $x \cdot 1 = x$ for every x of M.

Conversely, let $x \cdot 1 = x$ for all x of M. Then $x \in MR$ since $1 \in R$. Hence $M \subset MR$. Since $MR \subset M$, we have $M = MR$ and therefore M is unital.

It will be recalled that in the definition of a vector space, $x \cdot 1 = x$ for every vector x is an axiom of a vector space. Thus a vector space is unital. Note well however that a module can be unital without the ring being required to have an identity.

We now give some examples of modules.

Example 1. An additive commutative group A is a module over the ring Z of integers. If $n \in Z$ and $a \in A$ then $n \cdot a$, the sum of n terms each equal to a, is the scalar product. Thus A in this notation is a left Z-module. With the understanding that $1 \cdot a = a$, A is unital.

Example 2. Any vector space is a unital module (left or right, depending on how one writes the scalar product) over its field or division ring.

Example 3. A very important left or right module is the ring itself. This is the analogue of a field being a vector space over itself. A ring is both a left and a right module over itself. The scalar product in this case is merely the ordinary ring product. Thus the elements of the ring play a dual role, they are both elements of the module and scalars. If the ring has identity, then it is automatically a unital module. If the ring R does not have an identity, then $R^2 \subset R$ but R^2 may not equal R, in which case R is not unital. For example, if R is the ring of even integers 0, $\pm 2, \pm 4, \ldots$ then R^2 is the subring $0, \pm 4, \pm 8, \ldots$ and $R^2 \subset R$ but $R^2 \neq R$.

Example 4. The mappings M of any set S into a ring R will form an R-module under the following definitions: For $\alpha, \beta \in M$ define $\alpha + \beta$ by $x(\alpha + \beta) = x\alpha + x\beta$, $x \in S$. It is easy to show that this defines an associative and commutative addition in M. Let θ be the mapping of M that maps every element of S into the zero element 0 of R, and define $-\alpha$ by $x(-\alpha) = -(x\alpha)$ for all $x \in S$, then it follows that M is an additive commutative group with θ as the neutral element and $-\alpha$ as the inverse of α. A scalar product is obtained in M by defining αr, for $r \in R$, by $x(\alpha r) = (x\alpha)r$ for all $x \in S$, for it is easily checked that this product satisfies the conditions (i), (ii), (iii), above. Hence the mappings M constitute a right R-module.

Definition. A **submodule** N of a right R-module M is a subgroup of M such that $NR \subseteq N$. (Here NR stands for the set of all xr, $x \in N$, and $r \in R$.)

We describe $NR \subseteq N$ by saying that N is *stable* (or *closed*) under right multiplication by elements of R.

Note that the zero element alone $\{0\}$ and M itself are submodules of M. They are referred to as **improper** or **trivial** submodules.

The set MR of all finite sums of the form $\Sigma x_i r_i$, $x_i \in M$, $r_i \in R$, is easily seen to be a submodule of M. If S is a right ideal in the ring R, then the set MS of finite sums is a submodule of M.

If a ring R is regarded as a right (left) R-module then its submodules are its right (left) ideals. This follows at once from the definition of a submodule.

THEOREM 1. N is a submodule of a right R-module M if and only if $x, y \in N, r \in R$ imply that (i) $x - y \in N$, and (ii) $xr \in N$.

Proof: (i) makes N an abelian group and (ii) is equivalent to $NR \subseteq N$.

THEOREM 2. Any intersection of submodules of a module is a submodule.

Proof: This follows at once from the definitions of intersection and module.

Quotient Modules

Since a right R-module M is an additive abelian group, we can take any submodule N of M and form the factor group M/N consisting of all elements of the form $x + N$, $x \in M$. These are the cosets of N in M. However we can do more than this. We can define a right module (left module) structure on M/N by defining

$$(x + N)r = xr + N, \qquad r \in R.$$

Clearly this "scalar" product satisfies (i), (ii), (iii) above and so M/N becomes a right R-module, called a **quotient** or **factor module**.

Example 5. A ring R is a right (left) module over itself and its submodules are its right (left) ideals and its quotient modules are R/S, where S is a right (left) ideal in R. Recall that R/S is not a quotient ring unless S is a two-sided ideal.

Definition. A right R-module M is called **trivial** if $MR = \{0\}$, that is if the product $xr = 0$, for all $x \in M$ and all $r \in R$.

Example 6. We can turn any abelian additive group A into a trivial right module over any ring R, by defining $ar = 0$, for all $a \in A$ and $r \in R$.

Definition. A nontrivial right R-module M is called an **irreducible** right R-module if its only submodules are $\{0\}$ and M. Since $MR \neq \{0\}$ and is a submodule of M, $MR = M$ and hence an irreducible module is unital.

Example 7. A cyclic group of prime order is an irreducible module over the ring of integers.

Definition. A right R-module M is called **cyclic** if there exists a nonzero element $m \in M$ such that $mR = M$. Thus a cyclic module is unital.

THEOREM 3. An irreducible right R-module is cyclic.

Proof: Let M be an irreducible right R-module, and let N be the subset of M comprising all $x \in M$ such that $xR = 0$. Then clearly N is a submodule of M and hence $N = \{0\}$ or $N = M$. But $N = M$ implies $MR = \{0\}$, a contradiction of M being irreducible. Hence $N = \{0\}$, that is $xR = 0$ implies $x = 0$ and therefore any nonzero element of M generates M. For if $y \in M$ and $y \neq 0$, then yR is a nonzero submodule of M and therefore $yR = M$.

The irreducible R-modules of a ring are very important, that is if it has any. They determine a great many properties about the type of ring (semisimplicity, primitivity) and these in turn yield information about its structure (direct or subdirect sums of rings of simpler types).

EXERCISES

1. If R is a ring, show that the polynomial ring $R[x]$ in the variable x is an R-module.

2. Prove that $R[x, y]$ is a module over the ring $R[x]$ and also a module over the subring R of $R[x]$.

3. If S is a subring of a ring R, show that any R-module is also an S-module.

4. If R is a ring define an R-module structure on the set $R \times R$ of all ordered pairs of elements of R.

7-2 MODULE HOMOMORPHISMS

Definition. Let M and M' be two right R-modules. A mapping f of $M \to M'$ such that for all $m, m_1, m_2 \in M$ and $r \in R$,

$$(m_1 + m_2)f = m_1 f + m_2 f \quad \text{and} \quad (mr)f = (mf)r$$

is called a **module** or **R-homomorphism.** The *kernel* of f is the set of all $m \in M$ such that $mf = 0'$, where $0'$ is the zero element of M'. As before, if a module homomorphism is surjective, it is called an **epimorphism,** and if both surjective and injective then it is called a **module isomorphism.**

We now cite the following theorems on module homomorphisms. The proofs are so similar to the corresponding theorems on ring homomorphisms that they are left as exercises for the reader.

THEOREM 4. If f is a module homomorphism $M \to M'$ of two right (left) R-modules then $0f = 0'$ and $(-m)f = -(mf)$ for all $m \in M$.

THEOREM 5. A module homomorphism f is injective if and only if $\ker f = 0$.

THEOREM 6. The kernel of a module homomorphism f of $M \to M'$ is a submodule of M and the image im $f = Mf$ of M is a submodule of M'.

THEOREM 7. If f is a module homomorphism of $M \to M'$, then if N is a submodule of M, Nf is a submodule of M'. If N' is a submodule of M' then $f^{-1}N'$ is a submodule of M.

THEOREM 8. If N is any submodule of a module M, then the mapping f of $M \to M/N$, defined by $mf = m + N$ is an epimorphism. (It is called the **natural** or **canonical homomorphism** of M onto a quotient module of M.)

Proof: For $m_1, m_2 \in M$, we have $(m_1 + m_2)f = m_1 + m_2 + N = m_1 + N + m_2 + N = m_1 f + m_2 f$. Also, $(mr)f = mr + N = (m + N)r = (mf)r$. Hence f is a homomorphism. Since for any $m + N$ of M/N we have $mf = m + N$, clearly f is surjective.

Definition. An R-module M' is said to be a **homomorphic image** of an R-module M if there exists a module epimorphism f of $M \to M'$.

By the previous theorem we see at once that if N is a submodule of a module M then M/N is a homomorphic image of M. We now prove

THEOREM 9. Any homomorphic image of a module M is isomorphic to a quotient module of M.

Proof: Let $f: M \to M'$ be a module epimorphism and let ker $f = N$. Consider the mapping ϕ of $M/N \to M'$ defined by $(m + N)\phi = mf$. It is easy to see that ϕ is a module homomorphism, for

$$(m_1 + N + m_2 + N)\phi = (m_1 + m_2 + N)\phi = (m_1 + m_2)f$$
$$= m_1 f + m_2 f = (m_1 + N)\phi + (m_2 + N)\phi,$$

where m_1 and m_2 are elements of M; moreover, $((m + N)r)\phi = (mr + N)\phi = (mr)f = (mf)r = (m + N)\phi r$. The mapping ϕ is injective since ker $\phi = N$, the zero element of M/N, and ϕ is surjective since f is.

If a ring is commutative there is no need to distinguish between left and right R-modules, for it is easy to see that $rm \to mr$ is a module isomorphism of any left R-module with the corresponding right R-module. This identifies the two modules algebraically.

EXERCISES

1. If A and B are submodules of a module C, prove that $A + B$ is a submodule of C.

2. A, B, C are R-modules. If $A \supset B$, prove that

$$A \cap (B + C) = B + (A \cap C).$$

3. Let A be an ideal in a commutative ring R with 1. If M is an R-module, show that the set

$$S = \{am \mid a \in A, m \in M\}$$

is not in general an R-module. When is S an R-module?

4. Find the R-module generated by S.

5. If A is a right ideal in a ring R with 1, show that the set MA of all finite sums $\Sigma\, m_i a_i, m_i \in M, a_i \in A$ is a right R-module.

6. If M is an R-module and if $r \in R$, prove the set $rM = \{rm \mid m \in M\}$ is an R-module.

7. If R is a ring, define a new ring R' as being the set R with the addition as in R, but with a multiplication $r_1 \circ r_2$ defined by $r_1 \circ r_2 = r_2 r_1$ where $r_2 r_1$ is the product in R. Prove R' is a ring. If M is a right R-module, prove that M, with the same scalar multiplication, is a left R'-module. Thus properties of left modules can be derived from the theory for right modules.

8. Show that an additive abelian group G is a module over the ring of integers. What are its submodules? Prove that if A is the ideal $A = (m)$, $(m)G$ is a submodule.

9. A ring R has no proper zero-divisors. Regarded as an R-module, which of its ring homomorphisms are the same as its module homomorphisms?

10. Prove that the set M of all polynomials of degree ≤ 4 belonging to $R[x]$ is a left module over the ring R. If N is the submodule of all polynomials of degree ≤ 3, describe the quotient module M/N. Prove M/N is isomorphic to the left R-module R.

7-3 FINITELY GENERATED MODULES

A subset S of an R-module M is said to *generate* M if M is the least R-module that contains S, that is no proper submodule of M can contain S. In other words this means that every $m \in M$ has the form $m = n_1 s_1 + \cdots + n_k s_k + s_1 r_1 + \cdots + s_k r_k$, where the n_i are integers, $s_i \in S$ and $r_i \in R$. For clearly the set of all elements of M which have this form is a module and any module containing S would have to contain these. It is worth emphasizing that the same elements of S do not necessarily appear in the expressions for two distinct elements of M. The set S can be a finite set or an infinite set. In fact $S = M$ generates M. When S is a finite set we get the so-called **finitely generated modules,** which are the topic of this section.

Clearly if $f: M \to M'$ is an epimorphism of two R-modules and if S is a set of generators of M, then Sf is a set of generators of M'. In particular then, if M is a finitely generated module, a homomorphic image of M is likewise finitely generated.

We return now to finitely generated modules. Let M be a right R-module and let X be the finite subset of elements $x_1, x_2, \ldots, x_k$ of M. Clearly the set of all elements of M of the form

$$(1) \qquad n_1 x_1 + \cdots + n_k x_k + x_1 r_1 + \cdots + x_k r_k,$$

where the n_i are integers and the r_i are in R, forms a submodule $[X]$ of M. This submodule $[X]$ is called the submodule *generated by the set X* and is an example of what is called a **finitely generated module.** It is most important to observe that each element of X is required to be in the submodule generated by X. This is the reason for the presence in (1) above of the terms $n_1 x_1 + \cdots + n_k x_k$, where the n_i are integers. It may happen that M itself is a finitely generated module, in which case there would exist a finite set of elements $(y_1, y_2, \ldots, y_s)$ such that every $m \in M$ can be expressed in the form

$$m = n_1 y_1 + \cdots + n_s y_s + y_1 r_1 + \cdots + y_s r_s,$$

where the n_i are integers and the r_i are in R.

Example 8. A cyclic module is finitely generated, in fact generated by a single element.

If a finitely generated module over a ring R with identity 1 is unitary then its elements assume a simpler form. For example let M be unitary and generated by the elements $y_1, y_2, \ldots, y_n$. Then it is easy to see that every element $m \in M$ can be expressed in the simpler form

$$y_1 r_1 + y_2 r_2 + \cdots + y_n r_n,$$

where the r_i belong to R. For a term such as $n_i y_i$ where n_i is an integer can be written $n_i y_i = (n_i \cdot 1) y_i = y_i (n_i \cdot 1)$ and $n_i \cdot 1 \in R$.

If M is a finitely generated R-module, it does not follow that each submodule N of M is also finitely generated. For example, let M be a cyclic right R-module, that is $M = mR$ for some $m \neq 0$ of M. The right R-submodules of M clearly have the form mS, where S is a right ideal in the ring R. If S is a finitely generated right ideal, say $S = (a_1, a_2, \ldots, a_k)$, then clearly the submodule mS is generated by the elements $ma_1, \ldots, ma_k$ and so is a finitely generated R-module. (In fact, mS is a cyclic S-module.) We shall see later that in a Noetherian ring every ideal is finitely generated. However, if R is not a Noetherian ring, then the ideal S need not be finitely generated and hence the submodule mS of M would not be a finitely generated R-module. It is a theorem, which we shall not prove here, that states: a finitely generated R-module over a Noetherian ring R is itself a Noetherian module (that is if R has the ascending chain condition on right ideals, then M has the ascending chain condition on right submodules), hence every submodule of such a module is finitely generated.

7-4 DIRECT SUMS OF MODULES

Let M_i, $i \in I$, be a family of right R-modules indexed by the set I. Since modules are abelian groups, we can form the direct sum $\sum_{i \in I} \bigoplus M_i$ of the M_i regarded as abelian groups. Thus $\Sigma \bigoplus M_i$ is an abelian group whose elements are finite sums $x_{i_1} + x_{i_2} + \cdots + x_{i_k}$ of elements from the M_i. For $r \in R$, define

$$(x_{i_1} + x_{i_2} + \cdots + x_{i_k})r = x_{i_1}r + x_{i_2}r + \cdots + x_{i_k}r.$$

If $x_i \in M_i$, then $x_i r \in M_i$ and hence

$$(x_{i_1} + x_{i_2} + \cdots + x_{i_k})r \in \Sigma \oplus M_i.$$

It is easy to verify that this defines a scalar multiplication and with this agreement, $\Sigma \oplus M_i$ becomes a right R-module, called the (**external**) **direct sum** of the family of modules M_i, $i \in I$. The reader will recall from Chapter 4 that this direct sum is the *weak* direct product of the family of modules. Moreover it also follows therefore that each element of the direct sum has a unique expression as a sum of elements from the M_i.

Let $M_I, i \in I$, be a family of submodules of an R-module M which generates M. This means every $x \in M$ is a finite sum, with coefficients from the ring R, of elements from the M_i. If for each $i \in I$ the intersection of M_i with the submodule generated by all submodules M_j, $j \neq i$, is (0), then M is called the **direct sum** of the family M_i, $i \in I$, of submodules. We write

$$M = \sum_{i \in I} \oplus M_i.$$

As in the case of groups we can easily show that this intersection condition is equivalent to the condition that every element x of M has a unique representation in the form of a finite sum of elements of the M_i. As before this type of direct sum is called an **internal direct sum.**

We shall write a direct sum in the form $\Sigma \oplus M_i$ whether it is external or internal, for it will be clear from the context which is intended. Each term M_i of a direct sum is called a *direct summand*.

Let $M_\gamma, \gamma \in \Gamma$, be an arbitrary family of modules over the same ring R. We assume R has an identity 1 and that all the modules are unital.

Let $(m_\gamma)_{\gamma \in \Gamma}$ be a family of elements of the M_γ, $\gamma \in \Gamma$, such that almost all m_γ are 0 (that is all but a finite number) and let M be the set of all such elements $(m_\gamma)_{\gamma \in \Gamma}$. [Note that $(m_\gamma)_{\gamma \in \Gamma}$ is a generalization to an arbitrary index set Γ of the notation $(m_1, m_2, \ldots, m_k)$ when $\Gamma = \{1, 2, \ldots, k\}$.] Define

$$(m_\gamma)_{\gamma \in \Gamma} + (m'_\gamma)_{\gamma \in \Gamma} = (m_\gamma + m'_\gamma)_{\gamma \in \Gamma}$$

$$(m_\gamma)_{\gamma \in \Gamma} \cdot \gamma = (m_\gamma r)_{\gamma \in \Gamma}, \quad r \in R.$$

Then it can be verified that M is an R-module. M is called the **direct sum of the family** $M_\gamma, \gamma \in \Gamma$ of R-modules and we write

$$M = \sum_{\gamma \in \Gamma} \oplus M_\gamma.$$

Each M_γ is called a **direct summand of M**.

EXERCISES

1. If S is a subset of a module M, prove that an element $x \in M$ belongs to the submodule generated by S if and only if x is a linear combination of elements of S.

2. M is a module over a commutative ring with 1. Let R' be the R-module, $R' = R \oplus M$. If a multiplication defined by $(r_1 + m_1)(r_2 + m_2) = r_1 r_2 + r_1 m_2 + r_2 m_1$ is introduced in R', show that R' becomes a ring containing R and M. Prove M is an ideal in R'. Prove $M^2 = 0$.

3. Is M in Exercise 2 an R'-module?

7-5 REPRESENTATIONS OF RINGS

We saw earlier that any ring is isomorphic to a ring of endo-morphisms of an abelian group. A *representation* of a ring R is a homo-morphism of R into a ring of group endomorphisms of an abelian group, and the representation is called **faithful** if it is injective. Representations of rings (or for that matter of any algebraic system) are important in that they afford "pictures" of the rings. These pictures can be studied for properties of the rings, and often this can be done more easily than dealing with the rings themselves. Moreover, often the representations serve to uniformize the study of the rings themselves, and to give concrete mean-ings to abstract ideas. For instance, Cayley's theorem that every group is isomorphic to a permutation group is a representation theorem, since it says that every group can be represented as a permutation group. Linear transformations of a finite-dimensional vector space have a representation as matrices over the field of the vector space.

If R is any ring, then a representation of R determines a right R-module as follows: Let ρ be a representation of R into a ring of group endomorphisms of an abelian group M, that is ρ maps an element r of R into a group endomorphism $r\rho$ of M. We can use this fact then to define a module product mr in M, by putting $mr = m(r\rho)$, for $r\rho$ maps m into another element of M. It is easy to check that this qualifies as a module product by testing the three module axioms (i), (ii), and (iii). Hence the representation ρ of R determines a module structure on the abelian group M. Conversely if M is a given right R-module, then M is auto-matically an abelian group, and for each $r \in R$, a group endomorphism $\bar{r}$ of M is determined by the mapping $\bar{r}$ of $M \to M$ given by $m\bar{r} = mr$. This mapping $\bar{r}$ is easily verified to be a group endomorphism of M. Hence the mapping ρ of $r \to \bar{r}$ is a homomorphism of R into a ring of group endomorphisms of M, that is ρ is a representation of R. All this shows that the R-modules and representations of R are equivalent con-cepts, since one determines the other.

EXERCISES

1. Let M be a right module over the ring R. The ideal $(0; M) = \{r \in R \mid mr = 0, m \in M\}$ is called the **annihilator** of M. Prove it is an ideal.

2. Prove that the ring $R/(0:M)$ is isomorphic to a ring of endomorphisms of the abelian group M.

7-6 THE CHAIN CONDITIONS

Definition. A right R-module M is said to satisfy the *ascending-chain condition* for right submodules if every ascending chain of right submodules $N_1 \subseteq N_2 \subseteq N_3 \subseteq \cdots \subseteq N_i \subseteq \cdots$ terminates, that is if there is an integer n such that $N_n = N_{n+1} = \cdots$. The right R-module M is said to satisfy the *descending-chain condition* for submodules if every descending chain of submodules $N_1 \supseteq N_2 \supseteq N_3 \supseteq \cdots \supseteq N_i \supseteq \cdots$ terminates—that is, if there is an integer m such that $N_m = N_{m+1} = \cdots$.

Definition. A right R-module M is said to have or to satisfy the *maximum condition* on submodules if every nonempty family of submodules contains a (relatively) maximal submodule—that is, one that is not contained in any other submodule of the family. (This means it is maximal with respect to the family.) A right R-module is said to have or to satisfy the *minimum condition* on submodules if every nonempty family of submodules contains a (relatively) minimal submodule—that is, one that does not properly contain any other submodule of the family —i.e., minimal with respect to the family.

The axiom of choice (that is unlimited choice) is now used to prove the following theorem:

THEOREM 10.

(I) The ascending-chain condition and the maximum condition are equivalent.

(II) The descending-chain condition and the minimum condition are equivalent.

Proof: (I) Assume that the right R-module M satisfies the ascending-chain condition and let F be a non-empty family of submodules of M. Choose $N_1 \in F$. Then if N_1 is not maximal, choose $N_2 \in F$, where $N_1 \subset N_2$. Continuing in this way, since we assume unlimited choice we get an ascending chain $N_1 \subset N_2 \subset N_3 \subset \cdots$. This must lead in a finite number of steps to an $N_k \in F$ that is maximal, otherwise the ascending-chain condition would be violated. Hence F must contain a maximal submodule. Conversely assume every nonempty family of submodules contains a maximal submodule and let

$N_1 \subset N_2 \subset N_3 \subset \cdots$ be any ascending chain of submodules of M. Then the family of all submodules N_i of this chain must contain a maximal submodule N_k and hence $N_k = N_{k+1} = \cdots$, i.e., the chain must terminate.

(II) Assume that the right R-module M satisfies the descending-chain condition and let F be a nonempty family of submodules of M. Choose $N_1 \in F$. Then if N_1 is not minimal, choose $N_2 \in F$ such that $N_1 \supset N_2$. Continuing in this way, we get a descending chain $N_1 \supset N_2 \supset N_3 \supset \cdots$. This again must break off at a minimal submodule $N_k \in F$. Conversely, if $N_1 \supset N_2 \supset N_3 \supset \cdots$ is a descending chain of submodules of M, then the family F of all submodules N_i of this chain must contain a minimal submodule N_k and, as we have seen before, this means the descending chain terminates and $N_k = N_{k+1} = \cdots$.

It is most important to bear in mind that since a ring is both a right and a left modules over itself, these chain conditions and the maximum and minimum conditions apply to rings. If a ring is regarded as a right module over itself, then its submodules are its right ideals, so that this last theorem can be interpreted as applying to rings and to ascending and descending chains of right ideals of the ring. The same of course is true for left R-modules and for rings and their left ideals.

We can prove still one more equivalent condition to the ascending-chain condition.

THEOREM 11. A module M satisfies the ascending-chain condition for submodules if and only if every submodule of M is finitely generated.

Proof: Assume the ascending-chain condition. The zero submodule is generated by the zero element of the module. Let $N \neq \{0\}$ be a submodule of M. Choose $x_1 \in N$, where $x_1 \neq 0$. If $N = (x_1)$ then N is the cyclic submodule generated by x_1. If $(x_1) \subset N$, then choose $x_2 \in N$ where $x_2 \neq 0$ and $x_2 \notin (x_1)$. Either $N = (x_1, x_2)$ or if not $(x_1, x_2) \subset N$, and then choose an x_3 in a similar way. Continuing, we get an ascending chain of submodules $(x_1) \subset (x_1, x_2) \subset (x_1, x_2, x_3) \subset \cdots$ all contained in N. Thus we must reach some x_k where $(x_1, x_2, \ldots, x_k) = N$ or the chain condition is violated. Hence N is finitely generated. Conversely, assume that every submodule of M is finitely generated, and let $N_1 \subset N_2 \subset N_3 \subset \cdots$ be any ascending chain of submodules of M. Now it is easy to show that the union $\cup N_i$ of the N_i of this chain is a submodule of M and hence it is finitely generated. Let $\cup N_i = (x_1, x_2, \ldots, x_k)$. Now each $x_i \in \cup N_i$ (every generator of a submodule is in the submodule, by definition of a generator). Hence $x_i \in N_{m_i}$ for some m_i, and this is true of each $i = 1, 2, \ldots, k$. If we let

$$n = \max(m_1, m_2, \ldots, m_k),$$

then we see that $\bigcup N_i \subset N_n$ and hence $N_n = N_{n+1} = \cdots$, that is the chain breaks off.

Thus for a module M the following three conditions are equivalent:
1. M satisfies the ascending-chain condition for submodules.
2. M satisfies the maximum condition for submodules.
3. Every submodule of M is finitely generated.

Since any ring is a module over itself and its submodules are its right or left ideals as the case may be, we can restate the above three conditions for rings as follows:

In a ring R the following three conditions are equivalent:
1. R satisfies the ascending-chain condition for right (left) ideals.
2. R satisfies the maximum condition for right (left) ideals.
3. Every right (left) ideal in R is finitely generated.

We have defined earlier a *Noetherian ring*. It is a commutative ring which satisfies the ascending-chain condition for ideals. About such a ring we can now say that all its ideals are finitely generated and moreover that every nonempty family of ideals of a Noetherian ring contains a maximal ideal for the family (not necessarily however a maximal ideal in the ring).

It is now appropriate to give some examples of these concepts.

Example 9. The ring of integers Z is a principal ideal ring, and so every ideal in it is principal and therefore finitely generated. Hence Z satisfies the ascending-chain condition and therefore also the maximum condition. Z is a Noetherian ring. However, it does not satisfy the descending-chain condition, since in Z every ideal has the form (n) and $(m) \supset (n)$ means $m \mid n$. Thus $(2) \supset (4) \supset (8) \supset \cdots$ is a nonterminating descending chain.

Example 10. About the simplest example of a Noetherian ring is a principal ideal ring. If F is a field then we know that, if x is transcendental over F, $F[x]$ is a principal ideal domain and therefore a Noetherian ring. The Hilbert basis theorem also demonstrates that $F[x]$ has only finitely generated ideals. Of course, a field F has only the ideals (0) and F, and so trivially satisfies both chain conditions.

Example 11. The only integral domains that satisfy the descending-chain condition are fields. For let R be an integral domain and let $a \neq 0$ be an element of R. If R satisfies the descending-chain condition, then in the descending chain $Ra \supset Ra^2 \supset Ra^3 \supset \cdots$ we must have $Ra^n = Ra^{n+1}$

for some positive integer n. Hence there exists $b \in R$ such that $a^n = ba^{n+1}$; that is, $a^n(1 - ba) = 0$. Hence $ba = 1$, which means then every nonzero element of R has an inverse and therefore R is a field.

Example 12. If R is a commutative ring with identity that satisfies the descending-chain condition and P is a prime ideal in R, then the integral domain R/P satisfies the descending-chain condition. For the ideals of R/P have the form A/P, where A is an ideal of R containing P. Hence if $A_1/P \supseteq A_2/P \supseteq \cdots$ is a descending chain in R/P then $A_1 \supseteq A_2 \supseteq \cdots$ is a descending chain in R, and hence must terminate— that is, $A_n = A_{n+1} = \cdots$, and therefore $A_n/P = A_{n+1}/P = \cdots$ and thus R/P satisfies the descending-chain condition.

Example 13. In commutative rings without an identity maximal ideals need not be prime, and in commutative rings with an identity prime ideals do not have to be maximal. However, if a commutative ring with identity satisfies the descending-chain condition, then every prime ideal is maximal. For if P is a prime ideal in such a ring R then, as we saw above, the integral domain R/P satisfies the descending-chain condition and hence it is a field. Thus P is a maximal ideal.

Example 14. We leave as an exercise the proof of the following: a commutative ring R with identity satisfies the descending-chain condition if and only if (i) R satisfies the ascending-chain condition, and (ii) every prime ideal P of R, $P \neq R$, is maximal.

Example 15. So many examples of a Noetherian ring exist that it is interesting to study the following ring which we shall show is *not* a Noetherian ring. Let S be a given infinite set. Let B be the set of all sub-sets of S, including S itself and the empty set ϕ. Define an addition and multiplication in B by: for $\alpha, \beta \in B$, define $\alpha + \beta = \alpha \cup \beta - \alpha \cap \beta$ and $\alpha\beta = \alpha \cap \beta$. It is easy to verify that B is an additive abelian group with zero element ϕ and, since $\alpha + \alpha = \phi$ for every $\alpha \in B$, it follows that each element of B is its own inverse. Moreover the multiplication is seen to be both commutative and associative, and to be distributive with respect to addition. Hence B is a commutative ring with the identity S. (Note that B does have zero-divisors, since $\alpha\beta = \phi$ does not imply $\alpha = \phi$ or $\beta = \phi$.)

We prove that B is not a Noetherian ring by exhibiting an ideal in B that is not finitely generated. First observe that the subset C of B consisting of all finite subsets of S is an ideal in B. If C is finitely generated, then $C = (C_1, C_2, \ldots, C_m)$, where each C_i is a finite subset of S. Since B has an identity, every element D of C can be expressed in the form $D = \Sigma B_i C_i$, $B_i \in B$. Since $B_i C_i = B_i \cap C_i$, it is seen that each $B_i C_i$ is a finite set and thus each element D of C is a finite set. Moreover every element of D is an element of some $a \in C_i$. Hence there exist points in S that are not in any C_i and hence there exist finite subsets of S that are not in any C_i and therefore C, the ideal of all finite subsets of S, cannot be finitely generated.

1. Prove that the a.c.c. and d.c.c. are equivalent for a vector space (a) over a field (b) over a division ring.

2. N is a submodule of the module M. Prove that the a.c.c. (d.c.c.) holds for M if and only if it holds for both N and M/N.

3. If $M_1, M_2, \ldots, M_k$ are submodules of a module M and if

$$M = M_1 + M_2 + \cdots + M_k,$$

prove that if each M_i has the a.c.c. (d.c.c.), then so does M.

4. M is a finitely generated module over a ring R. If R satisfies the a.c.c. (d.c.c.) prove that M satisfies it also.

7-7 PRIMARY DECOMPOSITIONS IN NOETHERIAN RINGS

Noetherian rings form an important and common type of commutative ring. They are important in that they have the three desirable properties of satisfying the ascending chain condition, satisfying the maximum condition, and every ideal is finitely generated. They are common since (1) the ring Z of integers is Noetherian; (2) a P.I.D. is Noetherian; (3) all polynomial rings $F[x_1, \ldots, x_n]$ where F is a field are Noetherian (their polynomial ideals are of fundamental importance in algebraic geometry); (4) if R is a Noetherian ring then the Hilbert basis theorem asserts that the polynomial ring $R[x_1, \ldots, x_n]$ is likewise Noetherian.

Thus the study of Noetherian rings and their ideals constitutes a very important part of what is called commutative algebra. We are going to prove that every ideal in a Noetherian ring is an intersection of a finite number of primary ideals, which to some extent reduces the study of ideals in a Noetherian ring to the study of its primary ideals.

We remark that a module that satisfies the ascending chain condition for submodules is often called a **Noetherian module.**

Definition. Let R be a commutative ring with identity. A **primary ideal** Q in R is an ideal that has the property that whenever $xy \in Q$ and $x \notin Q$ then some power y^m of y is in Q.

Another way of describing a primary ideal is to say that whenever it contains a product xy, but contains no power of x, then y is in the ideal. Clearly a prime ideal is primary but not in general conversely.

Example 16. In the ring Z of integers the primary ideals are those of the form (p^m) where p is a prime and m is a positive integer. For if

$xy \in (p^m)$, then $xy = ap^m$ for some integer a, and if $x \notin (p^m)$, then $x = bp^k$ where $b > 1$, $p \nmid b$ and $k < m$. Hence $p \mid y$ and therefore $p^m \mid y^m$. Thus $y^m \in (p^m)$ and (p^m) is a primary ideal. Conversely, if an ideal (n) of Z is not of this form, then it is not primary. For let $n = ap^m$ where p is a prime, $m > 0$, $a > 1$, and $p \nmid a$. Then for $x = a$ and $y = p^m$, we see that $xy \in (ap^m)$, $x \notin (ap^m)$, yet ap^m cannot divide any power y^k of y. Hence no power of y is in (ap^m) and so it is not primary. Since every ideal in Z has one or the other of these two forms (p^m) and (ap^m) where $p \nmid a$, then every primary ideal in Z has the form (p^m).

Definition. An ideal A in a commutative ring is called **irreducible** if A is not the intersection of two ideals, properly containing A. This means if A is irreducible and $A = B \cap C$ then either $B = A$ or $C = A$.

THEOREM 12. A prime ideal is irreducible.

Proof: For if P is a prime ideal and if $P = A \cap B$ where $P \subset A$ and $P \subset B$, then there exist $a \in A$ and $b \in B$ such that $a \notin P$ and $b \notin P$, yet $ab \in P$, which contradicts P being a prime ideal.

A prime ideal is irreducible but a primary ideal need not be. For example, it is not hard to see that the ideal $B = (x^2, xy, y^2)$ is primary in the domain $F[x, y]$, where F is a field. A polynomial $f \in F[x, y]$ which is not in B would have to have a nonzero tail $ax + by + c$. Hence if $fg \in B$, then g would have to be either a polynomial without a constant term if $c = 0$, in which case g or g^2 is in B, or if $c \neq 0$ then $g \in B$. Thus B is a primary ideal. However,

$$B = (B + F[x,y]x) \cap (B + F[x,y]y)$$

and so is reducible. (For example, $F[x,y]x$ is the ideal of all elements of the form $f(x, y)x$).

Definition. Let A and B be ideals in a commutative ring R. The **quotient ideal** $(A : B)$ is defined as the set of all $r \in R$ such that $rB \subset A$.

Thus we see that $A \subset (A : B)$ and if $B \subset A$ then $(A : B) = R$.

THEOREM 13. In a ring R that has the ascending chain condition every ideal is the intersection of a finite number of irreducible ideals.

Proof: Let F be the family of ideals in R which are not finite intersections of irreducible ideals. Then F contains a maximal ideal A, and A cannot be irreducible. (For $A = A \cap A$ and if A were irreducible then $A \notin F$.) Hence $A = B \cap C$, where $A \subset B$ and $A \subset C$. Now $B, C \notin F$ (since A is maximal in F). Hence B and C must each be finite intersections of irreducible ideals, and this forces A to be. This contradiction implies F is empty and so we have the theorem.

THEOREM 14. In a commutative ring R with the ascending chain condition every irreducible ideal is primary.

Proof: Let B be an ideal in R that is not primary. The theorem will be proved if we show that B is reducible. There exists $a, b \in R$ such that $ab \in B$, $b \notin B$ and no power of a belongs to B. Now $b \in (B : (a))$ and hence B is properly contained in $(B : (a))$. Since $a^k \notin B$ for all $k = 1, 2, \ldots$ we see that $B \subset (a^k) + B$ for all $k = 1, 2, \ldots$. Now $(B : (a)) \subset (B : (a^2)) \subset (B : (a^3)) \subset \cdots$. Hence for some n, $(B : (a^n)) = (B : (a^{n+1}))$. Hence $B \subset (B : (a^n)) \cap (B + (a^{n+1}))$. If $x \in B + (a^{n+1})$, then $x = y + ma^{n+1} + za^{n+1}$, where $y \in B$, m is an integer, and $z \in R$. Hence if also $x \in (B : (a^n))$ then $ya^n + ma^{2n+1} + za^{2n+1} \in B$, and so $ma^{2n+1} + za^{2n+1} \in B$. But this means $ma + za \in (B : (a^{2n})) = (B : (a^n))$, and therefore $(ma + za)a^n \in B$. Since $y \in B$ this means $x \in B$. Thus

$$B = (B : (a^n)) \cap (B + (a^{n+1})),$$

and so B is reducible.

The two last theorems yield at once.

THEOREM 15. In a Noetherian ring every ideal is a finite intersection of primary ideals. (Such a finite intersection is called a **primary decomposition** of the ideal.)

Proof: If A is an ideal in a Noetherian ring, then

$$A = Q_1 \cap Q_2 \cap \cdots \cap Q_k,$$

where the Q_i are primary ideals.

A primary decomposition is called **irredundant** if no factor Q_i can be omitted, that is for example, if Q_1 were omitted, then $A \subset Q_2 \cap \cdots \cap Q_k$, but $A \neq Q_2 \cap \cdots \cap Q_k$. Clearly any primary decomposition can be made irredundant by omitting enough of its factors.

A primary decomposition is thus another type of factoring of an ideal, in contrast to factorization of an ideal into the product of prime ideals. There is a certain degree of uniqueness about an irredundant primary decomposition. The primary ideals themselves are not unique, but it can be proved that their associated prime ideals are. (If Q is a primary ideal, the set of all elements $r \in R$, such that $r^n \in Q$ for some positive integer n, is a prime ideal, called its **associated prime ideal**.)

If R is a commutative ring with identity and A is not a prime ideal in R then there exist ideals B and C such that $A \subset B$ and $A \subset C$ and $BC \subset A$. For A is not maximal, and since there exist $xy \in A$ such that $x \notin A$ and $y \notin A$ we can take $B = (x) + A$ and $C = (y) + A$. Bearing this in mind we can prove

THEOREM 16. In a Noetherian ring with identity every ideal contains a product of prime ideals.

We outline the proof and leave the details for the reader. Apply the maximum condition to the family F of all ideals in R that do not contain any product of prime ideals, to obtain a contradiction which implies that F is empty. For F contains a maximal ideal A and A is therefore not a prime. Making use of the above remark we get $BC \subset A$, $A \subset B$ and $A \subset C$. Since B and C do not belong to F, we can reach our contradiction.

We conclude this topic with this remark. We have seen that there are domains in which every ideal is a product of prime ideals. In general this is not possible in Noetherian rings. For example, consider the Noetherian ring $Z[x]$. The ideal $(4, x)$ in $Z[x]$ is not a prime ideal for $x^2 + 2$ and $x + 6$ do not belong to it, but their product does. In fact, a polynomial is not in $(4, x)$ if and only if it has a constant term not divisible by 4. However $(2, x)$ is a prime ideal and is in fact the only proper ideal containing $(4, x)$. Moreover no power of $(2, x)$ can equal $(4, x)$ and so $(4, x)$ cannot be a product of prime ideals. If an ideal is the product of prime ideals then it has to be contained in each of the prime ideals.

Of course the Noetherian rings possess another type of factorization of an ideal, namely *primary decomposition*. Some Noetherian rings have both. In the ring Z of integers if $n > 1$, then $n = p_1^{a_1} p_2^{a_2} \cdots p_k^{a_k}$, and therefore $(n) = (p_1)^{a_1}(p_2)^{a_2} \cdots (p_k)^{a_k} = (p_1^{a_1}) \cdots (p_k^{a_k})$. This is also true of a principal ideal domain which is obviously a Noetherian ring. For its prime ideals are those generated by its prime (irreducible) elements. Every non-unit element x is a unique product of prime elements, and so, as in the case of the integers, we get both types of factorization. As an interesting example, the reader should find the primary decomposition of the ideal $(4, x)$ in the domain $Z[x]$.

EXERCISES

1. Prove that a prime ideal in a commutative ring is irreducible.

2. If Q is a primary ideal in a commutative ring R with 1 and if $a \in R$, $a \notin Q$, prove $\sqrt{Q}$ is the radical of the ideal

$$(Q : aR) = \{r \in R \mid raR \subsetneqq Q\}.$$

3. Show that (x^2, y), (x), and (x^2, y^3) are primary ideals in $F[x, y]$. (F is a field.)

4. Prove that the radical of the ideal (x^2, xy) in $F[x, y]$ is (x). Is (x^2, xy) a primary ideal?

5. If Z is the domain of integers, show that $(4, x)$ and $(4, 2x, x^2)$ are primary ideals in $Z[x]$.

6. In $Z[x]$ prove the primary decompositions

$$(4, 2x, x^2) = (4, x) \cap (2, x^2)$$
$$(9, 3x + 3) = (3) \cap (9, x + 1).$$

7-8 FREE MODULES

We assume throughout the discussion of free modules that the ring R has an identity 1 and that all modules are unital.

Definition. Let S be any nonempty set. A **free R-module on the set** S is the pair (F, γ), where F is an R-module and γ is a mapping from $S \to F$ such that for each choice of an R-module X and of a mapping ϕ of $S \to X$, there exists a *unique* module homomorphism h of $F \to X$ for which $\gamma h = \phi$. The accompanying diagram depicts the property.

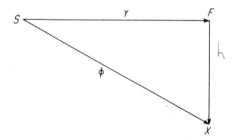

This property of a free R-module is called the **universal factorization property** (U.F.P.) and the couple (F, γ) is said to have the U.F.P. on the set S, universal for all R-modules X and all mappings ϕ of $S \to X$. This is to say that any mapping ϕ of $S \to X$ can be factored into $S \xrightarrow{\gamma} F \xrightarrow{h} X$, where h is a module homomorphism which is unique for each particular choice of X and ϕ. The U.F.P. is a type of what is often called a **universal mapping property.**

In the above diagram we have $\phi = \gamma h$ and, as before, this is described by saying that the diagram is *commutative*.

As usual, a definition does not guarantee the existence or uniqueness of such an object as a free module, so this will have to be substantiated. Before doing so however, we shall discover some of its properties.

We emphasize a property of module homomorphisms which is of great use to us now.

Lemma 1. A module homomorphism f of $M \to M'$ is completely determined by its values on a set of generators of M.

Proof: Suppose the subset S generates M. Then every element x of M has the form

$$x = \Sigma \ (n_i s_i + s_i r_i)$$

where the n_i are integers and $s_i \in S, r_i \in R$. Since f is a module homomorphism, we have

$$xf = \Sigma \left[n_i(s_i f) + (s_i f) r_i \right]$$

which expresses the value xf, for any $x \in M$, in terms of the values of f on S.

Returning now to free modules, the following three theorems are proved in exactly the same ways as for the case of the free abelian group. The reader is referred to Chapter 4 for the details.

THEOREM 17. If (F, γ) is a free R-module on the set S, then the mapping γ of $S \to F$ is injective.

THEOREM 18. The image $S\gamma$ of γ is a set of generators of the free module F.

THEOREM 19. If (F, γ) and (F', γ') are free R-modules on the same set S, then F and F' are isomorphic modules and there exists a unique isomorphism h of $F \to F'$ such that $\gamma' = \gamma h$.

Thus if the free R-module (F, γ) on S exists then it is unique up to isomorphism. In other words, the universal factorization property of a free R-module (F, γ) characterizes it up to an isomorphism.

THEOREM 20. There exists a free R-module on any set S.

Proof: Define F to be the set of all functions (mappings) f from S to the ring R, such that almost all values $xf, x \in S$, of f are 0 (that is all except a finite number are zero). For $f, g \in F$ define $f + g$ by $x(f + g) = xf + xg$ for $x \in S$. Clearly $f + g \in F$. (Check this out!) Let Θ be that mapping in F which takes every element of S into the 0 of the ring R. If we define $-f$ by $x(-f) = -(xf)$ for all $x \in S$, we see that $f + (-f) = f - f = \Theta$. Since the addition is both commutative and associative, we see that F is an additive abelian group. Now define $fr, f \in F$, and $r \in R$ by $x(fr) = (xf)r$ for all $x \in S$. Then it is easy to verify that this qualifies as a module product (scalar product) and hence F has now become an R-module.

For each $x \in S$ define $\bar{x}$ as that mapping of $S \to R$ which carries x into 1 and every other element of S into 0. Clearly $\bar{x} \in F$. Now define a mapping γ of $S \to F$ by $x\gamma = \bar{x}, x \in S$. We shall now show that the pair (F, γ) constitute a free R-module on the set S.

Let $f \in F$. Then $xf = 0$ except at a finite number $x_1, \ldots, x_n$ of points of S. Thus we can express f in the form

$$f = \bar{x}_1(x_1 f) + \cdots + \bar{x}_n(x_n f)$$

where each $\bar{x}_i \in F$ and $\bar{x}_i$ is that mapping of $S \to R$ which maps x_i into 1 and every other element of S into 0. We now show that (F, γ), as defined, has the U.F.P. Let X be any R-module and let ϕ be any mapping of $S \to X$. Define a mapping h of $F \to X$ by

$$fh = (x_1 \phi)(x_1 f) + \cdots + (x_n \phi)(x_n f).$$

In particular note that $\bar{x}h = x\phi$. If $g \in F$ and $xg = 0$ except at the points $y_1, \ldots, y_k$ of S, then the values of $f + g$ are 0 except, at most, at these $n + k$ points $x_1, \ldots, x_n, \ y_1, \ldots, y_k$. If we designate these points by $z_1, z_2, \ldots, z_{n+k}$ then we see that

$$(f + g)h = (z_1\phi)(z_1 f + z_1 g) + \cdots + (z_{n+k}\phi)(z_{n+k} f + z_{n+k}g),$$

since $z_i f = 0$ if z_i is not an x_j and $z_i g = 0$ if z_i is not a y_j. Hence $(f + g)h = fh + gh$. Also, $(fr)h = (fh)r, r \in R$, and hence h is a module homomorphism of $F \to X$. Since $\bar{x} = x\gamma$ and $\bar{x}h = x\phi$, we have $x\gamma h = x\phi$ for all $x \in S$, and therefore $\gamma h = \phi$.

Now all that remains to prove is that h is unique and then we shall have shown that (F, γ) has the U.F.P. on S and hence is a free R-module on S. Let h' be another module homomorphism from $F \to X$ such that $\gamma h' = \phi$. Then for any $f \in F$ we have

$$fh' = (\bar{x}_1 h')(x_1 f) + \cdots + (\bar{x}_n h')(x_n f).$$

Now $\gamma h' = \phi$ implies $\bar{x}_i h' = x_i \phi$ for each $i = 1, 2, \ldots, n$, and therefore $fh' = (x_1 \phi)(x_1 f) + \cdots + (x_n \phi)(x_n f) = fh$. Thus $fh' = fh$ for every $f \in F$, and hence $h' = h$.

Corollary. There exists a free abelian group on any set S.

Proof: Simply specialize the ring R in the theorem to be the ring Z of integers.

Since the mapping γ in the free R-module (F, γ) is injective, we can identify S with $S\gamma$ and regard S as a subset of F. If then we make this identification, γ becomes simply the inclusion $S \to F$ and we speak of F as the *free R-module generated by* S. In this case the module homomorphism h of $F \to X$ is simply the extension of the mapping ϕ of $S \to X$.

By this we mean that, since $S \subset F$, then $h = \phi$ when h is restricted to S. We call an R-module *free* if and only if it is isomorphic to the free R-module generated by some set S.

THEOREM 21. Every R-module is isomorphic to a quotient module of a free R-module. [By what is popularly known as an abuse of the language we often speak of F itself, instead of properly (F, γ), as the free module.]

Proof: Let M be an R-module and let S be a subset of M that generates M. (For instance we can take $S = M$.) There exists a free R-module (F, γ) on S. The U.F.P. with $X = M$ and ϕ the inclusion $S \to M$ yields $\phi = \gamma h$, where h is a unique module homomorphism of $F \to M$. Now S generates M, so that every element $m \in M$ has the form $m = \Sigma x_i r_i$, $x_i \in S$, $r_i \in R$. (Since R is assumed to have an identity, the element m takes this simpler form.)   Here $x_i = x_i \phi (\phi$ is the inclusion). Hence $m = \Sigma (x_i \phi) r_i = \Sigma (x_i \gamma h) r_i = \Sigma ((x_i \gamma) r_i) h$. This proves h to be surjective, since ϕ generates M and $\gamma h = \phi$. Hence M is a homomorphic image of F and therefore M is isomorphic to the quotient module F/N of F, where $N = \ker h$.

Definition. A **basis** of an R-module M is a subset S of M that has the following properties: (i) S is a set of generators of M, (ii) if a linear combination $\sum_{i=1}^{n} s_i r_i$, $s_i \in S$, $r_i \in R$, of elements of S is 0, then each $r_i = 0$.

If S is a basis of M then every element $x \in M$ can be expressed uniquely in the form

$$x = \sum_{i=1}^{k} s_i r_i, \quad s_i \in S, \quad r_i \in R.$$

For if

$$x = \sum_{i=1}^{k} s_i r_i',$$

$r_i' \in R$ (there is no loss of generality in assuming the same s_i occur in both expressions for x, since we can always add terms with zero coefficients) then

$$\sum_{i=1}^{k} s_i r_i - \sum_{i=1}^{k} s_i r_i' = \sum_{i=1}^{k} s_i (r_i - r_i') = 0$$

and hence $r_i = r_i'$, for all i from 1 to k. Thus *a basis of M is a set of linearly independent generators of M*.

THEOREM 22. An R-module M is a free R-module if and only if it has a basis.

Proof: Let $S \subset M$ be a basis of M. Let j be the inclusion $S \to M$. We show (M, j) is the free R-module generated by S. Let (F, γ) be the free R-module generated by S. The U.F.P. of (F, γ) with $X = M$ and $\phi = j$, yields the unique module homomorphism h of $F \to M$ such that $\gamma h = j$. Thus h is the extension of j to F. Since j generates M and γ generates F, it follows from $\gamma h = j$ that h is surjective. Let $x \in \ker h$. Then

$$x = \sum_{i=1}^{n} (s_i \gamma) r_i, \quad s_i \in S, \quad r_i \in R,$$

and

$$0 = xh = \sum_{i=1}^{n} (s_i(\gamma h)) r = \sum_{i=1}^{n} (s_i j) r_i = \sum_{i=1}^{n} s_i r_i.$$

Since S is a basis for M, each $r_i = 0$. Hence $x = 0$ and therefore h is injective. Thus h is an isomorphism and $M = F$. Thus M is the free R-module generated by S.

Conversely, let (M, γ) be a free R-module on S. We show that in $\gamma = S\gamma$ is a basis for M. We know $S\gamma$ generates M and hence all we need to prove is that the set $S\gamma$ is linearly independent. Suppose

$$\sum_{i=1}^{k} (s_i \gamma) r_i = 0, s_i \in S, r_i \in R.$$

Now use the U.F.P. of (M, γ). Take $X = R$ (the ring R is a module over itself) and define a mapping ϕ_i of $S \to R$ by $s_i \phi_i = 1$, $s\phi_i = 0$ for $s \neq s_i$. Then $\gamma h_i = \phi_i$, where h_i is the unique module homomorphism from $M \to R$. Now

$$0 = \sum_{i=1}^{k} (s_i \gamma) r_i) h_i = \sum_{i=1}^{k} (s_i(\gamma h)) r_i = \sum_{i=1}^{k} (s_i \phi_i) r_i = r_i.$$

This is true for each choice of ϕ_i, $i = 1, 2, \ldots, k$ and so each $r_i = 0$, $i = 1, 2, \ldots, k$. Hence $S\gamma$ is a linearly independent set of generators and is therefore a basis of M.

Corollary. A subset S of an R-module M is a basis of M if and only if the inclusion j of $S \to M$ extends to an isomorphism g of $F \to M$, where F is the free R-module generated by S.

Proof: (g is an extension of j if $g = j$ on S.) In the proof of the first part of the theorem, we have seen that this is true if S is a basis.

For the converse, let S be a subset of M and let the inclusion j of $S \to M$ extend to the isomorphism g of $F \to M$, where (F, γ) is the free R-module generated by S. We want to show that S is a basis of M. The U.F.P. of (F, γ) with $X = M$ and $\phi = j$ yields $\gamma h = j$, where, as usual, h is the unique module homomorphism of $F \to M$. But $\gamma g = j$, and hence $h = g$. Thus h is the isomorphism extending j. Since $S\gamma = S$ generates F, $j = \gamma h$ is surjective and hence S generates M. Since M is isomorphic to F, M is free on S and hence S is a basis of M.

We make a few emphases about the facts we have proved for free modules. There exists a free R-module on any set S, whether S is a finite or an infinite set. An R-module is free if and only if it has a basis, and if (F, γ) is a free module on the set S, then $S\gamma$ is a basis of F. As with vector spaces this is a highly useful property of a free module, for it brings an organization to the elements of the module in that it determines their form. Every element of a free module has a unique expression as a finite linear sum of the basis elements with coefficients from the ring. For modules with a finite basis this is proved as easily as in the case of finite-dimensional vector spaces. The generalization to an infinite basis is scarcely more difficult if it is kept in mind that linear independence still means that the vanishing of a finite sum of basis elements implies each ring coefficient is zero.

7-9 MODULES OVER A PRINCIPAL IDEAL DOMAIN

A finitely generated module is not in general a free module, for its generators are not necessarily linearly independent. For example, a cyclic R-module M is generated by a single element $m \in M$—that is, $M = mR$—but it is not a free module unless $mr = 0$ implies $r = 0$.

The direct sum of free modules over R is a free module over R, its basis being the union of the bases of the direct summands.

A submodule of a free module over a ring R is not necessarily a free module. However we shall prove that every submodule of a free module over a principal ideal domain (P.I.D.) is free.

Let M be a module over a P.I.D. D. An element $x \in M$ is said to be of *finite order* or a *torsion element* if $ax = 0$ for some $a \neq 0$ in D. (Since D is commutative it is immaterial whether the scalar product is written as ax or xa.) Since D has an identity 1, this includes the elements which are of finite order as group elements of the additive abelian group M. For if n is an integer and $nx = 0$, $x \in M$, then $(n \cdot 1)x = 0$ and $n \cdot 1 \in D$.

If $x \in M$ is of finite order, then so is ax, where $a \neq 0$ is in D. For if $bx = 0$, $b \neq 0$, then $bax = abx = 0$. If $x, y \in M$ are of finite order,

$ax = 0, a \neq 0$ and $by = 0, b \neq 0$ then $ab(x - y) = 0$ and $ab \neq 0$. Hence $x - y$ is of finite order. This proves that all the elements of finite order of a module M over a P.I.D. form a submodule N of M, called the **torsion submodule** of M.

It is worth noting that if x is an element of a module over a P.I.D. D and if x is not of finite order then, for any $a \neq 0$ in D, ax is not of finite order.

THEOREM 23. Let M be a free module over a P.I.D. D with a finite basis $x_1, x_2, \ldots, x_n$. Then every submodule N of M is free and has a basis of $\leq n$ elements.

Proof: For $i = 1, 2, \ldots, n$ let S_i be the set of all elements of R that occur as coefficients of x_i in the expressions for the elements of N in terms of the basis elements of M. Clearly each S_i is an ideal and hence a principal ideal in D. Let $S_i = (a_i)$, $a_i \in D$. Of course some of the a_i may be 0. Each $S_i x_i$ is a submodule of M (but of course not necessarily of N).

Now $N \subseteq S_1 x_1 + \cdots + S_n x_n$. For $i = 1, 2, \ldots, n$ put

$$N_i = N \cap (S_1 x_1 + \cdots + S_i x_i).$$

$N_1 = N \cap S_1 x_1 = (0)$ or is a submodule generated by an element of the form $bx_1, b \neq 0$, which cannot be of finite order. [If $N \neq (0)$ we can assume that at least one $S_i x_i$ has a zero intersection with N and so by renumbering the basis, if necessary, we are entitled to assume $N_1 \neq (0)$.] Thus N_1 is a free module. Now there exists $z \in N_2$ such that $z = b_1 x_1 + a_2 x_2$, where recall we have $S_2 = (a_2)$. Hence for any $y \in N_2$, $y = c_1 x_1 + c_2 a_2 x_2$, and therefore $y - c_2 z \in N_1$.

Thus $y = (y - c_2 z) + c_2 z$ and therefore $N_2 = N_1 + (z)$. Note (z) is a free module, since it is generated by an element that is not of finite order. If $a_2 = 0$, $N_2 = N_1$ and if $a_2 \neq 0$ then $N_1 \cap (z) = (0)$, and we have $N_2 = N_1 \oplus (z)$. If $N_2 = N_1$, then N_2 is free and if

$$N_2 = N_1 \oplus (z),$$

then N_2 is also free, since it is the direct sum of free modules.

All this suggests that, since N_1 is free, we use induction on i, that is assume N_i is free and show N_{i+1} is free. The method of doing this is precisely the same as above and the details are left as an exercise for the reader.

To conclude, we can therefore prove by induction that N_n is free and, since $N_n = N$, the theorem is proved.

Corollary. A submodule N of a finitely generated module M over a P.I.D. is finitely generated.

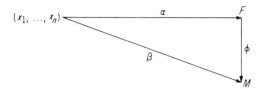

Proof: Let $M = (x_1, x_2, \ldots, x_n)$, and let F be the free module generated by the set $\{x_1, x_2, \ldots, x_n\}$.

The U.F.P., with α and β the inclusion mappings, yields a unique module homomorphism ϕ which is clearly surjective. Now $\phi^{-1}N$ is a submodule of F and hence is free and finitely generated. Since ϕ is an epimorphism, this forces N to be finitely generated.

We state here (without the evidence) that it can be proved that a finitely generated module M over a P.I.D. is the direct sum of its torsion submodule N and a free submodule F, $M = N \oplus F$. This theorem can be refined to the main theorem of elementary divisor theory which states (again presented here without proof): if $M = (x_1, \ldots, x_n)$ is a finitely generated module over a P.I.D., then $M = M_1 \oplus M_2 \oplus \cdots \oplus M_k$, where the M_i are cyclic submodules of M and $k \leq n$. If for

$$i = 1, 2, \ldots, k,$$

we denote by y_i the generator of M_i, then those y_i which are of finite order generate cyclic submodules that are not free (for these y_i are not bases), while those y_i that are not of finite order generate free cyclic submodules. The direct sum of the free cyclic submodules is the free submodule F above and the direct sum of the cyclic submodules generated by elements of finite order is the torsion submodule N.

The properties which a finitely generated R-module may have are seen to depend not only on the nature of the module as a group (for instance that it may be periodic, torsion-free or mixed or even a finite group) but on the kind of ring R is.

It is a theorem in elementary linear algebra that any finitely generated vector space over a field or over a division ring has a basis. In fact it has been proved that any vector space over a field or over a division ring has a basis. Consequently all vector spaces are free modules over a field or over a division ring.

EXERCISE

1. Let V be a vector space over a field F and let R be the P.I.D., $R = F[x]$. If T is an endomorphism of V, then V, with the use of T, can be regarded as a left R-module in the following way.

Define a scalar multiplication in V by

$$f(x)\alpha = \alpha f(T), \quad f(x) \in F[x], \quad \alpha \in V.$$

For example, if $f(x) = a_n x^n + a_{n-1} x^{n-1} + \cdots + a_0$, then

$$\alpha f(T) = \alpha[a_n T^n + a_{n-1} T^{n-1} + \cdots + a_0 I]$$

where I is the identity endomorphism of V, $T^n = T(T^{n-1})$, $\alpha(cT) = c(\alpha T)$ for $c \in F$. Prove V is an R-module.

This device of combining a vector space with an endomorphism and creating a module over a P.I.D. is very important in group theory. An infinite abelian group is a module over the P.I.D. of the integers. It can be proved that many structure theorems for infinite abelian groups, notably primary groups (each element of the group has an order which is a power of the same prime p), are valid for modules over an arbitrary P.I.D.

Naturally the scalar multiplication in the module over $F[x]$ varies with the choice of T. Look at the cases where (a) $T = I$, (b) T is a projection, $T^2 = T$, and (c) T is an endomorphism such that $T^2 = 0$.

7-10 PROJECTIVE MODULES

We next prove another important property of a free module. In homology a type of module called a projective module is of great importance.

Definition. An R-module P is called **projective** if and only if for any homomorphism α of P into an R-module B and for any epimorphism (surjective homomorphism) β of an R-module A onto B, there exists a homomorphism h of P into A such that $\alpha = h\beta$. This is expressed by saying that the diagram is commutative. The homomorphism α is described as being "lifted" to the homomorphism h.

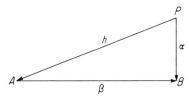

While a projective module is not necessarily free, we can prove

THEOREM 24. A free module is projective.

Proof: Let (F, γ) be a free R-module on a set S. Let α be a homomorphism of F into a module B and β an epimorphism of a module A onto B. Now the composite map $\gamma\alpha$ is a mapping of $S \to B$. Since β is surjective, for each $x \in S$ we can determine an $a \in A$ such that $a\beta = x(\gamma\alpha)$. Define a mapping τ of $S \to A$ by $x\tau = a$. The U.F.P. of (F, γ) with $X = A$ and $\phi = \tau$ yields a unique homomorphism h of $F \to A$ such that $\gamma h = \tau$. Hence we have

$$x(\gamma h\beta) = x(\tau\beta) = (x\tau)\beta = a\beta = x(\gamma\alpha).$$

Therefore $(\chi\gamma)(h\beta) = (x\gamma)\alpha$ for every $x\gamma$ of F. Now $h\beta$ is a homomorphism and the $x\gamma$ generate F, hence $h\beta = \alpha$ on all of F. Hence h is the homomorphism from $F \to A$ whose existence proves F to be a projective module.

7-11 THE TENSOR PRODUCT

The tensor product of two R-modules is an R-module constructed from the two given modules. It is of interest as an example of a new module created out of two given ones, but it is also very important for its many applications, including assigning a satisfactory definition to a tensor. We shall make extensive use of the tensor product in the next chapter and so perhaps it is unnecessary at this time to further motivate its study.

Definition. Let R be a commutative ring with the identity 1. Let A, B, C be unital right R-modules. A mapping f of the Cartesian product $A \times B \to C$ is called **bilinear** if it has the following two properties:

(i) For any $a_1, a_2 \in A$, $b_1, b_2 \in B$, and $r_1, r_2 \in R$

$$(a_1 r_1 + a_2 r_2, b)f = ((a_1, b)f)r_1 + ((a_2, b)f)r_2,$$

(ii) $(a, b_1 r_1 + b_2 r_2)f = ((a, b_1)f)r_1 + ((a, b_2)f)r_2,$

that is, if and only if $(a, b)f$, $a \in A$, $b \in B$, is linear in a when b is fixed and linear in b when a is fixed. Since for $r, s \in R$ we have

$$(ar, bs)f = ((a, bs)f)r = (((a, b)f)s)r = ((a, b)f)sr,$$
$$(ar, bs)f = ((ar, b)f)s = (((a, b)f)r)s = ((a, b)f)rs,$$

we see that in general R needs to be a commutative ring for a bilinear mapping. The image im $f = (A \times B)f$ of a bilinear mapping is not, in general, an R-module. A bilinear mapping is the simplest example of what are called **multilinear mappings.**

We assume throughout the rest of the chapter that the ring R is commutative, has an identity, and that all modules are unital.

Definition. Let A and B be two modules over a commutative ring R with identity. The **tensor product** of A and B is the pair (T, α) where T is an R-module and α is a bilinear mapping of $A \times B \to T$ such that for each choice of an R-module X and of a bilinear mapping ϕ of $A \times B \to X$, there exists a *unique* module homomorphism h of $T \to X$ for which $\alpha h = \phi$.

This means we are requiring the accompanying diagram to be commutative, and we can describe this by saying that the pair (T, α) has the universal factorization property (U.F.P.) on $A \times B$, universal for all

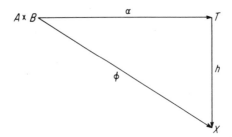

R-modules X and all bilinear mappings of $A \times B \to X$. This is to be understood to mean that every bilinear mapping ϕ from $A \times B \to X$ factors or splits into αh and this factoring is unique for each X and each ϕ. We mention here that T itself is often referred to as the tensor product.

Again we shall need to demonstrate the existence of a tensor product and this is again done by actual construction of it, as was done in the case of a free module. There are similarities in the concepts of free module and tensor product (T, α) and one would look for some similarity in their properties.

THEOREM 25. im $\alpha = (A \times B)\alpha$ generates the module T of the tensor product (T, α) of A and B.

> **Proof:** Let T' be the submodule of T generated by im α. The U.F.P. of (T, α) with $X = T'$ and $g = \alpha$ yields a unique module homomorphism h of $T \to T'$ such that $\alpha = \alpha h$. The U.F.P. of (T, α) with $X = T$ and $g = \alpha$ yields the unique module homomorphism I_T (the identity mapping of T on itself) such that $\alpha = \alpha I_T$. Since T' is a submodule of T, the homomorphism h can be regarded as a module homomorphism of $T \to T$ such that $\alpha = \alpha h$. From the uniqueness property therefore $h = I_T$ and hence $T' = T$.

The condition that each bilinear mapping ϕ of $A \times B \to X$ can be factored uniquely into $\phi = \alpha h$ is equivalent to the condition that im α generates T. The last theorem shows that the uniqueness condition implies that im α generates T. For the converse, assume that im α generates T, then from $\phi = \alpha h$, we see that the homomorphism h is defined by means of ϕ on all the generators of T. Hence h is uniquely defined on all of T, for a module homomorphism on T is uniquely determined by its values on a set of generators of T.

THEOREM 26. The tensor product (T, α) of A and B is unique up to isomorphism.

> **Proof:** Let (T', α') be a tensor product of A and B. We show there exists a unique module isomorphism h of $T \to T'$ such that

$\alpha' = \alpha h$. The U.F.P. of (T, α) with $\phi = \alpha'$ and $X = T'$ yields a unique module homomorphism of $T \to T'$ such that $\alpha h = \alpha'$. Similarly the U.F.P. of (T', α') with $\phi = \alpha$ and $X = T$ yields a unique module homomorphism h' such that $\alpha' h' = \alpha$. Hence $\alpha = \alpha h h'$ and therefore $h h' = I_T$. Similarly $h' h = I_T$. Hence h is an isomorphism and of course unique, since h is unique.

Thus if (T, α) exists it is essentially unique. Note that in contrast to the free module (F, γ) the mapping α in the tensor product (T, α) is not injective. This will become apparent when we construct (T, α).

THEOREM 27. The tensor product (T, α) of two modules A and B over a commutative ring R with identity exists.

Proof: Let (F, γ) be the free R-module on the set $A \times B$. Let G be the submodule of F generated by all elements of F of the form

$$g_1 = (a + a', b)\gamma - (a, b)\gamma - (a', b)\gamma$$
$$g_2 = (a, b + b')\gamma - (a, b)\gamma - (a, b')\gamma$$
$$g_3 = (ar, b)\gamma - ((a, b)\gamma)r$$
$$g_4 = (a, br)\gamma - ((a, b)\gamma)r.$$

Form the quotient module $T = F/G$ and define the mapping α of $A \times B \to T$ by $\alpha = \gamma j$, where j is the natural epimorphism of $F \to F/G = T$. Then (T, α) is the tensor product of A and B. To prove this we must show that (T, α) has the U.F.P. on $A \times B$.

We first point out that $g_i j = G$, the zero element of the module F/G, and so we write $g_i j = 0$, $i = 1, 2, 3, 4$. Now $\alpha = \gamma j$ is a bilinear mapping. This follows at once from the definition of the generators of G. For example,

$$0 = g_1 j = ((a + a', b)\gamma)j - ((a, b)\gamma)j - ((a', b)\gamma)j,$$

that is,

$$(a + a', b)\alpha = (a, b)\alpha + (a', b)\alpha.$$

Similarly,

$$(a, b + b')\alpha = (a, b)\alpha + (a, b')\alpha, \quad (ar, b)\alpha = ((a, b)\alpha)r$$

and

$$(a, br)\alpha = ((a, b)\alpha)r.$$

Thus α is bilinear.

Now let ϕ be any bilinear mapping of $A \times B$ into an arbitrary R-module X. We need to show there exists a unique homomorphism H from T to X such that $\alpha H = \phi$. The U.F.P. of the free R-module (F, γ) on $A \times B$ yields a unique module homomorphism h such that

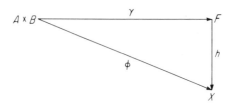

$\phi = \gamma h$. Define a mapping H from $T = F/G \rightarrow X$ by $jH = h$. Since j is surjective, this defines H on all of F/G. Clearly H is a module homomorphism. Also $\phi = \gamma jH = \alpha H$. To see this, take $x = (a, b)\gamma \in F$. Then $xh = (a, b)\gamma jH = (a, b)\alpha H$ and $xh = (a, b)\gamma h = (a, b)\phi$. Hence $(a, b)\alpha H = (a, b)\phi$, and this is true for all $(a, b) \in A \times B$. Hence $\alpha H = \phi$ on $A \times B$. Moreover, H is unique. For let $\alpha H' = \phi$, where H' is a module homomorphism from $T = F/G \rightarrow X$. Take $x = (a, b)\gamma \in F$. Then $(x + G)H' = (xj)H' = ((a, b)\gamma)jH' = ((a, b)\alpha)H' = (a, b)\phi = ((a, b)\alpha)H = (x + G)H$. This is true for all $x + G \in F/G$ and hence $H' = H$. This completes the proof that (T, α) is the tensor product of A and B.

Note that an element of the tensor product $T = F/G$ is an equivalence class (a coset of G in F) of elements of F, two elements being in the same coset if and only if their difference is an element of G. Thus under the natural homomorphism j of $F \rightarrow T = F/G$, two elements in the same coset map into the same element of T. The mapping j is surjective but is not injective. Hence the mapping $\alpha = \gamma j$ is not injective. This of course is true of the mapping α' in every tensor product (T', α') on the same set S, since $\alpha' = \alpha h$, where h is an isomorphism.

We now introduce a notation $\otimes$ for the tensor product. We write $a \otimes b = (a, b)\alpha$ and call it the **tensor product** of a and b. Write $A \otimes B = T$ for the tensor product of A and B, which will exhibit its dependence on A and B. As we have pointed out the elements $a \otimes b$ of $A \otimes B$ are actually equivalence classes of elements of the free R-module F on the set $A \times B$. The elements $(a, b)\gamma$ are known to be a basis of F, and since j is the natural epimorphism of $F \rightarrow A \otimes B$, it is surjective. Hence the elements $(a, b)\gamma j = a \otimes b$, $a \in A$ and $b \in B$, form a set of generators, but not a basis of $A \otimes B$ (unless $A = B = 0$, in which case j is an isomorphism). Thus an arbitrary element x of $A \otimes B$ has the form

$$x = \sum_{i=1}^{m} (a_i \otimes b_i) r_i$$

where $a_i \in A, b_i \in B, r_i \in R$.

The following rules of computation for tensor products of elements result from the expressions for the generators g_i of G. Apply $\alpha = \gamma j$ to

One can show that the Z-modules Z_n and Z_m that $Z_m \otimes Z_m \cong Z_\ell$

To see this show map $a \otimes b \xrightarrow{\quad} ab \pmod{\ell}$ is an isom. where $\ell = gcd(m,n)$

This follows (using cony next page) since $(a,b) \to (ab)\pmod{\ell}$ is bilinear and since if map $(ab)\pmod{\ell} \xrightarrow{\quad} 1 \otimes (ab)\pmod{\ell} = a \otimes b$ from $Z_\ell \to Z_n \times Z_m$ we get $L'L = 1_{Z_\ell}$ and $L'L = 1_{Z_n \otimes Z_m}$

g_1, g_2, g_3, g_4 in succession and we get

To see this note if $a \in Z_n$ and $a = g\ell + n$ then since $\ell = g_1 n + g_2 m$ we get using this that $a \otimes b = n \otimes b$

(i) $(a_1 + a_2) \otimes b = a_1 \otimes b + a_2 \otimes b$.

(ii) $a \otimes (b_1 + b_2) = a \otimes b_1 + a \otimes b_2$.

(iii) $(ar) \otimes b = (a \otimes b)r, r \in R$.

(iv) $a \otimes (br) = (a \otimes b)r$.

Hence $(na) \otimes b = n(a \otimes b) = a \otimes (nb)$, where n is an integer. In particular $0 \otimes b = 0 = a \otimes 0$ and $(-a) \otimes b = -(a \otimes b) = a \otimes (-b)$.

From the relations (iii) and (iv) we see that every element x of $A \otimes B$ can actually be put in the form

$$(1) \qquad x = \sum_{i=1}^{m} a_i \otimes b_i, \quad a_i \in A, \; b_i \in B,$$

where the elements r_i of the ring have been absorbed to form merely elements of A and B. Note also that the above relations plainly show that expression (1) for x is not unique.

7-12 MODULE HOMOMORPHISMS ON TENSOR PRODUCTS

A module homomorphism on $A \otimes B$ is uniquely determined by its values on the set $a \otimes b$ of generators of $A \otimes B$. We next derive a means of determining all module homomorphisms on the tensor product $A \otimes B$.

THEOREM 28. The set $\mathcal{L}$ of all module homomorphisms of $M \to M'$ where M and M' are R-modules is an R-module, under the usual definitions of addition and scalar multiplication of mappings.

Proof: This is so similar to previous proofs that it is left as an exercise.

THEOREM 29. With the same definitions of addition and scalar multiplication of mappings, the set $\mathcal{B}$ of all bilinear mappings of $A \times B \to X$, where A, B, X are R-modules, is an R-module.

Proof: This is again left as an exercise.

THEOREM 30. Let A, B, and X be R-modules. Let $\mathcal{L}$ be the R-module of all module homomorphisms of $A \otimes B \to X$ and let $\mathcal{B}$ be the R-module of all bilinear mappings of $A \times B \to X$. Then $\mathcal{L}$ and $\mathcal{B}$ are isomorphic.

Proof: Let the mapping f of $\mathcal{L} \to \mathcal{B}$ be defined by

$$Lf = \alpha L, \quad L \in \mathcal{L},$$

where $(A \otimes B, \alpha)$ is the tensor product of A and B. Since α is bilinear, αL is bilinear and therefore $\alpha L \in \mathcal{B}$. We show that f is a module isomorphism. If $\phi \in \mathcal{B}$ then the U.F.P. of $(A \otimes B, \alpha)$ on the set $A \times B$, with X and ϕ, yields a unique module homomorphism L of $A \otimes B \to X$ such that $\phi = \alpha L$. This proves f to be surjective. Next, if $L \in \ker f$, $\alpha L = 0$. Since $a \otimes b = (a, b)\alpha$ generates $A \otimes B$, $L = 0$ on the generators of $A \otimes B$ and hence, being a homomorphism, $L = 0$ on all of $A \otimes B$. Hence f is injective. That f is a module homomorphism is very easy to verify, and therefore f is a module isomorphism. Hence $\mathcal{L}$ and $\mathcal{B}$ are isomorphic.

Thus a bilinear mapping ϕ of $A \times B \to X$ determines a linear mapping L of $A \otimes B \to X$ defined by $(a \otimes b)L = (a, b)\phi$ and conversely. Moreover the linear mapping L is uniquely determined by ϕ. For the elements $a \otimes b$ generate $A \otimes B$ and a linear mapping is uniquely determined by its values on a set of generators.

An immediate consequence of this last theorem is the following simple and important test for a mapping on $A \otimes B$ to be a module homomorphism:

Corollary. There exists a module homomorphism L from $A \otimes B \to X$, having specified values $(a \otimes b)L$ on the generators $a \otimes b$ of $A \otimes B$ if and only if the function ψ of $A \times B \to X$ defined by $(a, b)\psi = (a \otimes b)L$ is a bilinear mapping.

Example. Let R be a commutative ring with identity, then R is an R-module. The mapping L of $A \otimes R \to A$ defined by $(a \otimes r)L = ar$ is a module homomorphism, since $(a, r)\psi = ar$ is a bilinear mapping from $A \times R \to A$.

THEOREM 31. If $A = (a_1, a_2, \ldots, a_n)$ and $B = (b_1, \ldots, b_m)$ are finitely generated modules over a commutative ring R with identity, then their tensor product $A \otimes B$ is a finitely generated R-module.

Proof: It is easy to verify that the mn elements $a_i \otimes b_j$, $i = 1, 2, \ldots, n$ and $j = 1, 2, \ldots, m$, of $A \otimes B$ generate $A \otimes B$.

Lemma. For any R-module M we have $R \otimes M \approx M \approx M \otimes R$.

Proof: For $r \in R$, $m \in M$, $(r \otimes m)L = rm$ is a module homomorphism of $R \otimes M \to M$. Define a mapping L' of $M \to R \otimes M$ by $mL' = 1 \otimes m$. Then $L'L = 1_M$ and also $LL' =$

$1_{R \otimes M}$. Hence L is both surjective and injective and is therefore an isomorphism. Similarly, we can show $M \approx M \otimes R$.

As is customary we can identify these modules and write the lemma as

$$R \otimes M = M = M \otimes R.$$

A word or two about such identifications. If an isomorphism of two modules M and M' is not dependent on some particular choice of bases, for instance, or sets of generators, but is unique in the sense that the definition of the isomorphism is independent of any free choices of objects, then M and M' are said to be canonically or naturally isomorphic. When two modules M and M' are canonically isomorphic, we frequently write $M = M'$.

7-13 TENSOR PRODUCT OF A FINITE NUMBER OF MODULES

We next consider the extension of the tensor product to three modules and then ultimately to any finite number of modules.

THEOREM 32. Let A, B, C be three unital modules over the same commutative ring R with identity element 1. Then

$$(A \otimes B) \otimes C = A \otimes (B \otimes C).$$

Proof: Let $A \times B \times C$ be the cartesian product of the sets A, B, and C. Define a mapping f_1 of $A \times B \times C \to A \otimes (B \otimes C)$ by $(x, y, z)f_1 = x \otimes (y \otimes z)$, where $x \in A$, $y \in B$, $z \in C$. Then f_1 is seen to be a trilinear mapping (an obvious extension of a bilinear mapping). For fixed $z \in C$, f_1 becomes a bilinear function on $A \times B$ and hence determines a unique linear mapping (that is, module homomorphism) on $(A \otimes B) \otimes C \to A \otimes (B \otimes C)$, where C is held to z. Now freeing z (that is allowing it to run through all of C) we have a bilinear function on $(A \otimes B) \times C$ and this in turn determines a unique linear mapping f of

$$(A \otimes B) \otimes C \to A \otimes (B \otimes C) \text{ with } [(x \otimes y) \otimes z]f = x \otimes (y \otimes z).$$

In the same way a linear mapping g of $A \otimes (B \otimes C) \to (A \otimes B) \otimes C$ is determined, such that $[x \otimes (y \otimes z)]g = (x \otimes y) \otimes z$. Now the composite mappings fg and gf are seen to be the identity mappings and hence f and g are both isomorphisms. We now identify the two modules $(A \otimes B) \otimes C$ and $A \otimes (B \otimes C)$, since these isomorphisms are canonical.

The Tensor Product of Three Modules

Let A, B, C be three given R-modules, and let W be any R-module. Let S be a trilinear mapping of $A \times B \times C \to W$. Let $x \in A$, $y \in B$,

$z \in C$. Then if z is held fixed, S is a bilinear mapping of $A \times B \rightarrow W$ and hence determines a unique linear mapping f_1 of $A \otimes B \rightarrow W$. This function f_1 is a bilinear mapping of $(A \otimes B) \times C \rightarrow W$ and hence again in turn determines a unique linear mapping f of $A \otimes B \otimes C \rightarrow W$ such that $(x \otimes y \otimes z)f = (x, y, z)S$. Thus if τ is the trilinear mapping of $A \times B \times C \rightarrow A \otimes B \otimes C$ for which $(x, y, z)\tau = x \otimes y \otimes z$, then the pair $(A \otimes B \otimes C, \tau)$ has the U.F.P. on $A \times B \times C$, universal for all trilinear mappings S of $A \times B \times C$ into the arbitrarily chosen module W. For we have seen that S determines a unique module homomorphism f of $A \otimes B \otimes C \rightarrow W$ such that $S = \tau f$. We state this as a theorem.

THEOREM 33. The tensor product of three modules exists.

An important and very useful consequence of this theorem is that if A, B, C, and W are modules, then a mapping f of $A \otimes B \otimes C \rightarrow W$ is linear if and only if the mapping g of $A \times B \times C \rightarrow W$ defined by $(x, y, z)g = (x \otimes y \otimes z)f$, $x \in A$, $y \in B$, $z \in C$, is trilinear. This is proved in the same way as in the corollary to Theorem 30 for the bilinear case. In applications the most useful form of this result is that if g is a trilinear mapping of $A \times B \times C \rightarrow W$, then g determines a unique linear mapping f of $A \otimes B \otimes C \rightarrow W$, where $(x \otimes y \otimes z)f = (x, y, z)g$, for $x \in A$, $y \in B$, $z \in C$. That f is unique follows from the fact that the elements $x \otimes y \otimes z$ generate $A \otimes B \otimes C$ and a linear mapping is uniquely determined by its values on a set of generators.

With this last theorem and the use of induction, the general associative law for tensor products can be proved.

THEOREM 34. The tensor product of any finite number n of modules, $n \geq 2$, exists.

Proof: We use induction on n and assume the theorem true for $n - 1$. Let $(x_1, \ldots, x_n)S = z$ be a multilinear mapping of $M_1 \times M_2 \times \cdots \times M_n$ into any module Z. Holding x_n fixed, S is multilinear on $M_1 \times \cdots \times M_{n-1}$ to Z and so, by the induction hypothesis, determines a unique linear mapping f_1 of

$$M_1 \otimes \cdots \otimes M_{n-1} \rightarrow Z$$

such that

$$(x_1 \otimes \cdots \otimes x_{n-1}, x_n)f_1 = (x_1, \cdots, x_{n-1}, x_n)S = z.$$

Now free x_n, then f_1 is a bilinear mapping of

$$(M_1 \otimes \cdots \otimes M_{n-1}) \times M_n \rightarrow Z$$

and determines in turn a unique linear mapping f of $M_1 \otimes \cdots \otimes M_n \to Z$ such that

$$(x_1 \otimes \cdots \otimes x_n)f = (x_1 \otimes \cdots \otimes x_{n-1}, x_n)f_1$$
$$= (x_1, \ldots, x_n)S = z.$$

Hence if τ is the mapping of $M_1 \times \cdots \times M_n \to M_1 \otimes \cdots \otimes M_n$ defined by $(x_1, \ldots, x_n)\tau = x_1 \otimes \cdots \otimes x_n$, then $S = \tau f$ and therefore the pair $(M_1 \otimes \cdots \otimes M_n, \tau)$ has the U.F.P. for multilinear mappings.

There follows in the same way from this theorem a result that is most useful to us in defining algebras in the next chapter. If g is a multilinear mapping of $M_1 \times M_2 \times \cdots \times M_n \to Z$ then g determines a unique linear mapping f of $M_1 \otimes \cdots \otimes M_n \to Z$ given by

$$(x_1 \otimes x_2 \otimes \cdots \otimes x_n)f = (x_1, x_2, \ldots, x_n)g.$$

For again the elements

$$x_1 \otimes \cdots \otimes x_n, x_i \in M_i,$$

generate the tensor product of the M_i.

7-14 TENSOR PRODUCT FOR A DIRECT SUM

THEOREM 35. If M_1, M_2, N are R-modules and if $M = M_1 \oplus M_2$ then

$$M \otimes N = M_1 \otimes N \oplus M_2 \otimes N.$$

Proof: Each $u \in M$ has a unique representation in the form

$$u = u_1 + u_2, u_1 \in M_1, u_2 \in M_2.$$

For $i = 1, 2$ let f_i be the inclusion homomorphism of $M_i \otimes N \to M \otimes N$ and p_i the projective homomorphism of $M \otimes N \to M_i \otimes N$. Thus for $v \in N$, $(u_i \otimes v)f_i = u_i \otimes v$ and $(u \otimes v)p_i = u_i \otimes v, u_i \in M_i$.

Put $S = M_1 \otimes N \oplus M_2 \otimes N$. Each $x \in S$ has a unique representation in the form $x = x_1 \oplus x_2$, $x_1 \in M_1 \otimes N$ and $x_2 \in M_2 \otimes N$.

Define a homomorphism h of $S \to M \otimes N$ by

$$xh = x_1 f_1 + x_2 f_2.$$

For $u = u_1 + u_2 \in M$ and $v \in N$, the mapping k of $M \otimes N \to S$ defined by

$$(u \otimes v)k = u_1 \otimes v + u_2 \otimes v$$

is a homomorphism.

We are going to prove that $kh = 1_{M \otimes N}$ and $hk = 1_S$. We have

$$(u \otimes v)kh = (u_1 \otimes v)h + (u_2 \otimes v)h$$
$$= (u_1 \otimes v)f_1 + (u_2 \otimes v)f_2$$
$$= u_1 \otimes v + u_2 \otimes v$$
$$= (u_1 + u_2) \otimes v = u \otimes v.$$

Hence $kh = 1_{M \otimes N}$. Next let $x = x_1 + x_2 \in S$. Then

$$xhk = x_1 f_1 k + x_2 f_2 k$$
$$= x_1 k + x_2 k = x_1 + x_2 = x.$$

Hence $kh = 1_S$.

This proves h and k are inverses and hence they are canonical isomorphisms.

Similarly we can prove

$$N \otimes M = N \otimes M_1 \oplus N \otimes M_2.$$

The next theorem is an important and interesting application of Theorem 35.

Lemma. If V and W are one-dimensional vector spaces over a field F, then $V \otimes W$ is a one-dimensional vector space.

Proof: Let α be a basis for V and β a basis for W. Every vector of $V \otimes W$ has the form $c(\alpha \otimes \beta)$ where $c \in F$. Hence $\dim V \otimes W$ is either 0 or 1. The mapping $c(\alpha \otimes \beta) \to c$ is a surjective linear transformation of $V \otimes W \to F$. Since $\dim F = 1$, it follows that $\dim V \otimes W = 1$ and $\alpha \otimes \beta$ is a basis for $V \otimes W$.

THEOREM 36. If V is an m-dimensional vector space with basis $\alpha_1, \ldots, \alpha_m$ and W is an n-dimensional vector space with basis $\beta_1, \ldots, \beta_n$, then the vectors $\alpha_i \otimes \beta_j$, $i = 1, 2, \ldots, m$, $j = 1, 2, \ldots, n$ form a basis for $V \otimes W$ and

$$\dim V \otimes W = mn.$$

Proof: For $i = 1, 2, \ldots, m$ let V_i be the one-dimensional vector space generated by α_i and for $j = 1, 2, \ldots, n$ let W_i be the one-dimensional vector space generated by β_i. Then

$$V = V_1 \oplus V_2 \oplus \cdots \oplus V_m$$
$$W = W_1 \oplus W_2 \oplus \cdots \oplus W_n.$$

By repeated use of Theorem 35, we have

$$V \otimes W = \sum_{j=1}^{n} \sum_{i=1}^{m} \oplus V_i \otimes W_j.$$

The sum is direct and, by the previous lemma, the dimension of each direct summand is 1. Hence

$$\dim V \otimes W = mn.$$

Since $\alpha_i \otimes \beta_j$ is a basis for each $V_i \otimes W_j$ it follows that this set of vectors is a basis for $V \otimes W$.

EXERCISES

1. Prove that the set A of all matrices of the form $\begin{bmatrix} x & 0 \\ 0 & 0 \end{bmatrix}$ over a field F is a noncommutative ring with no identity element. Show that A has infinitely many left identities but no right identity.

Prove that the set M of all matrices of the form $\begin{bmatrix} u & 0 \\ 0 & 0 \end{bmatrix}$ is a left A-module but not a right A-module.

2. Prove that the module endomorphisms of an irreducible R-module M form a division ring $\Delta(M)$.

3. Prove that M is a right vector space over $\Delta(M)$.

4. N is a submodule of a left R-module M and S is a subset of M. Define

$$(N:S) = \{r \in R \mid rs \in N \text{ for all } s \in S\}.$$

Prove $(N:S)$ is a left ideal in R. If $m \in M$, prove $(0:m)$ is a left ideal in R. Prove $(0:M) = \bigcap_{m \in M} (0:m)$.

5. If S and T are submodules of an R-module M, prove

$$S/S \cap T \approx S + T/T.$$

6. If N is a submodule of a right R-module M and if $m \in M$, prove

$$N + mR/N \approx R/(N:mR).$$

7. Let Hom (M, M') be the set of all module homomorphisms $M \rightarrow M'$ of two right R-modules M and M'. Show that under addition of mappings Hom (M, M') is an abelian group.

8. If $f \in$ Hom (M, M') and $r \in R$, define fr by $m(fr) = (mf)r$. Prove that this will convert Hom (M, M') into a right (or left) R-module, provided it is assumed that R is a commutative ring.

9. A linear form f on a right R-module M is a linear mapping of $M \rightarrow R$. If R is a commutative ring, prove the set Hom (M, R) is an R-module. It is called the **dual** of M and is usually denoted by M^*.

10. Show that an R-module M can be regarded as an abelian group with the ring R as its set of operators.

11. M is a right R-module where R is an arbitrary ring (not necessarily commutative) with identity. Prove that the set of all group endomorphisms T_r

of M, defined by $mT_r = mr$, $r \in R$, form a ring under map addition and composition; that is, first show $T_r + T_s = T_{r+s}$ and $T_r T_s = T_{rs}$.

12. A representation ρ of a ring R is a homomorphism of R into the ring of endomorphisms of an abelian group. Prove that if M is a right R-module $r\rho = T_r$, $r \in R$, (where $mT_r = mr$, $m \in M$) is a representation of R. Find the kernel of ρ.

13. If A is a left ideal in an arbitrary ring R, prove that R/A is a left R-module. Note that R/A is not a ring.

14. If A is a maximal left ideal (that is A is not properly contained in any other proper left ideal) prove that the left R-module R/A has no proper submodules.

15. A left ideal A in an arbitrary ring R is called modular if there exists $b \in R$ such that $xb - x \in A$ for all $x \in R$. (If R has an identity then all left (and right) ideals are modular.) If A is a maximal left ideal that is also modular prove that R/A is an irreducible R-module.

16. M is an irreducible left R-module. Hence for an $m \neq 0 \in M$, $M = Rm$. Fix this m. Let A be the left ideal in R for which $A_m = 0$—that is, $A = (0:m)$.
   (a) By assuming B is a left ideal properly containing A (that is, $B \supset A$ and $B \neq A$), show that $B = R$ and hence that A is a maximal left ideal.
   (b) For some $b \in R$, $m = bm$; hence prove that A is a modular left ideal.
   (c) Now prove that the mapping ψ of $R/A \to M$ defined by $(r + A)\psi = rm$, $r \in R$, is an isomorphism.
   This proves that if M is an irreducible R-module then there exists a modular maximal left ideal A such that $M \approx R/A$.

17. Let A be a modular maximal left ideal in R. Then if M is a left R-module which is isomorphic to the left R-module R/A, prove that M is an irreducible R-module. Combining this result with that of the previous exercise, it follows that an R-module M is irreducible if and only if there exists a modular maximal left ideal A in R such that $M \approx R/A$.

18. The *(Jacobson) radical $J(R)$* of an arbitrary associative ring R is defined by

$$J(R) = \cap (0:M),$$

where the intersection is taken over the annihilators (see Section 5 exercise 1) of all the irreducible R-modules M. If there are no irreducible R-modules $J(R)$ is defined by $J(R) = R$. Prove $J(R)$ is an ideal in R.

19. If R is a commutative ring with identity 1, then all left and right ideals are ideals. Moreover all ideals are now modular. Hence every irreducible R-module is isomorphic to an R/A for some maximal ideal A in R. Then $x \in J(R)$ if and only if $R/Ax = A$; that is, if and only if $Rx \subset A$, for every maximal ideal A in R. Prove $J(R) = \cap A$ where the intersection is taken over all maximal ideals A.

20. Prove by induction on the number of generators that a finitely generated module over a Noetherian ring is a Noetherian module (that is, every submodule is also finitely generated).

21. If $(M_1 \otimes M_2, \rho)$ is the tensor product of the modules M_1 and M_2, prove that the homomorphism h in the U.F.P. is unique if and only if im ρ is a set of generators of $M_1 \otimes M_2$.

22. Let M be a right module and L a left module over an arbitrary ring R with identity. Denote by A the free abelian group generated by the set $M \times L$ and let B be the subgroup of A generated by all elements of the form

$$(x_1 + x_2, y) - (x_1, y) - (x_2, y)$$

$$(x, y_1 + y_2) - (x, y_1) - (x, y_2)$$

$$(xr, y) - (x, ry),$$

where $x, x_1, x_2 \in M$, $y, y_1, y_2 \in L$ and $r \in R$. The tensor product $M \otimes L$ is defined as the abelian quotient group

$$M \otimes L = A/B.$$

Denote by $x \otimes y$ the coset $(x, y) + B$, $x \in M$, $y \in L$ and let j be the natural epimorphism of $A \to M \otimes L$.

If G is any abelian group and g is any bilinear mapping of $M \times L \to G$, prove the U.F.P. for $(M \otimes L, ij)$, where i is the inclusion mapping of $M \times L \to A$; that is, prove there exists a unique group homomorphism h of $M \otimes L \to G$ such that $g = ijh$.

23. Let R be a commutative ring with identity. The abelian group $M \otimes L$ in Exercise 21 can be given the structure of an R-module if a valid definition of the product $r(x \otimes y), r \in R$, can be given.

For fixed $r \in R$, define the mapping ρ_r of $M \times L \to M \otimes L$ by $(x, y)_{\rho_r} = xr \otimes y$. Show that ρ_r is bilinear. Then prove there exists a unique group endomorphism h_r of $M \otimes L$ such that $(x \otimes y)h_r = xr \otimes y$. If we now define

$$r(x \otimes y) = xr \otimes y = x \otimes ry,$$

prove that $M \otimes L$ becomes an R-module.

24. Let M and L be modules over a commutative ring R with identity. If $(M \otimes L, f)$ is the tensor product of M and L show $M \otimes L$ is not a free module. Note that it is a quotient module of a free module.

chapter 8

Algebras

The module, together with the group, constitute two of the most important and fundamental algebraic systems. We have seen enough evidence of this that it hardly needs any further substantiation. There are special kinds of modules called algebras, and it is the study of these that is our next concern.

Throughout this chapter R will always denote a commutative, associative ring with an identity 1. Since R is commutative, we can regard the modules over R to be either left or right modules. For the sake of definiteness, we shall write them as left R-modules. Moreover all modules in this chapter are assumed to be unital, that is $1x = x$ for each element x of the module.

8-1 ALGEBRAS AND IDEALS

Definition. An *R-algebra A* (or *algebra over R*) is a unital R-module and a bilinear mapping f of $A \times A \to A$ which defines a multiplication in A. That is, if $a_1, a_2 \in A$ then the product $a_1 a_2 = (a_1, a_2)f$.

Thus an R-algebra A is an R-module that is also a ring such that

$$(1) \qquad r(a_1 a_2) = r(a_1)a_2 = a_1(ra_2)$$

for all $a_1, a_2 \in A$ and all $r \in R$.

In particular, a finite-dimensional algebra A over a field F is a finite-dimensional vector space over F that is also a ring satisfying (1). This means there is a multiplication of vectors defined in the vector space satisfying (1) above and the two distributive laws of a ring.

If as a module an algebra is finitely generated, then it is called a **finitely generated algebra.**

If R is the ring Z of integers, then a Z-algebra is simply a ring.

Equivalently, an R-algebra can be defined as a unital R-module with a linear mapping (module homomorphism) ϕ of $A \otimes A \to A$ ($A \otimes A$ is the tensor product) defining a multiplication in A. In this definition the product $a_1 a_2 = (a_1 \otimes a_2)\phi$. This follows from Sec. 7-12 where we proved that a bilinear mapping f on the cartesian product $A \times A$ determines a unique linear mapping on the tensor product $A \otimes A$, and conversely.

Definitions. An R-algebra A is called *associative* if its multiplication is associative. If there is an element $e \in A$ such that $ae = a = ea$, for all $a \in A$, then A is called an **algebra with identity e.**

We assume for the rest of this chapter that the R-algebras are associative.

Since the mapping $A \times A \to A$ which defines the multiplication in A is bilinear, the multiplication itself in an algebra A is bilinear, that is for all $a_1, a_2, a_3 \in A$ and all $r \in R$,

$$a_1(a_2 + a_3) = a_1 a_2 + a_1 a_3,$$
$$(a_1 + a_2)a_3 = a_1 a_3 + a_2 a_3,$$
$$(ra_1)a_2 = r(a_1 a_2) = a_1(ra_2).$$

This is an important fact to bear in mind in the arithmetic of an algebra. Note that while the ring is commutative, the algebra itself is not in general commutative.

Frequently properties of algebraic systems are defined by diagrams involving modules and their homomorphisms. We give an example of this.

If an R-algebra A has an identity e, then it is easy to see that the mapping Ψ of $R \to A$ defined by $r\Psi = re$ is a module homomorphism. In particular then $1\Psi = 1e = e$, since A is a unital module.

If M_1, M_2, M_1', M_2' are unital R-modules, let T_1 and T_2 be module homomorphisms $M_1 \to M_1'$ and $M_2 \to M_2'$ respectively. Define a mapping $T_1 \otimes T_2$ of $M_1 \otimes M_2 \to M_1' \otimes M_2'$ by

$$(x_1 \otimes x_2)T_1 \otimes T_2 = x_1 T_1 \otimes x_2 T_2, \quad x_1 \in M_1, \quad x_2 \in M_2.$$

Now it is readily verified that the mapping

$$(x_1, x_2) \to x_1 T_1 \otimes x_2 T_2 \quad \text{of} \quad M_1 \times M_2 \to M_1' \otimes M_2'$$

is bilinear, and hence the mapping $T_1 \otimes T_2$ is linear—that is, it is a module homomorphism (Sec. 7-12).

We now show how an associative R-algebra A with identity e can be described by two diagrams.

Let A be a unital R-module which is equipped with (1) a module homomorphism ϕ of $A \otimes A \to A$ defining a bilinear multiplication in A, that is for $a_1, a_2 \in A$, $a_1 a_2 = (a_1 \otimes a_2)\Phi$, (2) a module homomorphism Ψ of $R \to A$.

Then A is an associative R-algebra with the identity $e = 1\Psi$ if the accompanying two diagrams are commutative.

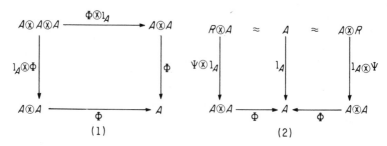

(1)   (2)

Let us explain what this means. First of all, as usual, let 1_A denote the identity mapping on A. We remind the reader that, in Sec. 7-12, it was proved that the mapping $r \otimes a \to ra$ is an isomorphism of $R \otimes A \to A$. Denote this isomorphism by f. Similarly let g denote the isomorphism of $A \otimes R \to A$, given by $(a \otimes r)g = ra$.

Then the commutativity of diagram (1) means that by map composition we have

$$(\Phi \otimes 1_A)\Phi = (1_A \otimes \Phi)\Phi.$$

Commutativity of the left side of diagram (2) yields

$$f \cdot 1_A = (\Psi \otimes 1_A)\Phi,$$

and for the right side of diagram (2),

$$g \cdot 1_A = (1_A \otimes \Psi)\Phi.$$

Consider diagram (1). Start with any element

$$a_1 \otimes a_2 \otimes a_3 \in A \otimes A \otimes A, \; a_1, a_2, a_3 \in A.$$

Then

$$(a_1 \otimes a_2 \otimes a_3)1_A \otimes \Phi = a_1 \otimes (a_2 a_3)$$

and

$$(a_1 \otimes a_2 a_3)\Phi = a_1(a_2 a_3).$$

Similarly, chasing along the other two arrows from $A \otimes A \otimes A$ to A, we obtain $(a_1 a_2)a_3$. Since the diagram is commutative, $a_1(a_2 a_3) = (a_1 a_2)a_3$. Thus the commutativity of diagram (1) implies the multiplication in A is associative.

Consider diagram (2). For $r \otimes a \in R \otimes A$ we have from the left half of the diagram

$$(r \otimes a)\Psi \otimes 1_A = r\Psi \otimes a \quad \text{and} \quad (r\Psi \otimes a)\Phi = (r\Psi)a.$$

Chasing the other path of the left half, we get $r \otimes a \to ra \to ra$. Hence $(r\Psi)a = ra$. The right half of diagram (2) yields $ra = a(r\Psi)$. For $r = 1$, these specialize to $(1\Psi)a = a = a(1\Psi)$, true for all $a \in A$. Hence $1\Psi = e$ is a multiplication identity for A. The commutativity of diagram (2) implies 1Ψ is an identity for A. It also yields $(r\Psi \otimes a)\Phi = ra = (a \otimes r\Psi)\Phi$.

Definitions. An R-submodule B of an R-algebra A is called an **R-subalgebra** of A if B is stable (closed) under multiplication. An *R-subalgebra* B of an R-algebra A is called a **left ideal** of A if $AB \subset B$, and is called a **right ideal** if $BA \subset B$. Here AB denotes the set of all elements ab, $a \in A$, $b \in B$. If B is both a left and a right ideal of A, then B is called an **ideal** of A.

If B is an ideal of A, then the quotient module A/B is an R-algebra under the following definitions of multiplication:

$$a_1(a + B) = a_1 a + B$$
$$(a + B)a_1 = aa_1 + B.$$

This can easily be verified as an acceptable definition of the product, and this multiplication is easily proved bilinear.

Thus if B is an ideal of A then A/B is an R-algebra. It is called the **quotient algebra** of A with respect to B.

If S is a subset of A, the ideal of A that is generated by S is clearly seen to be the set of all finite sums of elements of the form $a_1 s a_2$, where $a_1, a_2 \in A$ and $s \in S$. (For any ideal containing S would have to contain all such elements and moreover the set of all such elements is seen to be an ideal.)

Definition. Let A and A' be two R-algebras. A module homomorphism f of $A \to A'$ is called an **algebra homomorphism** if

$$(a_1 a_2)f = (a_1 f)(a_2 f), a_1, a_2 \in A.$$

If A and A' have identities e and e' we shall always assume an algebra homomorphism f of $A \to A'$ preserves identities—that is, $ef = e'$.

A bijective homomorphism is called an **isomorphism.** If f is an isomorphism then it has an inverse f^{-1} which is an isomorphism of $A' \to A$. This means $ff^{-1} = 1_A$ and $f^{-1}f = 1_{A'}$.

Example. Any commutative ring with identity 1 is an algebra over itself, the algebra multiplication being the multiplication of the ring.

Example. An arbitrary ring R is an algebra over the ring Z of integers. For R is a Z-module and the algebra multiplication is again the multiplication of the ring.

Example. The $n \times n$ matrices over a commutative ring R with identity 1 form an algebra over R under matrix addition and multiplication. The multiplication of a matrix (a_{ij}) by $r \in R$ gives the matrix (ra_{ij}).

Example. The endomorphisms $E(V)$ of a vector space V over a field F form an F-algebra whose multiplication is map composition. For $f, g \in E(V)$, $f + g$ is defined by $x(f + g) = xf + xg$, $x \in V$, and cf, for $c \in F$, is defined by $x(cf) = c(xf)$, $x \in V$. The endomorphisms of a module over a commutative ring with identity form an algebra over this ring.

THEOREM 1. If f is an algebra homomorphism $A \to A'$, where A and A' are R-algebras, then ker f is an ideal of A and im f is a subalgebra of A'.

Proof: ker $f = \{a \in A \mid af = 0'\}$. From Chapter 7 we know that ker f is an R-submodule of A. If $a_1, a_2 \in$ ker f, then $(a_1 a_2)f = (a_1 f)(a_2 f) = 0'$. Hence $a_1 a_2 \in$ ker f, and so ker f is a subalgebra of A. (ker $f)A$ is the set of all $a_1 a_2$, $a_1 \in$ ker f, $a_2 \in A$. Hence $(a_1 a_2)f = (a_1 f)(a_2 f) = 0'$. Therefore $a_1 a_2 \in$ ker f and (ker $f)A \subset$ ker f. Similarly, $A($ker $f) \subset$ ker f. Hence ker f is an ideal.

Again from Chapter 7 we know im $f = Af$ is a submodule of A'. If $a_1', a_2' \in$ im f, then $a_1' = a_1 f$, $a_2' = a_2 f$, $a_1, a_2 \in A$. Hence $a_1' a_2' = (a_1 f)(a_2 f) = (a_1 a_2)f \in$ im f. Hence im f is a subalgebra of A'.

Note that im f is not in general an ideal of A'. It is, for instance, if f is surjective.

THEOREM 2. If $F: A \to A'$ is an algebra epimorphism then the algebras $A/$ker f and A' are isomorphic.

Proof: Let ϕ be the mapping of $A/$ker $f \to A'$ defined by

$$(a + \text{ker } f)\phi = af.$$

The proof that ϕ is an isomorphism is so similar to analogous theorems for groups, rings and modules that it is left to the reader.

EXERCISES

1. If B is an ideal in an algebra A, prove that the mapping j of $A \to A/B$ defined by $aj = a + B$, $a \in A$, is an algebra epimorphism.

2. Prove that the intersection of a family of subalgebras of an algebra A is a subalgebra of A.

3. If f is an algebra isomorphism $A \to A'$ of two algebras, prove f^{-1} defined by $a'f^{-1} = a$, where $af = a'$, is an algebra isomorphism of $A' \to A$.

4. Prove that the set A of real functions of period 2π is a commutative algebra over the real field. Show that the set B of real functions of period π is a subalgebra of A. Is B an ideal in A?

5. Prove the set P of polynomials in one variable x over a field F is a commutative algebra over F.

Show that the subset Q of P consisting of polynomials with zero constant term is an ideal in P.

Prove P/Q is isomorphic to the algebra F.

6. If S is a subset of an algebra A, the intersection of all subalgebras of A that contain S is called the **subalgebra generated** by S.

If $S^2 \subseteq S$ (that is, S is closed under multiplication) prove that the submodule of A generated by S is a subalgebra of A.

7. If S is any subset of an algebra A, describe the form of the elements of the subalgebra generated by S.

8. Determine the subalgebras of the algebra P of exercise (5) generated by (a) the element $x \in P$, (b) the element $x^2 \in P$, (c) the elements 1 and x of P, and (d) the element $x + 1$ of P.

8-2 GRADED MODULES

We next introduce another special type of module, called a **graded module**.

A unital R-module M is said to be a *graded R-module* if M is the direct sum of a family $M_d, d \in D$, of submodules of M,

$$M = \sum_{d \in D} \oplus \ M_d.$$

The index set D of the family is called the *set of degrees* of the graded module M. The elements x_d of the direct summand M_d are called the **homogeneous elements of degree** d of M. Since the zero element of M is in every M_d, it can be regarded as homogeneous of any degree. It is the only element of M that the M_d have in common.

We shall also use the notation $(M)_d$ for M_d.

From the properties of a direct sum we know that every element $x \in M$ has a unique representation as a finite sum of elements from the M_d,

$$x = \sum_{d \in D} x_d, \quad x_d \in M_d, \quad d \in D,$$

and almost all $x_d = 0$. The $x_d, d \in D$ are called the **homogeneous components of x.**

Any module M can be regarded as graded, by simply assigning $(M)_{d_o} = M$ and $(M)_d = 0, d \neq d_o$.

A submodule N of a graded R-module $M = \sum_{d \in D} \oplus \ M_d$ is called a

graded submodule of M if $N = \sum_{d \in D} \oplus \ N_d$ where, for each $d \in D$, N_d is a submodule of M_d. Thus we can write the graded submodule N as

$$N = \sum_{d \in D} \oplus (M_d \cap N).$$

The graded submodule has the same set of degrees as the graded module.

A graded submodule is often called a **homogeneous** or **admissible submodule**.

THEOREM 3. N is a graded submodule of the graded module M if and only if, for each $x \in N$, all the homogeneous components of x are in N.

Proof: If N is a graded submodule $N = \sum_{d \in D} \oplus (M_d \cap N)$, and clearly $x \in N$ has all its homogeneous components in N.

Conversely assume for any $x \in N$ that each homogeneous component x_d of x is in N. Then $x_d \in M_d \cap N$. Hence $x \in \Sigma \oplus (N \cap M_d)$, and therefore $N \subset \Sigma \oplus (N \cap M_d)$. Since $N \cap M_d \subset N$ for each $d \in D$, $\Sigma \oplus (N \cap M_d) \subset N$. Thus $N = \Sigma \oplus (N \cap M_d)$, and therefore N is a graded submodule of M.

THEOREM 4. N is a graded submodule of the graded module M if and only if N is generated by a set S of homogeneous elements of M.

Proof: If N is a graded submodule, then, by the previous theorem, all the homogeneous components of its elements are in N, and clearly N is generated by these homogeneous elements of M.

Conversely, assume N is generated by a set S of homogeneous elements of M. Then for any $x \in N$,

$$x = s_1 r_1 + s_2 r_2 + \cdots + s_k r_k$$

where the $r_i \in R$, $s_i \in S$. Each s_i is a homogeneous element of M. For each $d \in D$, the homogeneous component x_d of x is the sum of all terms $s_i r_i$ of degree d in the above expression for x. Hence $x_d \in N$. This proves all homogeneous components of an element x of N are in N. Hence N is a graded submodule of M.

THEOREM 5. If N is a graded submodule of a graded module M, then the quotient module M/N is a graded module with the same set of degrees as M and N, and

$$M/N = \Sigma \oplus (M_d f)$$

where f is the natural epimorphism of $M \to M/N$.

Proof: For $x \in M$, $xf = x + N$.

$$M = \sum_{d \in D} \oplus \; M_d$$

and clearly the $M_d, d \in D$ are a set of generators of the module M. Since f is surjective, it follows that the $M_d f, d \in D$ form a set of generators of M/N. We want to prove that M/N is the direct sum of this family $M_d f, d \in D$ of generators. To do this we must show that for any $d_o \in D$, the intersection of $M_{d_o} f$ with the submodule S generated by all $M_d f, d \neq d_o$, is N (the zero element of M/N).

An element of $M_{d_o} f$ has the form $x_{d_o} f$, where $x_{d_o} \in M_{d_o}$. If $x_{d_o} f \in S$ then $x_{d_o} f = xf$ where $x \in \bigcup M_d$. Hence $(x_{d_o} - x)f = N$ and therefore $x_{d_o} - x \in N$. Since N is a graded submodule it must contain the homogeneous component x_{d_o} of $x_{d_o} - x \in N$—that is, $x_{d_o} \in N$. Hence $M_{d_o} f \cap S = N$ and therefore

$$M/N = \sum_{d \in D} \oplus \; (M_d f).$$

Corollary. If $N = \displaystyle\sum_{d \in D} \oplus \; N_d$ is a graded submodule of the graded module

$$M = \sum_{d \in D} \oplus \; M_d,$$

then

$$M/N \approx \sum_{d \in D} \oplus \; M_d/N_d.$$

Proof: $M_d f = M_d + N/N$. Let g be the mapping

$$(x_d + N)g = x_d + (N \cap M_d) \text{ of } M_d + N/N \to M_d N \cap M_d.$$

(This is actually nothing but the analogue for modules of the second isomorphism theorem for groups.) Clearly g is a surjective homomorphism of the two quotient modules. The kernel of g is seen at once to be N. Hence g is a module isomorphism. Writing N_d for $N \cap M_d$ we have our result.

Definition. Let M and M' be graded R-modules with the same set D of degrees, where now D is taken to be an additive group.

A module homomorphism f of $M \to M'$ is called **homogeneous of degree r, $r \in D$**, if and only if $M_d f \subseteqq M'_{d+r}$ for every $d \in D$.

We shall be dealing principally with module homomorphisms of degree zero, so that

$$M_d f \subsetneq M_d'.$$

Unless otherwise specified, module homomorphisms are assumed to have degree zero.

8-3 GRADED ALGEBRAS

Definition. An associative R-algebra A with an identity e is said to be a **graded R-algebra** if

(1) A is a graded R-module, $A = \sum_{d \in D} \oplus A_d$.

(2)   The set D of degrees is an abelian group.

(3)   The associative multiplication satisfies the condition

$$A_d A_{d'} \subsetneq A_{d+d'}, \quad d, d' \in D.$$

According to the definition then, the A_d, $d \neq 0$, are submodules, but they are not subalgebras, since $(A_d)^2 = A_{2d}$, and so the A_d are not closed under multiplication. However A_o is closed under multiplication and is a subalgebra. Condition (3) asserts that if a homogeneous element of degree d is multiplied by one of degree d', then the product is a homogeneous element of degree $d + d'$.

We next prove that the identity element e is in A_o. Let $e = \Sigma e_d$ and let $x \in A_r$. Then $x = xe = \Sigma x e_d$. Now $x e_d \in A_{r+d}$, by condition (3). Hence $e_d = 0$ for all $d \neq 0$, and therefore $x = xe_o$ for all homogeneous elements x. If $x = \Sigma x_d$, then $xe_o = \Sigma (x_d e_o) = \Sigma x_d = x$, for all elements x. Similarly, $e_o x = x$ for all elements $x \in A$. Hence $e = e_o \in A_o$.

Definition. A graded R-algebra A is called a **regularly graded R-algebra** if

(1) The set D of degrees is the additive abelian group of integers.

(2) $A_n = 0$ for $n < 0$.

(3) The elements of A_o have the form re, $r \in R$ and $re = 0$ implies $r = 0$.

Condition (3) states that the subalgebra $A_o = Re$. Since R is itself an R-module, it is easy to see that $r \rightarrow re$ is an algebra isomorphism of $R \rightarrow A_o$. Thus $A_o \approx R$. If we identify the element e of A with the identity element 1 of R, that is take $e = 1$, then $A_o = R$.

Example. The set $F[x]$ of all polynomials in a variable x over a field F forms a regularly graded algebra A, in which A_n is the module of all polynomials of degree n. This particular algebra is called the **graded polynomial algebra of degree 0.**

A *graded polynomial algebra* A of degree 2 is the graded module over a field F, defined by $A = \Sigma \oplus A_n$, where $A_n = 0$, $n \neq 0$ (modulo 2). $A = A_o \oplus 0 \oplus A_2 \oplus 0 \oplus A_4 \oplus \cdots$. Here A_2 is the free F-module on the generator x^2, and so on.

A *graded polynomial algebra* A *of degree m* is defined by

$$A = \Sigma \oplus A_n, \qquad A_n = 0, \qquad n \neq 0 \ (\text{mod } m).$$

Definition. A graded R-algebra A is called an **anticommutative graded algebra** if its multiplication has the property $a_p a_q = (-1)^{pq} a_q a_p$, for $a_p \in A_p, a_q \in A_q$.

For example any graded polynomial algebra of even degree can be seen to be anticommutative.

If in addition the multiplication has the property $a_p^2 = 0$ whenever p is odd, the graded algebra is said to be **strictly anticommutative.**

Definition. Let A and A' be two graded R-algebras with the same set of degrees. A module homomorphism of $A \rightarrow A'$ is called an **algebra homomorphism** if it preserves multiplication and units.

This means if $x \in A_p, y \in A_q$ then

$$(xy)f = (xf)(yf)$$

where $xy \in A_{p+q}$ and $xf \in A_p'$, $yf \in A_q'$, so that $(xf)(yf) \in A_{p+q}'$. Moreover units are preserved, so that $ef = e'$.

Let A be a graded R-algebra. A graded R-submodule B of A is called a **graded ideal of A** if for each $p = 0, 1, 2, \ldots$ and $b \in B_p$, the elements ab and ba, for all $a \in A_q, q = 0, 1, 2, \ldots$, are in B_{p+q}.

Let S be a set of homogeneous elements of a graded R-algebra A, that is $S \subset \cup s_n, S_n \subset A_n$. The **graded ideal B generated by S** is the least graded ideal of A such that $S_n \subset B_n, n = 0, 1, 2, \ldots$. B consists of all finite sums of elements of the form asa', where $a, a' \in A$ and $s \in S$. It follows that for $a \in A_p, s \in S_q, a' \in A_r, asa' \in B_{p+q+r}$.

Let $A = \displaystyle\sum_{n=0}^{\infty} \oplus A_n$ be a graded R-algebra, and let $B = \displaystyle\sum_{n=0}^{\infty} \oplus B_n$ be a graded ideal of A. Then the quotient module $A/B = \displaystyle\sum_{0}^{\infty} \oplus A_n/B_n$ is a graded R-algebra.

Multiplication in the algebra A/B can be illustrated by: For $a_n \in A_n, a_m \in A_m, (a_n + B_n)(a_m + B_m) = a_n a_m + a_n B_m + B_n a_m + B_n B_m$. Now $a_n a_m \in A_{n+m}$. Since each of the other three terms on the right is contained in B_{n+m}, then so is their sum. Hence we can define

$$(a_n + B_n)(a_m + B_m) = a_n a_m + B_{n+m}.$$

Thus $(A_n/B_n)(A_m/B_m) \subset A_{n+m}/B_{n+m}$, as it should be for a graded algebra.

8-4 TENSOR ALGEBRAS

We again assume that R is a commutative ring with 1 and that all R-modules are unital.

Definition. A **tensor algebra** on an R-module M is the pair (T, f), where T is an associative R-algebra with $e = 1$ and f is a module homomorphism of $M \rightarrow T$, such that (T, f) has the U.F.P. on M, universal for module homomorphisms.

This means that for any associative R-algebra A with unit element e and for any module homomorphism g of $M \rightarrow A$, there exists a *unique* algebra homomorphism h of $T \rightarrow A$ for which $g = fh$, and for which $1h = e$. The diagram is commutative, $g = fh$.

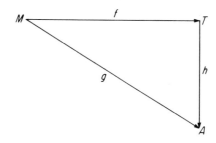

By proofs similar to those for the analogous theorems for a free module and the tensor product of modules, the following two theorems follow.

THEOREM 6. The elements of Mf and the element $1 \in T$ generate the tensor algebra T.

THEOREM 7. If (T, f) and $T'/f')$ are tensor algebras on the same R-module M, then $T \approx T'$, and there is a unique algebra isomorphism j of $T \rightarrow T'$ for which $f' = fj$.

This last theorem allows us to speak of *the* tensor algebra on an R-module M, since it is essentially unique.

We prove the existence of the tensor algebra by actually constructing it.

Let M be a given unital R-module where, as usual R is a commutative ring with 1.

Let T be the direct sum of the family of R-modules $T_0 = R$, $T_1 = M$,

$$T_2 = M \otimes M, \ldots, T_n = M \otimes M \otimes \cdots \otimes M \ (n \text{ factors}), \ldots$$

$$T = \sum_{n=0}^{\infty} \oplus \ T_n$$

We remind the reader that, by definition, the elements of T have therefore the form $\sum_i t_i$, $t_i \in T_i$, where almost all (all but a finite number) $t_i = 0$. Thus each element of T is a finite sum of terms from the T_i.

A multiplication is defined in the graded R-module T by

(2) $1(x_1 \otimes \cdots \otimes x_p) = x_1 \otimes \cdots \otimes x_p = (x_1 \otimes \cdots \otimes x_p)1$

(3) $(x_1 \otimes \cdots \otimes x_p)(y_1 \otimes \cdots \otimes y_q)$

$$= x_1 \otimes \cdots \otimes x_p \otimes y_1 \otimes \cdots \otimes y_q$$

where $x_i, i = 2, 2, \ldots, p$, and $y_i, i = 1, 2, \ldots, q$, are elements in M.

$1 \in T_0 = R$ is the multiplication identity for this multiplication in T. The tensor product of elements is associative and bilinear, and so this multiplication is associative and bilinear. It inherits these properties. Moreover, by (3),

$$T_p T_q \subseteq T_{p+q}.$$

Thus T has become a regularly graded R-algebra.

Now define the mapping f of $M \to T$ as the inclusion module homomorphism of $M \to T$ given by $f = 1_M$, the identity mapping of $M \to T_1 = M$.

We now show the pair (T, f) is the tensor algebra on M by proving (T, f) has the U.F.P. on M.

Let A be any R-algebra with the identity e, and let g be any module homomorphism of $M \to A$. Define a mapping h of $T \to A$ by

(4) $\qquad (x_1 \otimes \cdots \otimes x_p)h = (x_1 g)(x_2 g) \ldots (x_p g)$

for $x_1, x_2, \ldots, x_p \in M$,

(5) $\qquad rh = re, \quad r \in R.$

Thus for $x \in M$, $xh = xg$, and since $xf = x$, we have $x(fh) = xg$, for all $x \in M$. Hence $fh = g$.

To complete the proof we need to show that h is an algebra homomorphism and that h is unique.

Now A may or may not be a graded R-algebra. If it is not we can convert it into one by choosing $A_0 = A$ and $A_n = 0$, $n > 0$. Then $A = \Sigma \oplus A_n$.

It is easy to see that h is a homogeneous algebra homomorphism of degree 0. For the mapping $(x_1, x_2, \ldots, x_p) \to (x_1 g)(x_2 g) \ldots (x_p g)$ is clearly a multilinear mapping (since g is a homomorphism) and this implies (Sec. 7-12) that the mapping $x_1 \otimes \cdots \otimes x_p \to (x_1g)\ldots(x_pg)$ is linear, hence h is a module homomorphism. Moreover, from (4) we see that h preserves multiplication, and from (5) that h preserves units. Hence h is an algebra homomorphism of $T \to A$.

The elements Mf and 1 generate T, and h is defined on these generators. Since a module homomorphism (linear mapping) is uniquely determined by its values on a set of generators, it follows that h is unique.

The pair (T, f) is therefore the tensor algebra on M. We shall denote this tensor algebra by $T(M)$.

We have proved

THEOREM 8. For any given R-module M, there exists a tensor algebra on M.

Let M be a unital module over a commutative ring R with 1. Clearly R is a module over itself. Let $\mathrm{Hom}_R(M, R) = M^*$ be the set of all module homomorphisms of $M \to R$.

For $f, g \in M^*$ define an addition in M^* by $x(f + g) = xf + xg$, $x \in M$ and a scalar multiplication rf by $x(rf) = r(xf)$, $x \in M$. It is easily verified that with these definitions M^* becomes a unital R-module. It is called the **dual of M.** Similarly we can form the *bidual* $M^{**} = \mathrm{Hom}_R(M^*, R)$ of M, also a unital R-module.

Thus we can form the tensor algebra $T(M^*)$ on the module M^*. It is called the **covariant tensor algebra on M,** and $T(M)$ is called the **contravariant tensor algebra on M.** Both are regularly graded R-algebras.

8-5 EXTERIOR ALGEBRAS

An important type of graded algebra with a particularly simple structure (that is, simple multiplication rules) is the exterior algebra on a unital R-module M. Again R is a commutative ring with 1.

Definition. An **exterior algebra** E on an R-module M is a tensor algebra (E, f) on M such that

  (i) $(xf)^2 = 0$, for all $x \in M$.
  (ii) (E, f) has the U.F.P. on M, universal for all associative algebras A with identity e, and for all module homomorphisms g of $M \to A$ for which $(xg)^2 = 0$, for all $x \in M$.

The accompanying diagram illustrates the situation.

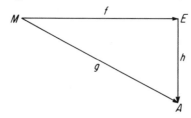

Here h is the unique algebra homomorphism of $E \rightarrow A$ such that $g = fh$ and $1h = e$. (That is, $1h$ is not just the unit element of $im\ h = Eh$ but the unit element of A.)

It follows, as for tensor algebras that

(a) The elements Mf and 1 generate the algebra E.

(b) If (E,f) and (E',f') are two exterior algebras on the same module M, then there exists a unique algebra homomorphism j of $E \rightarrow E'$ such that $fj = f'$.

The existence of the exterior algebra (E, f) on an R-module M is proved by constructing it.

Let (T,f') be the (contravariant) tensor algebra on M. Let

$$B = \sum_{n=0}^{\infty} \oplus B_n$$

be the graded ideal of T generated by the set of homogeneous elements $x \otimes x, x \in M$. Then T/B is a graded R-algebra. Note that $B_0 = B_1 = 0$. Let j denote the natural algebra epimorphism of $T \rightarrow T/B$— that is, j is the algebra epimorphism (homogeneous of degree zero) where $T_n j \subseteqq T_n/B_n$, defined by $t_n j = t_n + B_n$, $t_n \in T_n$. Now from $M \xrightarrow{f'} T \xrightarrow{j} T/B$ we see that $f = f'j$ is a mapping of $M \rightarrow T/B$. We are going to show that $(T/B,f)$ is the exterior algebra on M by proving it satisfies the conditions stipulated in the definition.

First of all, we have for all $x \in M$, $xf = x(f'j) = xj$ (since f' is the inclusion mapping of $M \rightarrow T$ and so $xf' = x$). Hence $(xf)^2 = (xj)^2 = (x \otimes x)j = 0$ [$x \otimes x \in B$ and $(x \otimes x)j = B$, the zero element of T/B]. Next let A be an algebra of the type specified and let g be any module homomorphism of $M \rightarrow A$. The U.F.P. for the tensor algebra (T,f') on M yields a unique algebra homomorphism k of $T \rightarrow A$ such that $g = f'k$ and $1k = e$. Hence $xk = xg$ for all $x \in M$.

Now $(x \otimes x)k = (xk)^2 = (xg)^2 = 0$. Hence k maps the generators $x \otimes x$ of B into 0 and, k being a homomorphism, we have $Bk = 0$. For $t \in T$, $tj = t + B$. Define the mapping h of $T/B \rightarrow A$ by $(t + B)h = tk$. Then h is an algebra homomorphism induced by k and clearly $jh = k$. Hence $jh = g$; that is, $fh = f'jh = g$. Since $1k = e$ for $1 \in T$, it follows that $(1 + B)h = e$ for $1 + B \in T/B$.

Now Mf' and 1 generate T. Since j is surjective, $(Mf')j = M(f'j) = Mf$ and $1j$ generate T/B. Thus h is an algebra homomorphism defined on the generators of T/B. Since an algebra homomorphism is uniquely determined by its values on a set of generators of its domain, h is unique. This completes the proof that $(T/B,f)$ is the exterior algebra on M.

We have proved

THEOREM 9. For any given R-module M there exists an exterior algebra on M.

We have shown that the exterior algebra on M is the quotient algebra of the tensor algebra on M with respect to the graded ideal generated by all elements $x \otimes x$ of the tensor product $M \otimes M$.

The exterior algebra

$$E = T/B = R \oplus M \oplus M \otimes M/B_2 \oplus \cdots$$

on M is commonly denoted by $\wedge M$. Also we write $\wedge^n M$ for E_n. Also for any $x_1, x_2, \ldots, x_n$ of M the product in T is $x_1 \otimes x_2 \otimes \cdots \otimes x_n$ and the product in $\wedge M$ is $(x_1 \otimes x_2 \otimes \cdots \otimes x_n)j$, which is denoted by $x_1 \wedge x_2 \wedge \cdots \wedge x_n$ and called the **exterior product**. The module

$$T_n/B_n = \wedge^n M$$

is generated by these exterior products and $1j$.

Observe that the mapping f of $M \to \wedge M$ is now proved to be the inclusion mapping, since $\wedge^1 M = M$. Multiplication in an exterior algebra has the following properties.

1. Multiplication in $\wedge M$ is bilinear. This follows from the fact that multiplication is bilinear in T and this bilinearity is passed on to $\wedge M$ by the homomorphism j, $\wedge M = Tj$.

2. $x \wedge x = 0$, for all $x \in M$. For $x \wedge x = (x \otimes x)j = 0$.

3. $x \wedge y = -y \wedge x$ for all $x, y \in M$. This follows from (2), for
$0 = (x + y) \wedge (x + y) = x \wedge x + x \wedge y + y \wedge x + y \wedge y = x \wedge y + y \wedge x$.

4. $x_1 \wedge x_2 \wedge \cdots \wedge x_n = 0$, $x_1, x_2, \ldots, x_n \in M$, if $x_i = x_j$ for some i and j. This follows at once from (2) and (3).

5. A permutation of the elements in the exterior product $x_1 \wedge x_2 \wedge \cdots \wedge x_n$ either (a) leaves the product the same if the permutation is even, or (b) changes its sign if the permutation is odd. This follows at once from (3).

6. $\wedge M$ is an anticommutative algebra. Let $x \in \wedge M$ be homogeneous of degree m—that is, $x \in (\wedge M)_m$. If y is homogeneous of degree n, then

$$xy = (-1)^{mn} yx.$$

To see this let $x = x_1 \wedge x_2 \wedge \cdots \wedge x_m$ and $y = y_1 \wedge y_2 \wedge \cdots \wedge y_n$, where the x_i and y_i belong to M. Then $xy = x_1 \wedge \cdots \wedge x_m \wedge y_1 \wedge \cdots \wedge y_n$, and we see that $yx = y_1 \wedge \cdots \wedge y_n \wedge x_1 \wedge \cdots \wedge x_m$ is a permutation of the elements in the product xy, causing mn changes of sign.

Definition. Elements of the form $x_1 \wedge \cdots \wedge x_n$ are called **decomposable elements**. The **indecomposable elements** are finite sums of decomposable elements. Rules (2) to (6) above apply to the decomposable elements. As we have seen the decomposable elements and $1j$ are a set of generators for $\wedge M$.

For example if M is a finitely generated module, generated by m elements, then by reason of (4), $\wedge^n M = 0, n > m$.

EXERCISE

1. If M is a cyclic R-module describe the elements of the tensor algebra on M and the elements of the exterior algebra of M.

8-6 TENSOR PRODUCTS OF ALGEBRAS

If A and B are R-algebras then we can form the R-module $A \otimes B$. Let a multiplication be defined in $A \otimes B$ by

$$(6) \qquad (a_1 \otimes b_1)(a_2 \otimes b_2) = a_1 a_2 \otimes b_1 b_2$$

for $a_1, a_2 \in A$ and $b_1, b_2 \in B$.

Since tensor and algebra multiplications are bilinear, we find for fixed a_2 and b_2 that $(a_1, b_1) \to a_1 a_2 \otimes b_1 b_2$ is a bilinear mapping of $A \times B \to A \otimes B$. Hence $a_1 \otimes b_1 \to a_1 a_2 \otimes b_1 b_2$ is a linear mapping of $A \otimes B \to A \otimes B$. Similarly, $a_2 \otimes b_2 \to a_1 a_2 \otimes b_1 b_2$ is also a linear mapping of $A \otimes B \to A \otimes B$. Thus

$$(a_1 \otimes b_1, \quad a_2 \otimes b_2) \to a_1 a_2 \otimes b_1 b_2$$

is a bilinear mapping of $(A \otimes B) \times (A \otimes B) \to A \otimes B$. Hence (6) defines a bilinear multiplication in $A \otimes B$. With this definition (6), $A \otimes B$ becomes an R-algebra called the **tensor product** of the algebras A and B.

We can easily extend the definition above to the tensor product of two graded R-algebras A and B.

First we define the tensor product of two graded R-modules

$$A = \sum_0^\infty \oplus A_n \quad \text{and} \quad B = \sum_0^\infty \oplus B_n.$$

Let $(A \otimes B)_n$ denote $\sum_{p+q=n} \oplus (A_p \otimes A_q)$, and let C denote the graded R-module $\sum_0^\infty \oplus (A \otimes B)_n$. For $a \in A, b \in B, (a, b) \to a \otimes b$ is a bilinear mapping of $A \times B \to C$ and hence it defines a unique linear mapping h of $A \otimes B \to C$, given by $(a \otimes b)h = a \otimes b$. This mapping h is seen to be an isomorphism. Hence we can define

$$A \otimes B = \sum_0^\infty \oplus (A \otimes B)_n.$$

This is called the *tensor product* of the two graded R-modules A and B. It is a graded R-module.

Now let A and B be two graded associative R-algebras. Form the graded R-module $A \otimes B$. Define a multiplication in $A \otimes B$ by

$$(7) \qquad (a_p \otimes b_r)(a_q \otimes b_s) = a_p a_q \otimes b_r b_s,$$

where $a_p \in A_p, a_q \in A_q, b_r \in B_r, b_s \in B_s$.

The multiplication is associative since A and B are associative algebras. Reasoning in the same way as for (6) it follows that (7) defines a bilinear multiplication in $A \otimes B$. Hence with this definition (7), the graded R-module $A \otimes B$ becomes a graded associative R-algebra.

Instead of (7) we can choose the following definition of multiplication in $A \otimes B$,

$$(8) \qquad (a \otimes b)(a' \otimes b') = (-1)^{\deg b)(\deg a')} aa' \otimes bb',$$

where $\deg b = m$ if and only if $b \in B_m$. This multiplication can readily be shown to be bilinear and so it is an acceptable definition. It has this advantage over definition (7), that if A and B are anticommutative algebras then so is $A \otimes B$. This is not true of (7). The product (8) is used in the definitions of graded coalgebras and graded Hopf algebras.

8-7 APPLICATIONS

Besides its intrinsic interest the following section should serve to clarify some of the previous theory. In this respect the reader can regard it as a series of illustrative examples.

Let M and M' be two R-modules, and let $\wedge M$ and $\wedge M'$ be their exterior algebras. We form the tensor product of these exterior algebras and obtain the algebra $\wedge M \otimes \wedge M'$. In forming this tensor product we shall assume its multiplication to be defined by (8) rather than (7), in order that $\wedge M \otimes \wedge M'$ shall be an anticommutative algebra.

With this understanding we can prove

THEOREM 10. $\wedge(M \oplus M') = \wedge M \otimes \wedge M'$.

Proof: Define the mapping ρ of $M \oplus M' \to \wedge M \otimes \wedge M'$ by

$$(x \oplus x')\rho = x \otimes 1 + 1 \otimes x', \quad x \in M, \quad x' \in M'.$$

We are going to show that $(\wedge M \otimes \wedge M', \rho)$ meets the requirements of the definition (Sec. 8-5) for the exterior algebra on $M \oplus M'$.

First we must prove $((x + x')\rho)^2 = 0$ for all $x \oplus x' \in M \oplus M'$. $((x + x')\rho)^2 = (x \otimes 1)^2 + (x \otimes 1)(1 \otimes x') + (1 \otimes x')(x \otimes 1) + (1 \otimes x')^2$. Now $x \otimes 1 \in (\wedge M)_1 = M$, and by multiplication rule (2) for exterior algebras, $(x \otimes 1)^2 = 0$. Similarly, $1 \otimes x' \in M'$ and for the same reason $(1 \otimes x')^2 = 0$. From (8) we have $(x \otimes 1)(1 \otimes x') =$

$x \otimes x'$ and $(1 \otimes x')(x \otimes 1) = (-1)^{\deg x \, \deg x'} x \otimes x' = -x \otimes x'$, since $\deg x = \deg x' = 1$. Thus $((x + x')\rho)^2 = 0$.

Let ϕ be a module homomorphism of $M \oplus M'$ into an associative algebra A with identity e, for which $((x + x')\phi)^2 = 0$ for all $x + x' \in M \oplus M'$.

The U.F.P. for the exterior algebra $(\wedge M, f)$ on M with the module homomorphism g of $M \to A$ defined by $xg = x\phi$, $x \in M$, yields a unique algebra homomorphism h of $\wedge M \to A$ and $g = fh$. Similarly the U.F.P. for $(\wedge M', f')$ with g' defined by $x'g = x'\phi$, $x' \in M'$, yields h' from $\wedge M' \to A$ and $g' = f'h'$. Thus $xg = (xf)h = xh$ and $x'g = (x'f')h' = x'h'$, since f and f' are inclusion mappings.

Now define the algebra homomorphism H of $\wedge M \otimes \wedge M' \to A$ by

$$(u \otimes u')H = (uh)(u'h'), \quad u \in (\wedge M)_p, \quad u' \in (\wedge M')_q$$

Here $(uh)(u'h')$ is a product in the algebra A. Thus H is uniquely determined by ϕ. Moreover, $(x + x')\rho H = (x \otimes 1 + 1 \otimes x')H = xh + x'h' = xg + x'g = (x + x')g$ for all $x + x' \in M \oplus M'$. Hence $\rho H = g$.

This completes the proof that $(\wedge M \otimes \wedge M', \rho)$ satisfies the U.F.P. as defined for an exterior algebra on $M \oplus M'$.

Lemma 1. Let M and M' be two R-modules. A module homomorphism T of $M \to M'$ can be extended to a unique algebra homomorphism $\wedge T$ of $\wedge M \to \wedge M'$ with $1 \wedge T = 1$. (This means that $\wedge T$ is an algebra homomorphism such that $\wedge T = T$ on M.)

Proof: The U.F.P. for the exterior algebra $(\wedge M, f)$ with $A = \wedge M'$ and $g = T$ yields the unique algebra homomorphism $\wedge T$ of $\wedge M \to \wedge M'$ such that $T = f(\wedge T)$ with $1(\wedge T) = 1$. For all $x \in M$, $xT = x(f \wedge T) = (xf) \wedge T = x(\wedge T)$. Thus $\wedge T$ is the unique algebra homomorphism defined on the set of generators M and 1 of $\wedge M$.

We consider next the exterior algebra of a finite-dimensional vector space over a field.

THEOREM 11. Let V be an n-dimensional vector space over a field F. Then

$$\wedge^p V = 0, \quad p > n$$

and

$$\dim \wedge^p V = \binom{n}{p}, \quad p \leqq n.$$

Proof: By definition $\binom{n}{p} = \dfrac{n!}{p!(n-p)!}$. If $\dim V = 1$, then $\wedge V = F \oplus V$. For if α is a basis for V, then for $c, c' \in F$, $c\alpha \wedge c'\alpha =$

$cc'(\alpha \wedge \alpha) = 0.$ Hence $\wedge^p V = 0,\ p > 1$ and dim $\wedge^0 V = 1,$ dim $\wedge^1 V = 1.$

Now use induction on the dimension n. Assume the theorem is true for all n-dimensional vector spaces. Let dim $V = n + 1$ and let $\alpha_1,$ $\ldots, \alpha_{n+1}$ be a basis for V. If $U = L(\alpha_1, \ldots, \alpha_n)$ and $W = L(\alpha_{n+1})$, then $V = U \oplus W$ and dim $U = n$, dim $W = 1$.

By Theorem 10 we have $\wedge V = \wedge U \otimes \wedge W$. Moreover $\wedge W = F \oplus W$. Hence

$$\wedge^k V = \sum_{i+j=k} \oplus \wedge^i U \otimes \wedge^j W$$

$$= \wedge^k U \otimes F \oplus \wedge^{k-1} U \otimes W.$$

By the induction hypothesis and by Theorem 36 (Chapter 7),

$$\dim \wedge^k U \otimes F = \binom{n}{k} \cdot 1 = \binom{n}{k},$$

$$\dim \wedge^{k-1} U \otimes W = \binom{n}{k-1} \cdot 1 = \binom{n}{k-1}.$$

As a direct sum, therefore,

$$\dim \wedge^k V = \binom{n}{k} + \binom{n}{k-1} = \binom{n+1}{k}.$$

Moreover, if $k > n + 1$, the induction hypothesis gives $\wedge^k U = 0$ and $\wedge^{k-1} U = 0$. Hence $\wedge^k V = 0$. This completes the induction and the theorem is proved.

Since

$$\wedge V = \sum_{p=0}^{n} \oplus \wedge^p V,$$

$$\dim \wedge V = \sum_{p=0}^{n} \dim \wedge^p V = \sum_{p=0}^{n} \binom{n}{p} = 2^n.$$

We close this section with an interesting and surprising definition.

Let V be an n-dimensional vector space and let T be an endomorphism of V. By Lemma 1, T extends to a unique algebra endomorphism $\wedge T$ of $\wedge V$. Now dim $\wedge^n V = 1$ by Theorem 11. Hence, since an endomorphism of a 1-dimensional vector space with basis, say, α, has to be of the form $\alpha \to c\alpha$, for some scalar c, it follows that for all $\alpha \in \wedge^n V$, $\alpha(\wedge T) = c\alpha,\ c \in F.$

Definition. If dim $V = n$ and if T is an endomorphism of V, then the *determinant of T* is the scalar det T, such that

$$\alpha(\wedge T) = (\det T)\alpha, \quad \text{for all } \alpha \in \wedge^n V.$$

All the well-known properties of a determinant can be derived from this definition; for instance, if T_1 and T_2 are endomorphisms of V, then

$$\det(T_1 T_2) = (\det T_1)(\det T_2).$$

This property we used once in Chapter 3. It follows directly from the key fact that $\wedge(T_1 T_2) = \wedge T_1 \wedge T_2$. The reader should have no difficulty in proving this and that $\det 1_V = 1$, $\det T = 0$ if and only if T is not an automorphism, and if T is an automorphism then $\det T^{-1} = 1/\det T$.

We illustrate the above definition for dim $V = 2$. Let α_1, α_2 be a basis for V and let T be an endomorphism of V. Then $\alpha_i T = \sum_{j=1}^{2} a_{ij}\alpha_j$, $a_{ij} \in F$ and $i = 1, 2$. For $\alpha, \beta \in V$, let $\alpha \wedge \beta$ be any element of $\wedge^2 V$. Then

$$\alpha = \sum_{1}^{2} x_i \alpha_i, \quad \beta = \sum_{1}^{2} y_i \alpha_i, \quad x_i, y_i \in F.$$

Hence $\alpha \wedge \beta = (x_1 y_2 - x_2 y_1)\alpha_1 \wedge \alpha_2$. Moreover,

$$\alpha T \wedge \beta T = (\alpha \wedge \beta) \wedge T \in \wedge^2 V.$$

A simple calculation yields

$$\alpha T \wedge \beta T = (a_{11}a_{22} - a_{12}a_{21})(x_1 y_2 - x_2 y_1)\alpha_1 \wedge \alpha_2.$$

Hence

$$(\alpha \wedge \beta) \wedge T = \begin{vmatrix} a_{11} & a_{12} \\ a_{21} & a_{22} \end{vmatrix} (\alpha \wedge \beta).$$

8-8 GRADED TENSOR ALGEBRAS

As a generalization of the tensor algebra on a module M we have the graded tensor algebra on a graded module M. We give here an outline of this concept.

Definition. A **graded tensor algebra** on a graded R-module M is a graded algebra Υ and a module homomorphism ρ of $M \to \Upsilon$, homogeneous of degree 0, such that (Υ, ρ) has the U.F.P. on M for all graded algebras A and all homogeneous module homomorphisms of $M \to A$ of degree 0.

THEOREM 12. For any graded R-module M the graded tensor algebra $\Upsilon(M)$ on

$$M = \sum_{0}^{\infty} \oplus M_n$$

exists, and

$$\Upsilon(M) = \sum_{0}^{\infty} \oplus \Upsilon_n,$$

where

$$\Upsilon_0 = R \oplus M_0 \oplus M_0 \otimes M_0 \oplus \cdots,$$
$$\Upsilon_1 = (M)_1 \oplus (M \otimes M)_1 \oplus (M \otimes M \times M)_1 \oplus \cdots$$

and for $n > 1$

$$\Upsilon_n = (M)_n \oplus (M \otimes M)_n \oplus (M \otimes M \otimes M)_n \oplus \cdots.$$

The proof is very similar to that for tensor algebras. An important detail is

$$(M \otimes M \otimes M)_n = \sum_{i+j+k=n} \oplus M_i \otimes M_j \otimes M_k,$$

where $(M)_n = M_n$. The bookkeeping has become a little more involved.

The multiplication in Υ is defined analogously to (2) and (3) in Sec. 8-4 for the tensor algebra T. For example in (3) if $x_i \in M_{h_i}$, $y_j \in M_{k_j}$, where $\Sigma h_i = h$, $\Sigma k_j = k$, then (3) is a product of an element of Υ_h and an element of Υ_k, the product being an element of Υ_{h+k}.

The mapping ρ of $M \to \Upsilon$ is defined to be the inclusion mapping and hence ρ is injective. Moreover it can be proved that im $\rho = M\rho$ and the element 1 form a set of generators of $\Upsilon(M)$. Hence all elements of $\Upsilon(M)$ are linear combinations of elements of M and R.

Given a module M it can be "graded" in many ways and different possibilities occur for the resulting graded module and its graded tensor algebra. For instance, if we take the graded module M to be the one defined by $(M)_1 = M, (M)_n = 0, n \neq 1$, then it is readily seen that the graded tensor algebra on this M is precisely the tensor algebra T of M.

Returning now to the graded tensor algebra Υ on M, let B be the graded ideal generated by all elements of the form

(9) $\qquad xy - (-1)^{pq} yx, \quad x \in (M)_p, \quad y \in (M)_q$

then Υ/B is an anticommutative graded algebra.

Let j be the natural epimorphism of $\Upsilon \to \Upsilon/B$. For $y = x$ and p odd (9) becomes $2x^2$. Hence in Υ/B we have $2\bar{x}^2 = 0$, where $\bar{x} = xj$. This does not assure $\bar{x}^2 = 0$. (For instance if the characteristic of R is 2, then $2r = 0, r \in R$, does not imply $r = 0$.)

However if (i) we take B to be the graded ideal generated by all elements (9) and all elements x^2, where x is a homogeneous element of Υ of odd degree; and (ii) specialize M to be the graded module for which $(M)_1 = M, (M)_n = 0, n \neq 1$, where M is an R-module, then it can be seen that Υ/B becomes the exterior algebra $\wedge M$ of the module M. $\wedge M$ is an anticommutative algebra.

The type of grading used here is often called **internal grading.** It requires a graded module to be the direct sum of a family of submodules. In contrast to this, another type of grading, called **external grading,** is commonly used. In an external grading, a graded module M is a sequence $\{M_0, M_1, M_2, \ldots, M_n, \ldots\}$ of modules. The modules M_i are not yet assumed to be submodules of M, nor is M required to be a direct sum of submodules, as in internal grading. One type leads to the other; however there are many more elements in an internally graded module $M = \sum_0^\infty \oplus M_n$ than in the externally graded module $M = \{M_0, M_1, M_2, \ldots\}$.

EXERCISE

1. M is a module. Find the graded tensor algebra of the graded module $M = \sum_{n=0}^\infty M_n$ when

(a) $M_n = 0$, $n \neq 0$ and $M_0 = M$.

(b) $M_n = 0$, $n \neq 1$ and $M_1 = M$.

8-9 LIE ALGEBRAS

We make an exception here and briefly mention a very useful type of nonassociative (i.e. not necessarily associative) algebra which, besides its own intrinsic interest, has widespread applications to topology and the theory of groups.

Let R be a commutative ring with unit element.

Definition. A **Lie algebra** L over R is a nonassociative R-algebra whose multiplication, denoted by $[x, y]$ for x and y in L, satisfies the two identities

(i) $[x, x] = 0$ for all $x \in L$

(ii) $[[x, y], z] + [[y, z], x] + [[z, x], y] = 0$ (Jacobi identity) for all $x, y, z \in L$.

From (i) we can deduce

$$0 = [x + y, x + y] = [x, x] + [x, y] + [y, x] + [y, y]$$
$$= [x, y] + [y, x].$$

Thus $[x, y] = -[y, x]$ for all $x, y \in L$.

Note the similarity in this multiplication to that of an exterior algebra.

Example. Let A be an associative R-algebra. Define a Lie product by $[x, y] = xy - yx$, $x, y \in A$. It is easy to prove that (i) and (ii) are satisfied and the resulting R-algebra (with this multiplication) is a Lie algebra. It is called the Lie algebra $L(A)$ of the associative algebra A. Thus every associative algebra defines a Lie algebra.

Conversely it can be proved that each Lie algebra over a field K defines an augmented associative algebra called its **enveloping algebra.** (An associative algebra A over a field K together with an algebra homomorphism of $A \rightarrow K$ is called an augmented algebra.)

The endomorphisms $E(V)$ of a vector space V over a field K form, as we have seen earlier, an associative algebra and hence the algebra $E(V)$ defines a Lie algebra $L(E(V))$.

A **representation** of an associative algebra over a field K is a homomorphism of the algebra into an algebra of endomorphisms of a vector space over K, the representation being called **faithful** if it is an isomorphism.

Every associative algebra with an identity e has a faithful representation. We see this as follows. Let A be the algebra. To every $a \in A$ associate the endomorphism T_a of A (where A is regarded as a vector space over K) defined by $xT_a = xa$, for all $x \in A$. Clearly $T_a + T_b = T_{a+b}$ and $T_a T_b = T_{ab}$, and it is not hard to see that these endomorphisms form a subalgebra B of the algebra of all endormorphisms of the vector space A. Moreover it also readily follows that $a \rightarrow T_a$ is a homomorphism of A into B. Now $T_a = T_b$ if and only if $xa = xb$ for all $x \in A$. In particular then $ea = a = eb = b$ and the homomorphism is therefore an isomorphism.

A **representation of a Lie algebra** L over K is defined as a homomorphism of L into a Lie algebra of endomorphisms of a vector space over K. It can be proved that every Lie algebra over K has a faithful representation, that is, every Lie algebra is isomorphic to a Lie algebra of endomorphisms of a vector space. It turns out that the enveloping algebra U of a Lie algebra serves to reduce the theory of representations of a Lie algebra to the theory of representations of the enveloping associative algebra itself and that the Lie algebra is isomorphic to a subalgebra of endomorphisms of U. By this means a faithful representation of every Lie algebra can be obtained.

EXERCISES

1. Prove that the condition $[x, x] = 0$ in a Lie algebra L can be described as a factorization of the linear mapping $L \otimes L \rightarrow L$ (defining the multiplication in L) into $L \otimes L \rightarrow \Lambda^2 L \rightarrow L$. (See Sec. 8-5.)

2. A is an associative algebra over a field K. A derivation D of A is defined as a linear mapping of $A \rightarrow A$ such that

$$(xy)D = (xD)y + x(yD).$$

Prove that the linear mappings $D + D'$, kD, where $k \in K$, and $DD' - D'D$ are derivations of A and hence prove that the set $D(A)$ of all derivations of A is a Lie algebra with the product $[D, D'] = DD' - D'D$. $D(A)$ is called the **derivation algebra** of A.

3. L is a Lie algebra. For any $x \in L$ a map, called the **adjoint map** ad x, of $L \to L$ is defined by

$$y(\mathrm{ad}\, x) = [x, y] \quad \text{for all } y \in L.$$

(a) Prove ad x is a derivation of L.

(b) Prove $x \to$ ad x is a homomorphism of the Lie algebra L into the Lie algebra $D(L)$. Note it is a Lie homomorphism.

8-10 TENSORS

Let M be a unital R-module and let M^* be the dual of M. Let $T(M)$ be the contravariant tensor algebra on M and let $T(M^*)$ be the covariant tensor algebra on M. $T(M)$ and $T(M^*)$ are graded R-algebras.

Definition. The graded R-algebra $\Upsilon(M)$ defined by

$$\Upsilon(M) = T(M) \otimes T(M^*)$$

is called the **tensor algebra on M.** The elements of $\Upsilon(M)$ are called tensors. Write

$$\Upsilon(M) = \Sigma \oplus T_q^p$$

where

$$T_q^p = \underbrace{M \otimes M \otimes \cdots \otimes M}_{p} \times \underbrace{M^* \otimes \cdots \otimes M^*}_{q},$$

and $T_0^0 = R$. The elements of T_q^p are called **mixed tensors, contravariant of rank p and covariant of rank q.**

Explicit Form for the Multiplication in $\Upsilon(M)$

First note $T_q^p = T_0^p \otimes T_q^0$. Also note that for $x \in T_q^p$ and $y \in T_s^r$ the mapping ϕ of $T_q^p \otimes T_s^r \to T_q^p \times T_s^r$ given by $(x, y) \to x \otimes y$ is bilinear. From the fact that there exists a unique isomorphism of $A \otimes B$ and $B \otimes A$ given by $a \otimes b \to b \otimes a$ we know there exists a unique isomorphism j of

$$T_q^p \otimes T_s^r =$$

$$T_0^p \otimes T_q^0 \otimes T_0^r \otimes T_s^0 \to T_0^p \otimes T_0^r \otimes T_q^0 \otimes T_s^0 = T_{q+s}^{p+r}$$

such that

$$(x \otimes y^* \otimes z \otimes w^*)j = x \otimes z \otimes y^* \otimes w^*$$

for $x \in T_o^p$, $y^* \in T_q^o$, $z \in T_o^r$, $w^* \in T_s^o$. Hence ϕj is a bilinear mapping of

$$T_q^p \times T_s^r \to T_{q+s}^{p+r}$$

which defines a product in $\Upsilon(M)$, and this product is given by

$$(x^p \otimes x_q^*) \otimes (y^r \otimes y_s^*) = x^p \otimes y^r \otimes x_q^* \otimes y_s^*,$$

where $x^p \in T_o^p$, $x_q^* \in T_q^o$, $y^r \in T_o^r$, $y_s^* \in T_s^o$.

The elements of T_q^p are linear combinations of tensor products of the form

$$x_1 \otimes x_2 \otimes \cdots \otimes x_p \otimes x_1^* \otimes \cdots \otimes x_q^*,$$

called *decomposable tensors*, where the x_i are any elements of M and the x_j^* are any elements of M^*.

8-11 TENSOR SPACES

We append a section on tensor spaces important in mathematical physics and differential geometry.

In this section we need the concept of a module with operators. It is merely an adaptation to modules of the concept of a group with operators in Chapter 4.

A *module with operators* is a module M, a set Σ, and a mapping $(x, \sigma) \to x\sigma$ of $M \times \Sigma \to M$, $x \in M$, $\sigma \in \Sigma$, satisfying

$$(x + y)\sigma = x\sigma + y\sigma, \quad x, y \in M$$
$$(rx)\sigma = r(x\sigma), \quad x \in M, \quad r \in R.$$

Σ is called the **set of operators.**

Thus a module with operators (see Sec. 4-11) can also be defined as a module M and a mapping of M into the ring of module endomorphisms of M.

Let R be a ring with unit element. Let $f: A \to B$ be a linear mapping of two left modules over R. For every $\phi \in B^*$ (the dual of B), $f\phi \in A^*$ as we see from $A \xrightarrow{f} B \xrightarrow{\phi} R$, and $\phi \to f\phi$ is a linear mapping of $B^* \to A^*$ called the **transpose f^T of f.**

If f is an isomorphism, (so that f^{-1} exists) then we can form the transpose of $f^{-1} : B \to A$. It is a mapping of $A^* \to B^*$ defined by $\psi \to f^{-1}\psi$, $\psi \in A^*$, $B \xrightarrow{f^{-1}} A \xrightarrow{\psi} R$, and is written f and called the **isomorphism contragredient to f.** Now let $\Upsilon(M)$ be the tensor algebra on M, that is $\Upsilon(M) = \Sigma \oplus T_q^p$ where

$$T_q^p = \underbrace{M \otimes \cdots \otimes M}_{p} \otimes \underbrace{M \otimes \cdots \otimes M}_{q}.$$

An automorphism α of M determines an automorphism α_q^p of T_q^p where

$$\alpha_q^p = \underbrace{\alpha \otimes \alpha \otimes \cdots \otimes \alpha}_{p} \otimes \underbrace{\breve{\alpha} \otimes \cdots \otimes \breve{\alpha}}_{q}$$

and where $\breve{\alpha}$ is the transpose of α^{-1}, i.e., the automorphism contragredient to α. $\breve{\alpha}$ is an automorphism of M^*. Note that $\alpha \to \alpha_q^p$ and $\beta \to \beta_q^p$ imply $\alpha\beta \to \alpha_q^p \beta_q^p = (\alpha\beta)_q^p$ and hence this is a representation of the multiplicative group of automorphisms of M in the set of automorphisms of T_q^p.

Every automorphism α of M determines an automorphism α_q^p of T_q^p defined by $\alpha_q^p = \overbrace{\alpha \otimes \cdots \otimes \alpha}^{p} \otimes \overbrace{\breve{\alpha} \otimes \cdots \otimes \breve{\alpha}}^{q}$, where $\breve{\alpha}$ is the contragredient automorphism of M^*. Thus the automorphisms of M constitute an operator set on the module T_q^p, in which the operator product is defined by $(x, \alpha) = x\alpha = x\alpha_q^p$, $x \in T_q^p$. Of course $x\alpha_q^p \in T_q^p$ and $x(\alpha_q^p \beta_q^p) = (x\alpha_q^p)\beta_q^p$ where α, β are any two automorphisms of M. We define α_o^o to be the identity map on $T_o^o = R$.

Now let T_q^p be made into a module with operators, the operator set being the multiplicative group of automorphisms of M and the operator product being defined as $x\alpha = x\alpha_q^p$, $\alpha \in T_q^p$. $[(x + y)\alpha = x \cdot \alpha + y \cdot \alpha, (rx)\alpha = r(x\alpha), (x\alpha)\beta = x(\alpha\beta).]$ A **tensor space** on a (unitary) R-module M is any submodule H of T_q^p which is stable (closed) under the operator product for every automorphism α of M. This means H is an operator submodule of the module with operators T_q^p.

Thus a *tensor space* H on M can be simply defined as any submodule of T_q^p which is closed under every automorphism of M.

If H and H' are two tensor spaces on an R-module M, a **tensor mapping** f on $H \to H'$ is a module homomorphism of $H \to H'$ such that $(\alpha \cdot x)f = \alpha \cdot (xf)$, where $x \in H$ and α is an automorphism of M.

Contraction of Tensors

Let H be a tensor space on M, $H \subset T_q^p$, and let H' be a tensor space on M, $H' \subset T_{q-1}^{p-1}$. The linear mapping c_j^i of $H \to H'$, defined on every decomposable tensor

$$z = x_1 \otimes \cdots \otimes x_p \otimes y_1^* \otimes \cdots \otimes y_q^*$$

by

$$c_j^i(z) = y_j^*(x_i)x_1 \otimes \cdots \otimes x_{i-1} \otimes x_{i+1} \otimes \cdots \otimes x_p \otimes y_1^* \otimes$$
$$\cdots \otimes y_{j-1}^* \otimes y_{j+1}^* \otimes \cdots \otimes y_q^*,$$

is known as **contraction** of the ith contravariant index and the jth covariant index, $(i \le p, j \le q)$. Note the first factor on the right is in R.

Clearly for every automorphism α of M

$$\bar{\alpha}(y_j^*)(\alpha x_i) = y_j^*(x_i).$$

Hence

$$c_j^i(\alpha z) =$$

$$y_j^*(x_i)\alpha(x_i) \otimes \cdots \otimes \alpha(x_p) \otimes \alpha(y_i) \otimes \cdots \otimes \alpha(y_q) = \alpha c_j^i(z)$$

(the $\alpha(x_i)$ and $\alpha(y_j)$ are deleted) which shows c is a tensor mapping. Similarly contractions p times contravariant and q times covariant can be defined.

EXERCISES

1. Show that the set of all complex numbers is a finite-dimensional algebra over the field of real numbers. What is its dimension?

2. If B is an ideal in an algebra A, define an algebra A/B and prove explicitly that the multiplication is bilinear. Prove the mapping j of $A \rightarrow A/B$ defined by $aj = a + B$, for all $a \in A$, is an algebra epimorphism.

3. Let A and A' be algebras over a field F and let B be an ideal in A. If T is an algebra homomorphism of $A \rightarrow B$ prove there exists a unique algebra homomorphism T' of $A/B \rightarrow B$ such that $jT' = T$ if and only if $B \subset \ker T$. Here j is the natural epimorphism of $A \rightarrow A/B$. Be sure to check that T' is well-defined.

4. If F is a field whose characteristic $\neq 2$, show that if M is a module over F we can define the exterior algebra $\wedge M$ by $\wedge M = T/B$ where T is the graded tensor algebra on M and B is the graded ideal generated by all elements of the form $xy + yx$, where $x, y \in M$.

5. If $x \in \wedge^n M$, where n is odd, prove $x \wedge x = 0$.

6. If $x \in \wedge^3 M, y \in \wedge^5 M, z \in \wedge^4 M$, prove

$$x \wedge (y + z) + (x + y - z) \wedge x = 0.$$

7. Describe the direct summands $\wedge^2 M$ and $\wedge^3 M$ of $\wedge M$ and write out their forms explicitly. Multiply an element of $\wedge^2 M$ by an element of $\wedge^3 M$ and check that the product is an element of $\wedge^5 M$.

8. If (T, f) is the tensor algebra of the R-module M, prove that im f and 1 generate T. Do this before constructing T.

9. If (T, f) is the tensor algebra of the R-module M, prove that f is injective. If (T, f) and (T', f') are tensor algebras of M, prove $T \approx T'$. Do this before constructing T.

10. Let A and B be graded algebras and let the linear mappings μ of $A \otimes A \rightarrow A$ and ν of $B \otimes B \rightarrow B$ define the bilinear multiplications of A and B respectively. In order to make the tensor product $A \otimes B$ a graded algebra a multiplication must be defined in $A \otimes B$ by a linear mapping of $(A \otimes B) \otimes (A \otimes B) \rightarrow A \otimes B$.

If ψ is an isomorphism of $B \otimes A \to A \otimes B$, show that

$$(A \otimes B) \otimes (A \otimes B) \xrightarrow{1_A \otimes \psi \otimes 1_B} A \otimes A \otimes B \otimes B \xrightarrow{\mu \otimes \nu} A \otimes B$$

defines a multiplication in $A \otimes B$, with the identity element of $A \otimes B$ to be given by the mapping $1_R \otimes 1_R$ of $R \otimes R = R \to A \otimes B$. Note that

$$(a \otimes (b \otimes a') \otimes b')(1_A \otimes \psi \otimes 1_B) = a \otimes (b \otimes a')\psi \otimes b',$$

$$a, a' \in A, \; b, b' \in B.$$

11. In defining tensors, the isomorphism ψ of $B \otimes A \to A \otimes B$ defined by $(b \otimes a)\psi = a \otimes b$ was used. Show that if the isomorphism $(b \otimes a)\psi = (-1)^{pq} a \otimes b$, $a \in A_p$, $b \in B_q$, is used, then if A and B are anticommutative algebras the tensor product $A \otimes B$ is an anticommutative algebra.

12. Show that the tensor product of two anticommutative algebras is not anticommutative if ψ is defined by $(b \otimes a)\psi = a \otimes b$ for $a \in A, b \in B$.

13. Let V be an n-dimensional vector space over a field F with the basis $\beta_1, \beta_2, \ldots, \beta_n$. Prove $f_1, f_2, \ldots, f_n$ is a basis of $V^* = \text{Hom } (V, F)$, where $(\Sigma c_i \beta_i) f_k = c_k$. Prove that $\beta_i \to f_i$, $i = 1, 2, \ldots, n$ is an isomorphism of $V \to V^*$. It is not natural (canonical) since it depends on the choice of a basis in V.

14. For each vector α of V define $\phi_\alpha \in V^{**} = \text{Hom } (V^*, F)$ by $f\phi_\alpha = \alpha f, f \in V^*$. Show that $\alpha \to \phi_\alpha$ is a natural isomorphism of $V \to V^{**}$.

15. If V and W are finite-dimensional vector spaces and $T = (V \otimes W)^*$ use this natural isomorphism to identify $V \otimes W$ and $(V \otimes W)^{**}$. Hence prove that $T^* \approx V \otimes W$.

16. Prove the set $B(V, W)$ of all bilinear mappings from $V \to W$ is a vector space under natural definitions of addition and scalar multiplication.

17. Prove that $B(V, W)^* = V \otimes W$.

18. Let R_n stand for the algebra of the $n \times n$ matrices over a commutative ring R with identity. Investigate the following theorem:

$$\text{The algebra } R_p \otimes R_q \approx R_{pq}.$$

chapter 9

Field Theory
and Galois Theory

A *field* F is a commutative ring with identity e in which every non-zero element $x \in F$ has a multiplication inverse $x^{-1} \in F$—that is, $xx^{-1} = e = x^{-1}x$.

If F and E are fields and $F \subset E$, F is called a subfield of E.

We remind the reader that we have proved in Chapter 5 that an integral domain can be embedded in a field, called its quotient field.

9-1 STRUCTURE OF FIELDS

Lemma 1. The intersection of any family F_i, $i \in I$, of subfields of a field F is a subfield of F.

> **Proof:** If $x, y \in \bigcap_{i \in I} F_i$, then $x + y$, xy, and x^{-1} belong to each F_i and hence to the intersection. It follows easily from this that the intersection is a subfield of F.

Let F and F' be two fields and let f be a ring homomorphism of $F \to F'$. A field F has only two ideals, the zero ideal (0) and F itself. Now ker f is an ideal in F and hence ker $f = (0)$ or ker $f = F$. If ker $f = F$ then f maps every $x \in F$ into the zero element $0'$ of F'. If ker $f = (0)$, then f is injective and, in this case, F is isomorphic to some subfield of F'. Consequently an endomorphism of a field F (that is a homomorphism of $F \to F$) is either the trivial endomorphism that maps every element of F into the zero element or if it is surjective it is an automorphism of F.

THEOREM 1. The automorphisms of a field F form a group under multiplication (map composition).

> **Proof:** If f and g are automorphisms, we show $f \circ g$ is an automorphism.

Let $x, y \in F$. Then

$$(x + y)fg = (xf + yf)g = (xf)g + (yf)g$$
$$= x(fg) + y(fg).$$

284

$$(xy)(fg) = ((xf)(yf))g = (xf)g \cdot (yf)g$$
$$= x(fg) \cdot y(fg).$$

Moreover, $x(fg) = 0$ implies $xf = 0$, and this in turn implies $x = 0$. Also for any $y \in F$, there exists a unique x such that $y = xg$ and a unique z such that $x = zf$. Hence $y = z(fg)$. Thus fg is a bijective endomorphism and hence an automorphism of F.

The identity map I_F on F is the neutral element and it is easy to verify that the inverse f^{-1} of an automorphism is an automorphism. For $xf^{-1} = y$ is equivalent to $x = yf$. We know map composition is associative and hence the automorphisms of F form a group.

A **prime field** is one that contains no proper subfields. Every field contains one and only one prime field, the intersection of all its subfields.

Being an integral domain, the characteristic of a field F is either 0 or a prime p. If the characteristic is 0, then the set of elements ne, n an integer, form an integral domain D. The quotient field of D is therefore the unique prime subfield of F. For any subfield of F must contain D and hence the quotient field of D. Clearly $m/n \rightarrow me/ne$, $n \neq 0$, is an isomorphism of the rational field with this prime subfield of F. Thus the prime subfield of a field of char 0 is isomorphic to the field of rational numbers.

On the other hand if char $F = p$, then the elements ne form a subfield of F, the inverse of me, $p \nmid m$, being ke where $mk \equiv 1 \pmod{p}$. This subfield obviously is the unique prime subfield of F and clearly it is isomorphic to the field of residue classes modulo p. Hence the prime subfield of a field of char p is isomorphic to the field of residue classes modulo p.

Thus the structure of the prime subfield of a field is completely determined by the characteristic of the field and, of course, conversely.

9-2 FIELD EXTENSIONS

If E and F are fields and $F \subset E$, then E is called an **extension field of F**. If multiplication in E is ignored, then E can be regarded as a vector space over the field F. If the dimension of the vector space E is finite, then E is called a **finite extension of F**. The dimension of the vector space E over F will be denoted by $E \mid F$.

Thus any field is an extension of a subfield. Any extension field of a field of char 0 is a field of char 0.

THEOREM 2. Let F, B, E be fields, and suppose $F \subset B \subset E$. (B is called an **intermediate field of F** and E.) If B is a finite extension of F and if E is a finite extension of B, then $E \mid F = (E \mid B)(B \mid F)$.

Proof: Let $E \mid B = m$ and let $\alpha_1, \ldots, \alpha_m$ be a basis of the vector space E over B. Let $B \mid F = n$ and let $\beta_1, \ldots, \beta_n$ be a basis of the vector space B over F.

If y is any element of E, then

$$y = b_1\alpha_i + b_2\alpha_2 + \cdots + b_m\alpha_m, \quad b_i \in B.$$

and for $i = 1, 2, \ldots, m$,

$$b_i = a_{i1}\beta_1 + a_{i2}\beta_2 + \cdots + a_{in}\beta_n, \quad a_{ij} \in F,$$

where $j = 1, 2, \ldots, n$.

Hence $y = \displaystyle\sum_{i=1}^{m} \sum_{j=1}^{n} a_{ij}\beta_j\alpha_i$

and the $\beta_j\alpha_i \in E$.

Now $y = 0$ if and only if each $b_i = 0$, and $b_i = 0$ if and only if each $a_{ij} = 0$. Hence $y = 0$ if and only if each $a_{ij} = 0$. Therefore the $\beta_j\alpha_i$ (there are mn of them) form a basis for E over F. Thus $E \mid F = mn = (E \mid B)(B \mid F)$.

We remark here, without proof, that Theorem 2 is also true if the field extensions are infinite.

Corollary. Let $F, F_1, F_2, \ldots, F_n$ be fields and let $F = F_0 \subset F_1 \subset \cdots \subset F_n$. (This is called a tower of fields.) If each F_{i+1} is a finite extension of F_i, $i = 0, 1, 2, \ldots, n - 1$, then F_n is a finite extension of F and

$$F_n \mid F = (F_1 \mid F)(F_2 \mid F_1) \cdots (F_n \mid F_{n-1}).$$

Proof: The corollary follows from the theorem by use of induction on n.

Definition. Let $f(x)$ be a polynomial over F, that is $f(x) \in F[x]$. An element α belonging to some extension field of F is called a **root** of **f(x)** if $f(\alpha) = 0$.

THEOREM 3. (Remainder Theorem). Let $f(x) \in F[x]$ where F is a field. Let E be an extension field of F. For any $b \in E$,

$$f(x) = (x - b)q(x) + f(b)$$

where $q(x) \in E[x]$ and where the degree of q is one less than the degree of $f(x)$.

Proof: Since $F \subset E$, $f(x) \in E[x]$. The division algorithm for $E[x]$ yields $f(x) = (x - b)q(x) + r$, where $q(x) \in E[x]$ and where $r \in E$. (For either $r = 0$ or degree $r = 0$.) Hence for $x = b$, $f(b) = r$ and the theorem is proved. It is easy to see that the degree of $q(x)$ is one less than that of $f(x)$.

Corollary. (Factor Theorem). If α is a root of $f(x)$, that is α belongs to some extension field K of F, then $f(x) = (x - \alpha)q(x)$ where $q(x) \in K[x]$. Hence $x - \alpha$ divides $f(x)$.

Proof: Put $b = \alpha$ in the theorem.

Definition. A root α of $f(x)$ is said to be of *multiplicity* m, if $(x - \alpha)^m \mid f(x)$ but $(x - \alpha)^{m+1} \nmid f(x)$. If $m \geq 2$, α is called a **multiple root of f(x)**.

A root α of $f(x)$ that is not a multiple root is called a **simple root of f(x)**. Thus α is a simple root of $f(x)$ if and only if $f(x) = (x - \alpha)g(x)$ and $g(\alpha) \neq 0$.

THEOREM 4. A polynomial $f(x)$ of degree n over a field F has at most n roots in any extension field. (A root of multiplicity m is to be counted as m roots.)

Proof: Use induction on n, that is assume the theorem true of all polynomials of degrees less than n. Let $F \subset E$. If $f(x)$ has a root $\alpha \in E$, suppose it is of multiplicity m. Then $f(x) = (x - \alpha)^m g(x)$ where degree $g(x) = n - m$. α cannot be a root of $g(x)$. If $\beta \neq \alpha$ is another root of $f(x)$ in E, then β is clearly a root of $g(x)$. Now by the induction hypothesis $g(x)$ has at most $n - m$ roots in E. Hence $f(x)$ has at most $n - m + m = n$ roots in E. (Observe the theorem is trivially true for $n = 1$.)

Definition. Let $f(x) = a_0 + a_1 x + \cdots + a_n x^n$ be a polynomial over F. The *formal derivative* $f'(x)$ of $f(x)$ is defined by

$$f'(x) = a_1 + 2a_2 x + 3a_3 x^2 + \cdots + na_n x^{n-1}$$

where $2a_2 = a_2 + a_2$, $3a_3 = a_3 + a_3 + a_3$, and so on. Thus $f'(x)$ is a polynomial over F.

The notion of limit is not involved in this definition of formal derivative. It is a simple, straightforward exercise to prove the following formulas for the formal derivative. For any $f(x), g(x) \in F[x]$,

$$(f(x) + g(x))' = f'(x) + g'(x)$$
$$(f(x) \cdot g(x))' = f'(x) \cdot g(x) + f(x) \cdot g'(x)$$
$$((f(x))^m)' = m(f(x))^{m-1} \cdot f'(x).$$

Use induction on m to prove the last one.

THEOREM 5. A polynomial $f(x)$ over F has a multiple root α if and only if α is a root of $f'(x)$.

Proof: Suppose α is a multiple root of $f(x)$. Then $f(x) = (x - \alpha)^m g(x)$ where $m \geq 2$ and $g(\alpha) \neq 0$. Then

$$f'(x) = (x - \alpha)^m g'(x) + m(x - \alpha)^{m-1} g(x).$$

Hence $f'(\alpha) = 0$ and α is a root of $f'(x)$.

Conversely, suppose α is a root of both $f(x)$ and $f'(x)$. Now $f(x) = (x - \alpha)g(x)$ and therefore $f'(x) = g(x) + (x - \alpha)g'(x)$. Since $f'(\alpha) = 0$, we have $g(\alpha) = 0$ and therefore α is a root of $g(x)$ and hence α is a multiple root of $f(x)$.

Definition. An irreducible polynomial over a field F is called **separable** if it does not have repeated (multiple) roots. An arbitrary polynomial over F is called **separable** if all its irreducible factors are separable.

Definition. A polynomial $f(x)$ over a field F is said to be **separable** if every irreducible factor of $f(x)$ in $F[x]$ has simple (nonrepeated) roots.

If E is an extension of the field F, an element $\alpha \in E$ is said to be **separable over F** if it is the root of a separable polynomial over F. If every element of E is separable over F, E is called a **separable extension of F.**

9-3 ALGEBRAIC EXTENSIONS

Let F be a field and let $F[x]$ be the ring of all polynomials in the variable x with coefficients in F. We know from our previous study that $F[x]$ is a principal ideal domain. Let E be an extension field of F. An element $\alpha \in E$ is called **algebraic over F** if α is a root of a polynomial belonging to $F[x]$, that is a polynomial in x with coefficients in F. Among all such polynomials with the root α we can select one of least degree (by the well-ordering property of the positive integers).

Let $f(x)$ be a polynomial of least degree with the root α and whose leading coefficient is 1. If $f(x)$ were reducible, that is $f(x) = g(x)h(x)$ where degree $g(x) > 0$ and degree $h(x) > 0$, then α would be a root of either $g(x)$ or of $h(x)$. But this contradicts the choice of $f(x)$, since the degrees of $g(x)$ and $h(x)$ are less than the degree of $f(x)$. Hence $f(x)$ is irreducible. Moreover, $f(x)$ is unique and is a divisor of every polynomial $g(x)$ that has α for a root. For, by the division algorithm,

$$g(x) = f(x)q(x) + r(x),$$

where $r(x) = 0$ or the degree of $r(x)$ is less than the degree of $f(x)$. Put $x = \alpha$ and we see that $0 = r(\alpha)$. Hence $r(x) = 0$, for otherwise $r(x)$ would be a polynomial having α for a root and degree $r(x) <$ degree $f(x)$. Therefore $f(x) \mid g(x)$ and this also shows $f(x)$ is unique.

THEOREM 6. An irreducible polynomial $f(x)$ over a field F of characteristic O is separable.

Proof: Let α be a root of $f(x)$. Since $f(x)$ is irreducible, $f(x)$ must be a polynomial over F of least degree with α as a root. Now $f'(x) \neq 0$, for its leading coefficient is an integral multiple of the leading coefficient of $f(x)$ and so cannot be zero, since char $F = 0$.

If α is a multiple root of $f(x)$ then α is a root of $f'(x)$. But the degree of $f'(x)$ is one less than the degree of $f(x)$ and this contradicts $f(x)$ being a polynomial of least degree with α as a root. Hence α must be a simple root of $f(x)$ and $f(x)$ is therefore separable.

Corollary. In a field of char O, every polynomial is separable.

Proof: Every polynomial is a product of irreducible polynomials.

Let E be an extension field of a field F, and let $\alpha \in E$. Denote by $F(\alpha)$ the smallest subfield of E that contains F and the element α. $F(\alpha)$ is the intersection of all subfields of E (and this includes E itself) that contain F and α. $F(\alpha)$ is said to be an extension of F obtained by the **adjunction** of α to F.

If E is an extension field of F then an element $\alpha \in E$ is said to be **transcendental over F** if it is not algebraic over F. Transcendental elements exist, for we constructed in Chapter 5 a polynomial ring in a transcendental element. It has been proved that π, the ratio of the circumference of a circle to its diameter, and e, the base of the natural logarithms, are transcendental elements over the field of rational numbers.

Definition. An extension field E of a field F is called **algebraic over F** if every element of E is algebraic over F.

THEOREM 7. A finite extension E of F is always algebraic.

Proof: Let α be any element of E, and let $E \mid F = n$. Then the $n + 1$ elements $1, \alpha, \alpha^2, \ldots, \alpha^n$ of E must be a linearly dependent set of vectors in the vector space E over F. Hence for scalars (that is elements of F) $b_0, b_1, \ldots, b_n$, not all zero, we have

$$b_0 + b_1\alpha + \cdots + b_n\alpha^n = 0.$$

Thus α is algebraic over F and hence E is algebraic over F.

Let E be a finite extension of F and let $\alpha \in E$. Form $F(\alpha)$. Let $\beta \in E$. Then β is algebraic over F and, since

$$E \mid F = (E \mid F(\alpha))(F(\alpha) \mid F)$$

we see that $E \mid F(\alpha)$ is finite and so β is an algebraic element over $F(\alpha)$.

Now form $F(\alpha)(\beta)$, that is adjoin β to $F(\alpha)$. Then clearly $F(\alpha)(\beta)$ is the least field containing F and α and β. Moreover $F(\alpha)(\beta) = F(\beta)(\alpha)$. We denote this **double extension** field by $F(\alpha, \beta)$ (or by $F(\beta, \alpha)$).

Clearly then we can form $F(\alpha_1, \alpha_2, \ldots, \alpha_n)$ where the $\alpha_i \in E$ by successive adjunctions of $\alpha_1, \alpha_2, \ldots, \alpha_n$ and we know that each α_{i+1} is algebraic over $F(\alpha_1, \alpha_2, \ldots, \alpha_i)$.

Definition. If E is an extension field of F and if $\alpha \in E$, then $F(\alpha)$ is called a **simple extension of F.**

THEOREM 8. If E is a separable extension of the infinite field F and if $\alpha_1, \beta_1 \in E$, then there exists an element $\theta \in E$ such that $F(\alpha_1, \beta_1) = F(\theta)$.

Proof: The element θ is called a **primitive** element and the theorem states that the extension $F(\alpha_1, \beta_1)$ is actually a simple extension. Let $f(x)$ be the irreducible polynomial with the distinct roots $\alpha_1, \alpha_2, \ldots, \alpha_m$, and let $g(x)$ be the irreducible polynomial with the distinct roots $\beta_1, \beta_2, \ldots, \beta_n$. Choose a $c \in F$ such that

$$(1) \qquad \alpha_i + c\beta_j \neq \alpha_r + c\beta_s, \quad \text{unless } i = r \text{ and } j = s.$$

This is possible, since F is an infinite field and there are only a finite number of elements $b \in F$ for which

$$(2) \qquad \alpha_i + b\beta_j = \alpha_r + b\beta_s.$$

To see this, observe that for each choice of i, j, r, s, there is either no element b that satisfies (2) or there is one and only one element b satisfying (2). If $s \neq j$, $b = \dfrac{\alpha_i - \alpha_r}{\beta_s - \beta_j}$. If $s = j$ there is no element b satisfying (2), unless $i = r$. Moreover there are only a finite number of distinct choices for i, j, r, s.

We now claim $\theta = \alpha_1 + c\beta_1$ where c is any element of F that satisfies (1).

Now $F(\alpha_1 + c\beta_1) \subset F(\alpha_1, \beta_1)$. We next show that $F(\alpha_1, \beta_1) \subset F(\alpha_1 + c\beta_1)$.

Form $f(\theta - cx)$. It is a polynomial over $F(\theta)$. It has the root $x = \beta_1$, since $\theta - c\beta_1 = \alpha_1$. No other $\beta_i, i = 2, \ldots, n$, is a root of $f(\theta - cx)$. For if $\theta - c\beta_i = \alpha_j$, say, then this would contradict the choice of c. Now $g(x)$ has the root β_1 and hence $f(\theta - cx)$ and $g(x)$ have the G.C.D. $x - \beta_1$. Thus there exist polynomials $h(x)$ and $k(x)$ over $F(\theta)$ such that $h(x)f(\theta - cx) + k(x)g(x) = x - \beta_1$. Hence $\beta_1 \in F(\theta)$, and therefore $\alpha_1 = \theta - c\beta_1 \in F(\theta)$. Thus $F(\alpha_1, \beta_1) \subset F(\theta)$ and so $F(\alpha_1, \beta_1) = F(\theta)$.

Corollary. If E is a separable extension of the infinite field F and if $\gamma_1, \gamma_2, \ldots, \gamma_n \in E$, then there exists a primitive element $\theta \in E$ such that

$$F(\gamma_1, \gamma_2, \ldots, \gamma_n) = F(\theta).$$

Proof: The corollary follows from the theorem by use of induction on n.

Corollary. If E is an algebraic extension of a field F of char 0 and if $\gamma_1, \ldots, \gamma_n \in E$ then there exists a primitive element $\theta \in E$ such that

$$F(\gamma_1, \ldots, \gamma_n) = F(\theta).$$

Proof: A field of char 0 is an infinite field and every polynomial in such a field is separable.

We have thus shown that every finite separable extension of an infinite field F is a simple extension of F.

Let E be an extension of the field F. If $\alpha \in E$ is transcendental (variable) over F, then we know from Chapter 5 that $F[\alpha]$ is an integral domain whose elements are polynomials in α over the field F. The quotient field of this integral domain consists of all elements of the form $f(\alpha)/g(\alpha)$, $g(\alpha) \neq 0$, where $f(\alpha)$ and $g(\alpha)$ are polynomials in α over F. Elements of this form are called **rational functions** of α over F. Now clearly all such rational functions would have to be contained in the field $F(\alpha)$, obtained by the adjunction to F of the element α. Since $F(\alpha)$ is the smallest subfield of E containing F and α, it follows at once that $F(\alpha)$ must be this quotient field of $F[\alpha]$. This affords us another description of the field $F(\alpha)$.

Let E be an extension of the field F. Let $\alpha \in E$ be algebraic over F, and let $f(x)$ be an irreducible polynomial over F with the root α. $F[x]$ is a principal ideal domain. Form the quotient domain $F[x]/(f(x))$, where $(f(x))$ is the principal ideal generated by $f(x)$. We are going to prove that $K = F[x]/(f(x))$ is a field by showing that $(f(x))$ is a maximal ideal in $F[x]$. (See Chapter 6, Theorem 17.) Let $g(x)$ be any polynomial in $F[x]$ which is not in the ideal $(f(x))$. Then $f(x) \nmid g(x)$. Hence, since $f(x)$ is irreducible, $f(x)$ and $g(x)$ are relatively prime and therefore there exist $s(x)$, $t(x) \in F[x]$ such that $s(x) f(x) + t(x)g(x) = 1$. Thus the ideal generated by $f(x)$ and $g(x)$ contains 1 and is therefore $F[x]$. Thus $(f(x))$ is maximal. Hence K is a field.

For $a \in F$, the mapping $a \to \bar{a} = a + (f(x))$ is clearly an isomorphism of F with a subfield $\bar{F}$ of K. Thus F is embedded in K. If we identify $\bar{F}$ with F, then we see that K is an extension of F.

Write $\bar{x} = x + (f(x))$ and $\overline{g(x)} = g(x) + (f(x))$, where $g(x) \in F[x]$. Then $\overline{g(x)} = g(\bar{x})$.

Hence $0 = \overline{f(x)} = f(\bar{x})$ and so K contains a root of the irreducible polynomial $f(x)$. This leads to

THEOREM 9. (Kronecker's theorem) If $f(x)$ is any polynomial over a field F, there exists an extension field of F in which $f(x)$ has a root.

Proof: Let $p(x)$ be an irreducible factor of $f(x)$. Then $p(x)$, and hence $f(x)$, has a root in the field $F[x]/(p(x))$.

Corollary. If $f(x)$ is an irreducible polynomial of degree n over F, then $K = F[x]/(f(x))$ is an extension field of F of degree or dimension n, that is $K \mid F = n$.

Proof: Now $1, \bar{x}, \bar{x}^2, \ldots, \bar{x}^{n-1}$, where $f(\bar{x}) = 0$, must be linearly independent vectors of K over F. For otherwise $\bar{x}$ would be a root of a polynomial over F of degree $< n$. This is impossible since $f(x)$ is irreducible. Hence $K \mid F \geq n$. On the other hand if $g(x) \in F[x]$, then $g(x) = q(x)f(x) + r(x)$, where $r(x) = 0$ or degree $r(x) <$ degree $f(x) = n$. Thus $g(\bar{x}) = r(\bar{x})$. (For $g(x) - r(x) \in f(x)$)). Hence $g(\bar{x}) = r(\bar{x})$ and so all elements of K are generated by $1, \bar{x}, \ldots, \bar{x}^{n-1}$. Hence these n elements form a basis of K over F. Hence $K \mid F = n$.

EXERCISES

1. Find a primitive element for the double extension $F(\sqrt{2}, i)$ of the rational field.

2. If a subfield F of a finite field K has h elements, prove that K has h^n elements, where $n = K \mid F$.

3. Prove that a finite field F of characteristic p has p^m elements, for some positive integer m.

9-4 PROPERTIES OF AN ALGEBRAIC EXTENSION

Let E be an extension of a field F, and let $\alpha \in E$ be algebraic over F, then α is the root of an irreducible polynomial over F. Let us discover some important facts about the elements of $F(\alpha)$. We know these elements are rational functions of α over F, that is quotients $h(\alpha) \mid g(\alpha) \neq 0$, of polynomials in α with coefficients in F. $g(\alpha) \neq 0$ implies $f(x) \nmid g(x)$ and hence that $f(x)$ and $g(x)$ are relatively prime over F. Therefore there exist polynomials $s(x)$ and $t(x)$ belonging to $F[x]$ such that $s(x)f(x) + t(x)g(x) = 1$. Putting $x = \alpha$, we see that $t(\alpha)g(\alpha) = 1$. *Thus the inverse of $g(\alpha) \neq 0$ is a polynomial $t(\alpha)$ over F.* Hence all the elements of $F(\alpha)$ can actually be expressed as polynomials in α. Moreover, if $h(x) \in F[x]$, then $h(x) = f(x)q(x) + r(x)$, where $r(x) = 0$ or degree $r(x) <$ degree $f(x)$. Again, for $x = \alpha$, we see that $h(\alpha) = r(\alpha)$. Thus any polynomial in α over F can be expressed as a polynomial in α of degree less than the degree of $f(x)$. This means that if $h(x)$ is any polynomial in $F[x]$ then we can reduce it modulo $f(x)$ to a polynomial $r(x)$ of degree less than the degree of $f(x)$, and hence $h(\alpha) = r(\alpha)$. All the elements of $F(\alpha)$ can be expressed as polynomials in α of degrees less than the degree of $f(x)$.

One more important fact that follows from this result is that if degree $f(x) = n$, then $F(\alpha) \mid F = $ degree $f(x) = n$. For every element of $F(\alpha)$, as a vector space over F, is a linear combination of the vectors $1, \alpha, \alpha^2, \ldots, \alpha^{n-1}$ and so these vectors span $F(\alpha)$. They are also independent vectors. For if

$$c_0 + c_1\alpha + \cdots + c_{n-1}\alpha^{n-1} = 0, \quad c_i \in F,$$

then all the $c_i = 0$. Otherwise α would be a root of a polynomial $c_0 + c_1 x + \cdots + c_{n-1}x^{n-1}$ of degree less than n, contradicting $f(x)$ as being the polynomial of least degree with α as a root. Hence the dimension of $F(\alpha)$ over F is n.

It is now easy to prove the following theorem.

THEOREM 10. If α is a root of the irreducible polynomial $f(x)$ over F, then
$$F(x)/(f(x)) \approx F(\alpha).$$

Proof: The mapping ϕ of $F(x)/(f(x)) \to F(\alpha)$ defined by

$$(g(x) + (f(x)))\phi = g(\alpha), \quad g(x) \in F[x]$$

preserves addition and multiplication and is therefore a homomorphism. Moreover clearly ϕ is surjective, and ker $\phi = (f(x))$, since $g(\alpha) = 0$ if and only if $f(x) \mid g(x)$. Hence the two fields are isomorphic.

Note in Theorem 10 that the element $\bar{x} = x + (f(x))$ of $F[x]/(f(x))$ corresponds to the element α of $F(\alpha)$. Hence $f(\bar{x}) = 0$. This provides the clue to the problem of finding an extension field of F in which a given polynomial has a root. As we did in proving Kronecker's theorem, we simply take an irreducible factor $f(x)$ of a given polynomial $h(x)$ and form the field $F[x]/(f(x))$. For example, if F is the field of real numbers, then $x^2 + 1$ is an irreducible polynomial over F. Hence the field $F[x]/(x^2 + 1)$ is the field of complex numbers. By Theorem 10, $F[x]/(x^2 + 1) = F(i)$ where $i^2 = -1$.

Corollary. If α and β are roots of the same irreducible polynomial $f(x)$ over F, then $F(\alpha) \approx F(\beta)$.

Proof: Each of them is isomorphic to $F[x]/(f(x))$ and hence to each other.

THEOREM 11. Let E be an algebraic extension of F. There exists a finite number of elements $\alpha_1, \alpha_2, \ldots, \alpha_n$ of E such that

$$E = F(\alpha_1, \ldots, \alpha_n)$$

if and only if E is a finite extension of F.

Proof: Suppose $E = F(\alpha_1, \ldots, \alpha_n)$. Then

$$E \mid F = (F(\alpha_1) \mid F)(F(\alpha_1, \alpha_2) \mid F(\alpha_1))\ldots(E \mid F(\alpha_1, \ldots, \alpha_n)).$$

Each factor on the right is finite and hence so is $E \mid F$.

Conversely let E be a finite extension of F. If $E \mid F = 1$, then $E = F$. If $F \subset E$, let $\alpha_1 \in E$, where $\alpha_1 \notin F$. Then if $F(\alpha_1) = E$ the theorem is proved. If there exists $\alpha_2 \in E$, $\alpha_2 \notin F(\alpha_1)$, form $F(\alpha_1, \alpha_2)$. If $F(\alpha_1, \alpha_2) \neq E$, select $\alpha_3 \in E$ where $\alpha_3 \notin F(\alpha_1, \alpha_2)$ and form $F(\alpha_1, \alpha_2, \alpha_3)$. Since E is a finite extension of F, this process must come to an end for some n, with

$$E = F(\alpha_1, \ldots, \alpha_n).$$

EXERCISE

1. Show that an extension field K of a field F is an algebra over F. If B is an intermediate field, $F \subset B \subset K$, show that B is a subalgebra of K. If K is a finite extension of F, then K is a finite-dimensional algebra over F.

9-5 SPLITTING FIELDS AND NORMAL EXTENSIONS

Definition. Let $f(x) \in F[x]$. An extension field E of a field F is called a **splitting field of f(x)** over F if $f(x)$ splits into linear factors in E, but does not so split in any proper intermediate field B, $F \subset B \subset E$.

Thus if E is a splitting field of $f(x)$, the roots $\alpha_1, \alpha_2, \ldots, \alpha_n$ of $f(x)$ generate E and $E = F(\alpha_1, \ldots, \alpha_n)$. For the field $F(\alpha_1, \ldots, \alpha_n)$ is contained in every field in which $f(x)$ splits.

If we can prove that there exists a field H in which $f(x)$ splits, then we have a field H among whose elements are all the roots $\alpha_1, \ldots, \alpha_n$ of $f(x)$, and $F \subset H$. Now form the field $F(\alpha_1, \alpha_2, \ldots, \alpha_n)$ generated by these roots. It is the intersection of all subfields of H containing F and the roots $\alpha_1, \ldots, \alpha_n$; that is, it is the least field containing F and the roots of $f(x)$. Then $F \subset F(\alpha_1, \ldots, \alpha_n) \subset H$.

THEOREM 12. If $f(x)$ is a polynomial over a field F, then there exists a splitting field E of $f(x)$.

Proof: We use induction on the degree n of $f(x)$. If $n = 1$, $E = F$. Assume the theorem is true for all fields and for all polynomials of degree $< n$. Let $p(x)$ be an irreducible factor of $f(x)$. Put $K = F[x]/(p(x))$. Then $p(x)$ has a root α_1 in K and $p(x) = (x - \alpha_1)q(x)$. Hence $f(x) = (x - \alpha_1)q(x)g(x)$ in $K[x]$. Now the degree of $q(x)g(x)$ is $n - 1$, and hence by the induction hypothesis there exists a splitting field $H \supset K$ of $q(x)g(x)$ containing the roots $\alpha_2, \ldots, \alpha_n$ of $q(x)g(x)$, and hence the roots of $f(x)$. Thus $f(x)$ splits in H. Hence $E = F(\alpha_1, \ldots, \alpha_n) \subset H$ is a splitting field of $f(x)$ over F.

Our curiosity would next lead us to ascertain whether or not a splitting field of a polynomial is unique.

THEOREM 13. Let σ be an isomorphism $F \to F'$ of two fields. Let $p(x)$ be an irreducible polynomial over F, where $p(x) = p_0 + p_1 x +$

$\cdots + p_n x^n$. Let $p'(x)$ be the polynomial $p'(x) = p_0\sigma + (p_1\sigma)x + \cdots + (p_n\sigma)x^n$ over F'. Let α be a root of $p(x)$ in an extension field of F and let α' be a root of $p'(x)$ in an extension field of F'. Then σ extends to an isomorphism $\bar{\sigma}$ of $F(\alpha) \to F'(\alpha')$.

Proof: First we observe that $p'(x)$ is also irreducible over F'.

We claim the mapping $\bar{\sigma}$ defined by

$$(3) \quad (a_0 + a_1\alpha + \cdots + a_m\alpha^m)\bar{\sigma} = a_0\sigma + (a_1\sigma)\alpha'$$
$$+ \cdots + (a_m\sigma)\alpha'^m$$

is an isomorphism of $F(\alpha) \to F'(\alpha')$.

It is a simple and straightforward calculation to verify that $\bar{\sigma}$ preserves both addition and multiplication. Moreover $\bar{\sigma}$ is surjective, since σ is an isomorphism. Also the right-hand side of (3) is 0 if and only if it is a multiple of $p'(\alpha') = 0$ and hence, under σ^{-1}, if and only if the polynomial on the left side of (3) is a multiple of $p(\alpha) = 0$. Hence $\bar{\sigma}$ is injective, and therefore $\bar{\sigma}$ is an isomorphism. Since $\bar{\sigma} = \sigma$ on F, we see that it is an extension of σ.

Corollary. If α and β are two roots of an irreducible polynomial over F, then $F(\alpha) \approx F(\beta)$.

Proof: Specialize the theorem to the case where $F' = F$. Take σ to be the identity automorphism of F and put $\alpha' = \beta$.

THEOREM 14. If E is an extension field of F then the set $G(E/F)$ of all automorphisms of E leaving F fixed is a group under multiplication.

Proof: Let $\sigma, \tau \in G(E/F)$. Then we have seen in Theorem 2, that $\sigma\tau$ is an automorphism, as well as σ^{-1}. It clearly follows that $\sigma\tau$ and σ^{-1} leave F fixed, since σ and τ do, and, of course, map composition is associative.

Definition. A finite extension E of a field F is called a **normal extension** of F if E is the splitting field of a separable polynomial over F.

Definition. Two elements of an extension field E of F are called **conjugates** over F if and only if they are roots of the same irreducible polynomial over F.

Lemma 2. Let E be a finite extension field of F. An automorphism σ of E over F maps each $\alpha \in E$ on to a conjugate element $\alpha\sigma$ of α over F.

Proof: α is algebraic over F and is therefore a root of an irreducible polynomial $p(x)$ over F. Since σ leaves each element of F fixed, then if

$$p(x) = p_0 + p_1 x + \cdots + p_n x^n, \quad p_i \in F,$$

$$p(\alpha) = p_0 + p_1\alpha + \cdots + p_n\alpha^n = 0$$

and therefore $[p(\alpha)]\sigma = p_0 + p_1(\alpha\sigma) + \cdots + p_n(\alpha\sigma)^n = 0$; that is, $p(\alpha\sigma) = 0$. Thus α and $\alpha\sigma$ are roots of the same irreducible polynomial over F.

THEOREM 15. Let σ be an isomorphism $F \to F'$ of two fields. Let $f(x)$ be a polynomial over F and $f'(x)$ the polynomial over F' corresponding to $f(x)$ under σ. Then σ extends to an isomorphism of $E \to E'$, where E is the splitting field of $f(x)$ and E' is the splitting field of F'.

Proof: If $f(x) = a_0 + a_1x + \cdots + a_nx^n$, then $f'(x) = a_0\sigma + (a_1\sigma)x + \cdots + (a_n\sigma)x^n$. Let $p(x)$ be an irreducible factor of $f(x)$ and let $p'(x)$ be the polynomial corresponding to $p(x)$ under σ. Then it is easy to see that $p'(x)$ is an irreducible factor of $f'(x)$. Let α be a root of $p(x)$ in E and α' a root of $p'(x)$ in E'. Then σ extends to an isomorphism $\bar\sigma$ of $F(\alpha) \to F'(\alpha')$, and $\alpha' = \alpha\bar\sigma$.

We use induction on n and assume the theorem is true for all fields and for all polynomials of degrees $< n$. Hence under the induction hypothesis $\bar\sigma$ extends to an isomorphism $\bar{\bar\sigma}$ of the splitting field of $q(x)g(x)$ over $F(\alpha)$ with the splitting field of $q'(x)g'(x)$ over $F'(\alpha')$. $[f(x) = p(x)g(x)$ and $p(x) = (x - \alpha)^m q(x)$, so that $f'(x) = p'(x)g'(x)$ and $p'(x) = (x - \alpha')^m q'(x)]$.

Now E is the splitting field of $q(x)g(x)$ over $F(\alpha)$, for $q(x)g(x)$ splits in E [E contains all roots of $f(x)$]. Moreover, $\alpha \in E$. Hence if $q(x)g(x)$ splits in B where $F(\alpha) \subset B \subset E$, then $f(x)$ would split in B, contradicting E as its splitting field.

Hence σ extends to an isomorphism

$$\bar{\bar\sigma} \text{ of } E \to E'.$$

Corollary. Any two splitting fields of the same polynomial over F are isomorphic.

Proof: Specialize the theorem to $F' = F$ and it follows that the two splitting fields E and E' of F are isomorphic.

Lemma 3. If the polynomial $f(x)$ in Theorem 15 is separable, then σ can be extended to $\bar\sigma$ in precisely $m = E \mid F$ distinct ways.

Proof: We use induction on m.

Any extension $\bar\sigma$ maps the root α of $p(x)$ into a root α' of $p'(x)$. Since $f(x)$ is separable, if $p(x)$ is of degree r, then $p(x)$ has r distinct roots. These r roots α yield r choices for $\bar\sigma$, where $\alpha' = \alpha\bar\sigma$. By the induction hypothesis each such σ can be extended to E in

$\dfrac{m}{r} = E \mid F(\alpha)$ distinct ways. Hence in all we get $r\left(\dfrac{m}{r}\right) = m$ distinct extensions of σ on F to $\bar{\sigma}$ on E.

THEOREM 16. If E is a normal extension of F, then $E \mid F = \#G(E/F)$, where $G(E/F)$ is the group of automorphisms of E that leave F fixed.

Proof: Let $E = E'$. E is the splitting field of a separable polynomial $f(x)$ over F. Then the identity automorphism I_F on $F \to F$ can be extended in $E \mid F$ different ways to an automorphism of E. These extensions are precisely the automorphisms belonging to $G(E/F)$. Hence $E \mid F = \#G(E/F)$.

THEOREM 17. If an irreducible polynomial $p(x)$ over F has one root in a splitting field of a polynomial $f(x)$ over F, then $p(x)$ splits in E.

Proof: Let $p(\theta) = 0$ where $\theta \in E$. Let $\beta \neq \theta$ be another root of $p(x)$ in some extension field of F. Then by the corollary to Theorem 10, $F(\theta) \approx F(\beta)$. This isomorphism extends to an isomorphism of the splitting field E of $f(x)$ over $F(\theta)$ with the splitting field K of $f(x)$ over $F(\beta)$. $E \approx K$.

Since $f(x)$ splits in $E(\beta)$, $K \subset E(\beta)$. But K contains all the roots of $f(x)$ and β. Hence $K \supset E(\beta)$. Therefore $K = E(\beta)$ and hence $E(\beta)$ is the splitting field of $f(x)$ over $F(\beta)$.

Now $E \mid F(\theta) = E(\beta) \mid F(\beta)$ and by the corollary to Theorem 10, $F(\theta) \mid F = F(\beta) \mid F$. Hence

$$E \mid F = (E \mid F(\theta))(F(\theta) \mid F) = (E(\beta) \mid F(\beta))(F(\beta) \mid F) = E(\beta) \mid F.$$

Thus $E \mid F = E(\beta) \mid F$. Since $E \subseteq E(\beta)$, $E = E(\beta)$; that is, $\beta \in E$. Hence all the roots of $p(x)$ are in E. Hence $p(x)$ splits in E.

Corollary. In a field of characteristic 0, an extension E of a field F is a normal extension if and only if it is the splitting field of an irreducible polynomial over F.

Proof: This follows at once from the theorem by use of the definition of a normal extension, and the corollary to Theorem 6.

Let E be an extension field of F. Let G be a group of automorphisms of E leaving the elements of F fixed. The set of all elements x of E such that $x\sigma = x$ for every $\sigma \in G$ forms a subfield of E that contains F. (This is very easy to verify.) It is called the **fixed field of G.**

If $f(x) = (x - \alpha_1), (x - \alpha_2), \ldots, (x - \alpha_n)$ then by multiplication we find that

$$f(x) = x^n - (\alpha_1 + \cdots + \alpha_n)x^{n-1}$$
$$+ (\alpha_1\alpha_2 + \cdots + \alpha_{n-1}\alpha_n)x^{n-2} \cdots (-1)^n \alpha_1\alpha_2, \cdots \alpha_n$$

and the coefficients of $f(x)$ turn out to be

$$- (\alpha_1 + \alpha_2 + \cdots + \alpha_n)$$

$$\sum_{i<j} \alpha_i \alpha_j = \alpha_1 \alpha_2 + \alpha_1 \alpha_3 + \cdots + \alpha_{n-1} \alpha_n$$

$$\sum_{i<j<k} \alpha_i \alpha_j \alpha_k = - (\alpha_1 \alpha_2 \alpha_3 + \cdots + \alpha_{n-2} \alpha_{n-1} \alpha_n)$$

$$\vdots$$

$$(-1)^n \alpha_1 \alpha_2 \cdots \alpha_n$$

The symmetry of these coefficients of $f(x)$ in the roots α_i of $f(x)$ is a property that is used in the next lemma.

Lemma 4. Let E be a finite extension of a field F. If F is the fixed field of $G(E/F)$, then any element $\alpha \in E$ is the root of an irreducible polynomial $f(x)$ over F which splits in E.

Proof: Let $1, \sigma_1, \ldots, \sigma_{n-1}$ be the automorphisms belonging to $G(E/F)$, the group of automorphisms of E leaving F fixed. Applying them successively to α, assume $\alpha, \alpha_2, \ldots, \alpha_r$ are the distinct elements of the set $\alpha, \alpha\sigma_1, \alpha\sigma_2, \ldots, \alpha\sigma_{n-1}$. If α_i is any one of the distinct elements, then for $\sigma_j \in G(E/F)$, $\alpha_i \sigma_j = (\alpha\sigma_k)\sigma_j$ for some σ_k, and hence $\alpha_i \sigma_j = \alpha\sigma_m$ (where $\sigma_k \sigma_j = \sigma_m$). Thus $\alpha_i \sigma_j$ is one of the set of distinct elements $\alpha, \alpha_2, \ldots, \alpha_r$. Moreover, $\alpha_i \neq \alpha_j$ implies $\alpha_i \sigma_k \neq \alpha_j \sigma_k$. This proves that $G(E/F)$ permutes the α_i among themselves. Hence the coefficients of the polynomial $f(x) = (x - \alpha)(x - \alpha_2) \cdots (x - \alpha_r)$ are left fixed by each automorphism in $G(E/F)$ and therefore, since F is the fixed field of $G(E/F)$ these coefficients belong to F. We have constructed a polynomial $f(x)$ over F, with α as a root, which splits in E.

If $g(x)$ is a polynomial over F with α as a root, then $g(\alpha) = 0$ and therefore $g(\alpha\sigma_i) = 0$ for $i = 2, 3, \ldots, r$. Hence degree of $g(x) \geq r$. Since degree $f(x) = r$, it follows that $f(x)$ is irreducible.

Corollary. If E is a finite extension of a field of characteristic 0 and F is the fixed field of $G(E/F)$, then E is a normal extension of F.

Proof: This follows at once from the corollary to Theorem 17.

In the next theorem we do not assume the characteristic of the field to be 0.

THEOREM 18. If E is a finite extension of F and if F is the fixed field of $G(E/F)$, then E is a normal extension of F.

Proof: Let $E \mid F = n$ and let $1, \beta_1, \ldots, \beta_{n-1}$ be a basis of E over F. Let $f_1(x)$ be the irreducible polynomial over F (constructed as in the previous lemma) having β_1 as a root, $f_2(x)$ the irreducible polynomial having β_2 as a root, and so on. Then E is the splitting field of the polynomial $p(x) = f_1(x)f_2(x) \cdots f_{n-1}(x)$. The polynomial $p(x)$ is separable since by construction each of its irreducible factors has simple roots. Hence E is a normal extension of F. We can now assert

THEOREM 19. If E is a finite extension of a field F such that $E \mid F = \#G(E/F)$, then F is the fixed field of $G(E/F)$ and E is a normal extension of F.

Proof: Assume F' to be the fixed field of $G(E/F)$. Then $F \subset F' \subset E$.

If $\sigma \in G(E/F)$, then σ leaves F' fixed and hence $\sigma \in G(E/F')$. Thus $G(E/F) \subset G(E/F')$. If $\tau \in G(E/F')$ then τ leaves F' fixed and therefore leaves F fixed. Hence $\tau \in G(E/F)$ and thus $G(E/F') \subset G(E/F)$. Therefore $G(E/F) = G(E/F')$.

It follows then that F' is the fixed field of $G(E/F')$ and so, by Theorem 18, we see that E is a normal extension of F'. Hence by Theorem 16 it follows that $E \mid F' = \#G(E/F') = \#G(E/F) = E \mid F$. Since $F \subset F'$, we have at once $F' = F$. Thus F is the fixed field of $G(E/F)$, and now, by Theorem 18, we see that E is a normal extension of F.

Let us now collect some results of our previous theorems, combine them, and restate them as a theorem in a more compact and efficient form. Because of their great importance, it is worth doing this.

THEOREM 20.
  (A) E is a normal extension of a field F if and only if $E \mid F = \#G(E/F)$.
  (B) E is a normal extension of a field F if and only if F is the fixed field of $G(E/F)$.

Proof: Part (A) is an immediate consequence of combining Theorems 16 and 19.

Part (B) follows from part (A) and Theorem 18.

9-6 ROOTS OF UNITY

Definition. For n a positive integer, an *nth root of unity* is a root of the polynomial $x^n - 1$.

Since the (formal) derivative nx^{n-1} of $x^n - 1$ has only the root 0, then by Theorem 5 we see that all the nth roots of unity are distinct. Note that $x^n - 1$ has n distinct roots, whatever the characteristic of F is.

THEOREM 21. In the splitting field of $x^n - 1$ the nth roots of unity form a cyclic group A of order n.

Proof: If α and β are nth roots of 1, then $\alpha^n = 1$, $\beta^n = 1$, and hence $(\alpha\beta)^n = 1$. Hence $\alpha\beta \in A$. Moreover if $\alpha^n = 1$, then $(\alpha^{-1})^n = 1$ and hence $\alpha^{-1} \in A$. Clearly $\alpha\beta = \beta\alpha$. Thus A is an abelian group of order n. Hence, by Theorem 29, Chapter 4, A is the direct product of cyclic groups; that is,

$$A = [y_1] X [y_2] X \cdots X [y_r].$$

Moreover, if t_i is the order of the cyclic group $[y_i], i = 1, 2, \ldots, r$ we know $t_i \mid t_{i+1}$ for $i = 1, 2, \ldots, r - 1$. But this means all the elements of A have the order t_r, and hence are t_rth roots of 1. Since there are only t_r distinct t_rth roots of 1, it follows that we must have $A = [y_r]$ and $t_r = n$. Thus A is a cyclic group.

Definition. An nth root of unity that is a generator of the cyclic group of nth roots of 1 is called a **primitive nth root of unity.**

Thus the element α is a primitive nth root of 1 if and only if its order in the cyclic group is n. This means α is a primitive nth root of 1 if and only if $\alpha^n = 1$ and $\alpha^m \neq 1$ for $m < n$.

In general, if α is a primitive nth root of 1, then α^k is a primitive nth root of 1 if and only if k and n are relatively prime. This is very easy to prove. Thus there are $\phi(n)$ primitive nth roots of 1, where $\phi(n)$ is the Euler ϕ-function, defined by $\phi(n)$ is the number of positive integers less than n that are relatively prime to n.

Example. The cube roots of unity are $1, \omega, \omega^2$, where

$$\omega^2 + \omega + 1 = 0. \quad \varphi(3) = 2,$$

and so there are two primitive cube roots of 1. They are ω and ω^2.

Example. The eighth roots of 1 are

$$\pm 1, \pm i, \pm \frac{1}{\sqrt{2}}(1 + i), \pm \frac{1}{\sqrt{2}}(-1 + i).$$

Here $i = \sqrt{-1}$. There are $\phi(8) = 4$ primitive roots, and it is easy to verify that they are the four last ones listed.

9-7 GALOIS THEORY

If E is an extension field of a field F the automorphisms of E that leave F fixed form a group G under map composition. The subfield F' of E which is invariant under this group is called the **fixed field** of G. F' is the totality of all elements of E that are mapped into themselves by the automorphisms belonging to G. Clearly $F \subset F' \subset E$. Of particular importance is the case where $F' = F$, in which case the extension E is a normal extension of F.

Galois theory relates the extensions of fields with the groups of automorphisms for which they are the fixed fields and establishes a one-to-one correspondence between the subgroups of the group of automorphisms of a splitting field with the subfields of the splitting field. Among its many applications, it yields the answer to the solvability of equations by radicals.

For example, the general quadratic equation $x^2 + bx + c = 0$ over the rational field F has the roots $\dfrac{-\alpha - b}{2}$ and $\dfrac{\alpha - b}{2}$, where $\alpha = \sqrt{b^2 - 4c}$, in the splitting field $F(\alpha)$. That is α is a root of $x^2 = \alpha^2$, $\alpha^2 \in F$. $F(\alpha)$ is known as an extension of F by a radical. If the student consults H. S. Hall and S. R. Knight, *Higher Algebra* (St. Martin's Press, New York, 1943), he will find that the general cubic equation and the general quartic equation can be solved by radicals—that is, the splitting fields of each can be obtained by adjoining a finite number of radicals to the rational field. However, this is not true of the general equation of degree $n > 4$. This we shall prove.

Definition. The group $G(E/F)$ of automorphisms of the splitting field E over F of a polynomial $f(x)$ over F is called the **Galois group** of the equation $f(x) = 0$.

Note that in general F is not the fixed field of the Galois group. It is if $f(x)$ is a separable polynomial.

THEOREM 22 (Fundamental Theorem). Let G be the Galois group and E the splitting field of a separable polynomial $f(x)$ over a field F. (Thus E is a normal extension of F.) Then

1. Each intermediate field B, $F \subset B \subset E$, is the fixed field for a subgroup G_B of G and distinct subgroups have distinct fixed fields.

2. For each B, $B \mid F$ = the index of G_B in G and $E \mid B$ = order of G_B. Thus G_B is called the *Galois group* of E over B. Since E is a normal extension of B, B is the fixed field of G_B. Thus G_B defines B uniquely.

3. B is a normal extension of F if and only if G_B is a normal subgroup of G. G/G_B is isomorphic to the group $G(B/F)$ of B over F, and so G/G_B is often referred to as the Galois group of B over F.

The simple diagram

$$
\begin{array}{ccc}
G & \supset\ G_B & \supset\ 1 \\
\updownarrow & \updownarrow & \updownarrow \\
F & \subset\ B & \subset\ E
\end{array}
$$

illustrates the situation, that is, to each subgroup of the Galois group G of $f(x)$ corresponds an intermediate field which is the fixed field of this subgroup and conversely. If the subgroup is normal its intermediate field is a normal extension and conversely.

Proof:

Part 1. E is the splitting field of $f(x)$ for any intermediate field. Hence E is a normal extension of B. Hence B is the fixed field of that subgroup $G_B = G(E/B)$ of G which leaves elements of B fixed. Distinct subgroups have distinct fixed fields, since each automorphism leaving a field fixed belongs to the subgroup for that field.

Part 2. This follows from the fact that $E \mid B$ = the order of G_B and $E \mid F$ = the order of G.   For $E \mid F = (E \mid B)(B \mid F)$ and hence $B \mid F$ must equal the index of G_B in G.

Part 3.   For each $\sigma \in G$, $B \to B\sigma$ is an isomorphism of two subfields of E which leaves F fixed.   These isomorphisms include all those in $G(B/F)$.

Now $\sigma^{-1}G_B\sigma$ is the subgroup of G corresponding to the intermediate field $B\sigma$, that is the automorphisms belonging to $\sigma^{-1}G_B\sigma$ have $B\sigma$ for a fixed field.   For if $b \in B$, $\sigma_1 \in G_B$, then $b\sigma(\sigma^{-1}\sigma_1\sigma) = b\sigma_1\sigma = b\sigma$. Hence $\sigma^{-1}\sigma_1\sigma$ leaves the elements of $B\sigma$ fixed and therefore $\sigma^{-1}G_B\sigma$ leaves $B\sigma$ fixed.   If $\tau \in G$ and if $(b\sigma)\tau = b\sigma$ for all $b \in B$, then $b(\sigma\tau\sigma^{-1}) = b$ and hence $\sigma\tau\sigma^{-1} \in G_B$, hence $\tau \in \sigma^{-1}G_B\sigma$. This proves $B\sigma$ is the fixed field of the subgroup $\sigma^{-1}G_B\sigma$.

Assume G_B is a normal subgroup of G. We wish to show B is a normal extension of F.

Then $\sigma^{-1}G_B\sigma = G_B$ for all $\sigma \in G$. Hence by Part 1, $B\sigma = B$ for all $\sigma \in G$. This implies that every $\sigma \in G$ is an automorphism of B over F and therefore $\sigma \in G(B/F)$. Moreover every automorphism of B extends to an automorphism of the splitting field E. The number of automorphisms $B \to B\sigma$ is the index of G_B in G (for $B\sigma = B\tau$ if and only if σ and τ belong to the same coset of G_B in G), and by Part 2 this equals $B \mid F$. Thus $B \mid F$ = the order of $G(B/F)$ and hence by Theorem 20, B is a normal extension of F.

Conversely, assume B is a normal extension of F. We show G_B is a normal subgroup of G, where G_B is that subgroup that leaves all elements of B fixed. Now $B \mid F$ = the order of $G(B/F)$. The number of distinct isomorphisms $B \to B\sigma$, $\sigma \in G$, is $B \mid F$.   Hence all isomorphisms $B \to B\sigma$ are automorphisms of B over F, and therefore $B\sigma = B$ for every $\sigma \in G$. If $\sigma \in G$, $\sigma_1 \in G_B$, form $\sigma^{-1}\sigma_1\sigma$. Now $b(\sigma^{-1}\sigma_1\sigma) = (b\sigma^{-1})\sigma_1\sigma = (b\sigma^{-1})\sigma = b$. Hence $\sigma^{-1}\sigma_1\sigma \in G_B$, and therefore G_B is a normal subgroup of G.

To complete the proof of Part 3 we now show that G/G_B is isomorphic to the group $G(B/F)$. Each automorphism of B over F, $B \approx B\sigma = B$, corresponds to a coset $G_B\sigma$; that is, to an element of G/G_B, and conversely. Hence $G(B/F) \approx G/G_B$.

Thus B is a normal extension of F if and only if all the isomorphisms $B \rightarrow B\sigma, \sigma \in G$, are automorphisms of B over F. A normal extension is often called **Galois** and the field B is said to be **Galois over F.** The $B\sigma$ are isomorphisms of B that leave F fixed and that are contained in the same overfield E. If B is a normal extension of F, then B must be the splitting field of one or more factors of some separable polynomial over F. Thus the condition that all isomorphisms $B \rightarrow B\sigma$ be automorphisms of B is equivalent to the condition that these isomorphisms permute the roots of each irreducible factor among themselves (in which case the isomorphisms are automorphisms of B).

Example 1. Let F be the field of rational numbers. Consider the polynomial $(x^2 + 1)(x^3 - 2)$ over F. Put $\alpha = \sqrt[3]{2}$. Let ω be an imaginary cube root of unity (ω is therefore a primitive cube root of unity), so that $\omega^2 + \omega + 1 = 0$. Let $i^2 = -1$. Then $F(\omega, i, \alpha)$ is the splitting field of $(x^2 + 1)(x^3 - 2)$. Now $F(\omega, i, \alpha) \,|\, F = 12$. A basis of $F(\omega, i, \alpha)$ over F is $1, i, \omega, \alpha, \alpha^2, i\omega, i\alpha, i\alpha^2, \omega\alpha, \omega\alpha^2, i\omega\alpha, i\omega a^2$.

An automorphism of $F(\omega, i, \alpha)$ over F is determined completely by its effect on ω, i and α. The 12 automorphisms are given by

$$
\begin{bmatrix} \omega \rightarrow \omega \\ i \rightarrow i \\ \alpha \rightarrow \alpha \end{bmatrix}, \;
\begin{bmatrix} \omega \rightarrow \omega \\ i \rightarrow i \\ \alpha \rightarrow \alpha\omega \end{bmatrix}, \;
\begin{bmatrix} \omega \rightarrow \omega \\ i \rightarrow i \\ \alpha \rightarrow \alpha\omega^2 \end{bmatrix}, \;
\begin{bmatrix} \omega \rightarrow \omega \\ i \rightarrow -i \\ \alpha \rightarrow \alpha \end{bmatrix},
$$

$$
\begin{bmatrix} \omega \rightarrow \omega \\ i \rightarrow -i \\ \alpha \rightarrow \alpha\omega \end{bmatrix}, \;
\begin{bmatrix} \omega \rightarrow \omega \\ i \rightarrow -i \\ \alpha \rightarrow \alpha\omega^2 \end{bmatrix}, \;
\begin{bmatrix} \omega \rightarrow \omega^2 \\ i \rightarrow i \\ \alpha \rightarrow \alpha \end{bmatrix}, \;
\begin{bmatrix} \omega \rightarrow \omega^2 \\ i \rightarrow -i \\ \alpha \rightarrow \alpha \end{bmatrix},
$$

$$
\begin{bmatrix} \omega \rightarrow \omega^2 \\ i \rightarrow i \\ \alpha \rightarrow \alpha\omega \end{bmatrix}, \;
\begin{bmatrix} \omega \rightarrow \omega^2 \\ i \rightarrow i \\ \alpha \rightarrow \alpha\omega^2 \end{bmatrix}, \;
\begin{bmatrix} \omega \rightarrow \omega^2 \\ i \rightarrow -i \\ \alpha \rightarrow \alpha\omega \end{bmatrix}, \;
\begin{bmatrix} \omega \rightarrow \omega^2 \\ i \rightarrow -i \\ \alpha \rightarrow \alpha\omega^2 \end{bmatrix}.
$$

These automorphisms form the Galois group G.

Observe that an automorphism is a permutation of the conjugate roots $\alpha, \alpha\omega, \alpha\omega^2$, of the conjugate roots $i, -i$, and of the conjugate roots ω, ω^2.

If B is the intermediate field $F(i)$, $F \subset B = F(i) \subset F(\omega, i, \alpha)$ then G_B is the subgroup of the 6 automorphisms for which $i \rightarrow i$. $B = F(i)$ is the splitting field of $x^2 + 1$ and is therefore a normal extension of F. Thus G_B is a normal subgroup of G. Note also that $F(i) \,|\, F = 2$ which is the index of G_B in G.

Example 2. $x^4 + 1$ is irreducible over the rational field F. However, $x^4 + 1 = (x^2 - \sqrt{2}x + 1)(x^2 + \sqrt{2}x + 1)$ over the real field. Let $\alpha = \dfrac{\sqrt{2}(1 + i)}{2}$, where $i^2 = -1$. Then α, $\sqrt{2} - \alpha$, $-\alpha$, $\alpha - \sqrt{2}$ are seen to be the roots of $x^4 + 1$. Now $F \subset F(\sqrt{2}) \subset F(\sqrt{2}, \alpha)$, where F is the rational field. $F(\sqrt{2}, \alpha)$ is the splitting field of $x^4 + 1$ and of $(x^2 - 2)(x^4 + 1)$.

Since $F(\sqrt{2}) \mid F = 2$ and $F(\sqrt{2}, \alpha) \mid F(\sqrt{2}) = 2$, we have $F(\sqrt{2}, \alpha) \mid F = 4$. $F(\sqrt{2})$ is a normal extension of F and $F(\sqrt{2}, \alpha)$ is a normal extension of F.

Since the order of $G(F(\sqrt{2}, \alpha)/F)$ is 4 it has to be an abelian group. Hence all its subgroups are normal and therefore all intermediate fields B, $F \subset B \subset F(\sqrt{2}, \alpha)$ are normal extensions of F. For example, $F(\sqrt{2} + i)$ is an intermediate field. Actually $F(\sqrt{2} + i) = F(\sqrt{2}, \alpha)$, for $F(\sqrt{2} + i)$ contains $\dfrac{1}{\sqrt{2} + i} = \dfrac{\sqrt{2} - i}{3}$. Thus $\sqrt{2} + i$ is a primitive element. It is a root of $x^4 - 2x^2 + 9 = 0$.

If α is a root of $f(x) = a_0 + a_1 x + \cdots + a_n x^n$, $a_0 \neq 0$, then $\dfrac{1}{\alpha} = -a_0^{-1}(a_1 + a_2\alpha + \cdots + a_n\alpha^{n-1})$. Thus

$$\frac{1}{\sqrt{2} + i} = \frac{2}{9}(\sqrt{2} + i) - \frac{1}{9}(\sqrt{2} + i)^3.$$

THEOREM 23. Let $\alpha_1, \alpha_2, \ldots, \alpha_k$ be the distinct roots of a polynomial $f(x)$ of degree n over a field F in a splitting field

$$E = F(\alpha_1, \alpha_2, \ldots, \alpha_k).$$

Then each automorphism σ of the Galois group $G(E/F)$ of $f(x)$ determines a permutation $\alpha_i \to \alpha_i \sigma$ of the distinct roots of $f(x)$, and conversely the automorphism σ is completely determined by a permutation of the roots.

Proof: σ maps $\alpha_1, \alpha_2, \ldots, \alpha_k$ into their conjugates $\alpha_1\sigma, \ldots, \alpha_k\sigma$— that is, into themselves. Moreover, $\alpha_i \neq \alpha_j$ implies $\alpha_i\sigma \neq \alpha_j\sigma$, since σ is an automorphism. Hence σ determines a unique permutation of the roots.

Conversely, every element $\beta \in E$ is a polynomial $\beta(\alpha_1, \alpha_2, \ldots, \alpha_k)$ in the α_i with coefficients in F. Hence

(4) $[\beta(\alpha_1, \ldots, \alpha_k)]\sigma = \beta(\alpha_1\sigma, \ldots, \alpha_k\sigma)$

and thus the effect of σ on β is completely determined by its effect on the roots α_i. Thus a permutation of the α_i determines a unique automorphism in the Galois group.

Since the product of two permutations of the roots corresponds to the product of the corresponding automorphisms, it follows that the Galois group of any polynomial is isomorphic to a group of permutations of its roots. If the degree of the polynomial is n then the order of the Galois group must be a divisor of n!

EXERCISES

1. Find the Galois group of the equation $x^3 - 2 = 0$ over the rational field F and write out the automorphisms explicitly. Find the splitting field E. Determine the intermediate fields corresponding to all subgroups of G and determine whether they are normal or not.

2. Do the same for the equation $x^4 - 2 = 0$ over the rational field.

3. Find the Galois group G of the equation $x^4 + 2x^2 - 5 = 0$ over the rational field. Find the splitting field and all intermediate fields corresponding to the subgroups of G.

4. Rework the previous problems when the base field is taken to be the real field in place of the rational field.

5. Prove that $F(i, \sqrt{\sqrt{2}+1}, \sqrt{\sqrt{2}-1})$ is a splitting field of the equation

$$(x^4 + 4x^2 + 2)^2 = 8x^4$$

over the rational field F. What is the order of the Galois group G of this equation? Write out the automorphisms and the intermediate fields corresponding to the subgroups of G.

9-8 EXTENSIONS BY RADICALS

Let F be a field and suppose F contains all roots of unity. Let $\alpha = a^{1/r}$ be a root of $x^r = a$, where $r > 1$ and $a \in F$. If ζ is a primitive rth root of unity, then $F(\alpha)$ contains all the roots $\alpha, \alpha\zeta, \ldots, \alpha\zeta^{r-1}$ of $x^r - a$. Clearly $F(\alpha)$ is the splitting field of $x^r - a$ and hence $F(\alpha)$ is a normal extension of F. $F(\alpha)$ is called an **extension of F by a radical,** the radical being $\alpha = a^{1/r}$. The Galois group $G = G(F(\alpha)/F)$ is of order r.

Lemma 5. If $F(\alpha)$ is an extension of F by a radical then $G(F(\alpha)/F)$ is an abelian group.

Proof: An automorphism $\sigma \in G(F(\alpha)/F)$ is completely determined by its effect on α. Let $\sigma, \tau \in G(F(\alpha)/F)$. Then $\alpha\sigma = \alpha\zeta^s$ for some s, and $\alpha\tau = \alpha\zeta^t$ for some t, where $0 \leq s, t \leq r - 1$. Hence $\alpha(\sigma\tau) = \alpha\zeta^{t+s} = \alpha(\tau\sigma)$. Therefore $\sigma\tau = \tau\sigma$ and $G(F(\alpha)/F)$ is abelian.

Definition. An extension field K of a field F is called an *extension of F by radicals* if

(5) $$F = K_0 \subset K_1 \subset K_2 \cdots \subset K_m = K$$

where each $K_i, i = 1, 2, \ldots, m$, is an extension of K_{i-1} by a radical.

This means that in an extension by radicals $K_i = K_{i-1}(\alpha_i)$ where $\alpha_i^{r_i} \in K_{i-1}, i = 1, 2, \ldots, m$. Each K_i is therefore a normal extension of K_{i-1} and each $G(K_i/K_{i-1})$ is an abelian group. However the definition does not require that K be a normal extension of F.

It is always possible to obtain an extension field K^* of F by radicals for which K^* is a normal extension of F. To see this, let (5) be an extension K of F by radicals. Let $K_1 = F(\alpha_1)$ where $\alpha_1^{r_1} \in F$. Then, as we have seen, K_1 is a normal extension of F. Next let $K_2 = K_1(\alpha_2)$, where $\alpha_2^{r_2} \in K_1$. Form the polynomial

$$f_1(x) = (x^{r_2} - \alpha_2^{r_2})(x^{r_2} - \alpha_2^{r_2}\sigma_2) \cdots (x^{r_2} - \alpha_2^{r_2}\sigma_{r_2})$$

where $1, \sigma_2, \ldots, \sigma_{r_2}$ are the automorphisms of K_1 over F. These automorphisms permute the factors of $f_1(x)$ and hence leave unchanged the coefficients of the polynomial $f_1(x)$. Since K_1 is a normal extension of F (and hence F is the fixed field of the σ_i), these coefficients must belong to F and hence $f_1(x)$ is a polynomial over F. If we adjoin successively the roots of $f_1(x)$ to K_1, we get an extension K_2^* of K_1 which is a normal extension of F. Continuing in this way we can reach an extension field K^* of F by radicals which is a normal extension of F.

THEOREM 24. If K is a normal extension of a field F by radicals then the Galois group $G(K/F)$ is a solvable group.

Proof: Let $F = K_0 \subset K_1 \subset \cdots \subset K_m = K$.

Now K is the splitting field of a separable polynomial over F and hence the splitting field of this same polynomial over K_1. Thus K is a normal extension of K_1, and in the same way we see that K is a normal extension of each K_i, $i = 0, 1, 2, \ldots, m - 1$. Moreover each K_i is a normal extension of F.

Put $G_i = G(K/K_i), i = 1, 2, \ldots, m$, and $G = G(K/F)$. Then

(6) $$G = G_0 \supset G_1 \supset G_2 \supset \cdots \supset G_m = 1$$

is a subnormal series for G, for by the (Fundamental) Theorem 22, each $G_i \triangle G_{i-1}$. Now

$$G_{i-1}/G_i = G(K/K_{i-1})/G(K/K_i)$$

and again by part (3) of Theorem 22, we have

$$G(K/K_{i-1})/G(K/K_i) \approx G(K_i/K_{i-1}).$$

Hence $G_{i-1}/G_i \approx G(K_i/K_{i-1})$. Since each $G(K_i/K_{i-1})$ is, by Lemma 5, an abelian group, G_{i-1}/G_i is an abelian group for each $i = 1, 2, \ldots, m$. Hence (6) has abelian factors and therefore G is a solvable group.

Definition. Again let F be a field and assume F contains all roots of unity. An equation $f(x) = 0$ over F is said to be **solvable by radicals** if its splitting field E lies in an extension K of F by radicals.

As we have remarked before there is no loss of generality in assuming K is a normal extension of F.

Lemma 6. The homomorphic image of a solvable group G is solvable.

Proof: Let $G = G_0 \supset G_1 \supset \cdots \supset G_m = 1$ be a solvable group, so that the factors G_{i-1}/G_i are abelian. Let f be an epimorphism of $G \to G'$. We want to prove that G' is solvable. Put $G_i' = G_i f$ then

$$G' = G_0' \supset G_1' \supset \cdots \supset G_m' = 1$$

is a subnormal series for G'. This follows from Theroem 15, Chapter 2. It is easy to verify that the mapping g of $G_{i-1}/G_i \to G_{i-1}'/G_i'$ defined by $(xG_i)g = (xf)G_i'$, $x \in G_{i-1}$, is an epimorphism for each $i = 1, 2, \ldots, m$. Since G_{i-1}/G_i is abelian, its homomorphic image G_{i-1}'/G_i' must be abelian. Hence G' is a solvable group.

THEOREM 25. If the equation $f(x) = 0$ over a field F is solvable by radicals, then its Galois group is a solvable group.

Proof: Let K be the normal extension by radicals in which the splitting field E of $f(x)$ lies. Then $G(K/F)$ is a solvable group.

By part (3) of the fundamental theorem 22,

$$G(E/F) \approx G(K/F)/G(K/E),$$

that is, $G(E/F)$ is a homomorphic image of the solvable group $G(K/F)$. Hence $G(E/F)$ is a solvable group.

9-9 THE GENERAL EQUATION OF DEGREE n

Definition. Let $x_1, x_2, \ldots, x_n$ be n variables (transcendental elements) over a field F. A polynomial $g(x_1, x_2, \ldots, x_n)$ is called **symmetric** if it is invariant under the symmetric group S_n of degree n—that is, if $g(x_1, \ldots, x_n)$ is left unchanged by all the permutations of the n subscripts of the n variables $x_1, x_2, \ldots, x_n$.

The n polynomials $s_1 = x_1 + x_2 + \cdots + x_n$, $s_2 = x_1 x_2 + x_1 x_3 + \cdots + x_{n-1} x_n, \ldots, s_n = x_1 x_2 \cdots x_n$ are symmetric polynomials. They are called the **elementary symmetric polynomials** (E.S.P.) in the n variables $x_1, x_2, \ldots, x_n$.

It is a fact (which we shall not prove here) that any symmetric polynomial can be expressed as a polynomial over F in the elementary symmetric polynomials.

Let F be a field and let $x_1, \ldots, x_n$ be elements of an extension field of F that are independent transcendental elements with respect to F. They are called **variables over** F. Now $F[x_1, \ldots, x_n]$ is an integral domain whose elements are polynomials in $x_1, x_2, \ldots, x_n$ with coefficients in F. We denote by $F(x_1, \ldots, x_n)$ the quotient field of $F[x_1, \ldots, x_n]$, that is the field of all rational functions in $x_1, \ldots, x_n$ with coefficients in the field F.

Definition. Let $F(u_1, \ldots, u_n)$ be the field of all rational functions in the variables $u_1, \ldots, u_n$. The equation

$$f(x) = x^n - u_1 x^{n-1} + u_2 x^{n-2} + \cdots + (-1)^n u_n = 0$$

is called the **general equation of degree** n over the field $F(u_1, \ldots, u_n)$.

Clearly $f(x)$ is a polynomial in the integral domain $F(u_1, \ldots, u_n)[x]$.

THEOREM 26. The Galois group of the general equation of degree n is the symmetric group of degree n.

Proof: Let E be the splitting field of the general equation $f(x) = 0$ of degree n. Let $v_1, \ldots, v_n$ be the roots of $f(x) = 0$ in E. Clearly the u_i are elementary symmetric polynomials of the v_i. We are out to prove that

$$G(E/F(u_1, \ldots, u_n)) = S_n.$$

Let $F(x_1, x_2, \ldots, x_n)$ be the field of rational functions in the n variables $x_1, \ldots, x_n$ over the field F. Let $\alpha_1, \ldots, \alpha_n$ be the n elementary symmetric polynomials in the variables $x_1, \ldots, x_n$. Form the polynomial

$$g(x) = (x - x_1)(x - x_2) \cdots (x - x_n)$$
$$= x^n - \alpha_1 x^{n-1} + \cdots + (-1)^n \alpha_n.$$

Then $g(x)$ is a polynomial over $F(\alpha_1, \ldots, \alpha_n)$, whose splitting field is $F(x_1, \ldots, x_n)$.

Consider the mapping σ of $F(u_1, \ldots, u_n) \to F(\alpha_1, \ldots, \alpha_n)$ defined by

$$\frac{g(u_1, \ldots, u_n)}{h(u_1, \ldots, u_n)} \sigma = \frac{g(\alpha_1, \ldots, \alpha_n)}{h(\alpha_1, \ldots, \alpha_n)}.$$

First, observe that $h(\alpha_1, \ldots, \alpha_n) = 0$ implies $h(u_1, \ldots, u_n) = 0$, since the α_i are E.S.P. of the x_i and the u_i are E.S.P. of the v_i. However the u_i are variables (transcendental elements) over F and hence a polynomial in the u_i is zero if and only if all the coefficients are 0—that is, if $h = 0$. This eliminates for the mapping σ the possibility that a denominator $h(\alpha_1, \ldots, \alpha_n)$ could vanish without the corresponding denominator $h(u_1, \ldots, u_n)$ vanishing, and conversely.

Second, observe that the mapping σ is surjective.

Third, it is easy to verify that the mapping σ is a homomorphism (it preserves sums and products).

Finally, ker σ is the set of all $g(u_1, \ldots, u_n)$ for which

$$g(\alpha_1, \ldots, \alpha_n) = 0.$$

But we have just seen that $g(\alpha_1, \ldots, \alpha_n) = 0$ implies $g(u_1, \ldots, u_n) = 0$ implies $g \equiv 0$. Hence ker $\sigma = \{0\}$, that is the mapping σ is injective. Hence σ is an isomorphism.

Now by Theorem 15 the isomorphism σ extends to an isomorphism of the splitting fields E of $f(x)$ and $F(x_1, \ldots, x_n)$ of $g(x)$.

$$E \approx F(x_1, \ldots, x_n).$$

Hence

$$G(E/F(u_1, \ldots, u_n)) \approx G(F(x_1, \ldots, x_n)/F(\alpha_1, \ldots, \alpha_n)).$$

Since the α_i are the E.S.P. in the x_i, each permutation of the symmetric group S_n leaves the α_i fixed and hence induces an automorphism of $F(x_1, \ldots, x_n)$ leaving its subfield $F(\alpha_1, \ldots, \alpha_n)$ fixed. And every automorphism of $F(x_1, \ldots, x_n)$ over $F(\alpha_1, \ldots, \alpha_n)$ permutes the roots of $g(x)$ among themselves and so it is completely determined by the permutation from S_n. Hence

$$G(F(x_1, \ldots, x_n)/F(\alpha_1, \ldots, \alpha_n)) = S_n,$$

and therefore

$$G(E/F(u_1, \ldots, u_n)) = S_n.$$

This completes the proof that the Galois group of the general equation of degree n is the symmetric group S_n of degree n.

For $n \geq 5$ we have proved (see Lemma 4, Sec. 2-11) that the alternating group A_n is simple. It is a normal subgroup of the symmetric group S_n. Hence

$$S_n \supset A_n \supset 1$$

is a composition series for the group S_n. Hence by Theorem 34, Chapter 4, S_n is not a solvable group when $n \geq 5$. This fact combined with Theorem 26 together prove

THEOREM 27. The general equation of degree n, $n \geq 5$, is not solvable by radicals.

For many years after the general biquadratic (4th degree) equation had been solved by radicals, it was believed that such solutions could be found for the general quintic (5th degree) equation, and possibly for the general equations of higher degrees. It was the work of Abel and Galois in the beginning of the nineteenth century that proved this to be impossible.

EXERCISES

1. Prove that any rational symmetric function in n variables (one that is invariant under the permutations belonging to S_n) is the quotient of two symmetric polynomials.

2. If $F(\alpha)$ is the set of all polynomials in α where F is the rational field and $\alpha^3 - \alpha^2 = 2$, prove directly that $F(\alpha)$ is a field. Show that every polynomial in $F(\alpha)$ can be expressed as a polynomial of degree ≤ 2 and find the inverses of $\alpha^2 + \alpha + 1$ and of $\alpha^5 + 3$.

3. Prove $F(\alpha) \approx F[x]/(x^3 - x^2 - 2)$, where $F(\alpha)$ is the field in Exercise 2.

4. Find the splitting fields of $x^4 - 2x^2 - 3 \in F[x]$ when (a) F is the rational field (b) F is the real field.

5. Show that $F[x]/(x^4 - 2x^2 - 3)$ is not a field.

6. Find the group of automorphisms of the splitting field of the polynomial $x^4 - 2x^2 - 3$ over the rational field.

7. Find a primitive element for the splitting field of $x^4 - 2x^2 - 3$ over the rational field.

8. Find the number of vectors in the vector space $V_n(Z_p)$ where Z_p is the field of residue classes modulo the prime p.

9. Write out the elements of the Galois group G of the equation

$$x^2 - 3x^2 - 10 = 0$$

over the field F of rational numbers.

10. Find all the sixth roots of unity. Do they form a cyclic group?

11. Find the number of primitive 8th roots of unity.

12. Find the dimension of the vector space E over the rational field F, where E is the splitting field of $x^4 - 5x^2 + 6$. Write down a basis for E.

13. If F is a field of characteristic p whose degree with respect to its prime field is m, find the number of elements in F.

14. F and K are fields and $F \subset K$. If $F(\theta)$ is a normal extension of F prove that the group of $F(\theta)$ over F is the same as the group of $K(\theta)$ over K if and only if $F(\theta) \cap K = F$.

15. $x^6 - x + 2$ is a polynomial over the rational field F. Find a field in which this polynomial has a zero.

16. Let F be a field of characteristic 0 that contains the 4th roots of 1. Prove that the group of the "pure" equation $x^4 - a = 0$, $a \in F$, $a \neq 0$, is cyclic over F.

Generalize this result to the pure equation $x^n - a = 0$ where $a \neq 0$, $a \in F$ and F contains the nth roots of 1.

17. If p is a prime prove that the polynomial $x^p - a$ over the field F where $a \neq 0, a \in F$ and F contains the pth roots of 1, either factors completely in F or is irreducible over F.

18. We know that a finite extension K of a field of characteristic 0 can be generated by a primitive element θ; that is, $K = F(\theta)$.

If $f(x)$ is the irreducible polynomial over F with the zero θ and if $\theta = \theta_1, \theta_2, \ldots, \theta_n$ are the zeros of $f(x)$ in some extension field of F, prove that K is identical with all the conjugate fields $F(\theta_i)$ if and only if K is the splitting field of $f(x)$.

19. G is a multiplicative group and Z is the ring of integers. Denote by $Z(G)$ the set of all functions of $G \to Z$ whose values are almost all zero. If an addition $f + g$ and a multiplication $f \cdot g$ for $f, g \in Z(G)$ are defined by

$$x(f + g) = xf + xg, \quad x(f \cdot g) = \sum_{y \in G} yf \cdot (y^{-1}x)g,$$

where $x, y \in G$, prove that $Z(G)$ is an associative ring. It is called the **integral group ring** of G. (In the same way this concept can be generalized to an arbitrary associative ring R in place of Z.)

20. If ψ is the mapping of $G \to Z(G)$ defined by $x\psi = \psi_x$, $x \in G$ where $x\psi_x = 1$ and $y\psi_x = 0$ for $y \neq x$, show that $(Z(G),\psi)$ is the free abelian group, under addition, on G.

21. Prove that $\psi_x \psi_y = \psi_{xy}$,

22. Show that the elements of the ring $Z(G)$ have the form $\sum_{x \in G} n(x)\psi_x$, where $n(x)$ are integers, almost all of which are 0.

23. Let G be a multiplicative group. Let K be an abelian group. Denote by $A(K)$ the (multiplicative) group of automorphisms of K and by Hom (K, K) the ring of group endomorphisms of K. Of course $A(K) \subset$ Hom (K, K).

(a) If a group homomorphism ϕ of $G \to A(K)$ is given, show that ϕ extends to a ring homomorphism τ of $Z(G) \to$ Hom (K, K) as follows:

For $f \in Z(G), f = \sum_{x \in G} n(x)\psi_x$, let τ be defined by $f\tau = \sum_{x \in G} n(x)(x\phi)$.

(b) Next for $k \in K$, $f \in Z(G)$, define kf by $kf = k(f\tau)$. Prove that this gives K the structure of a right $Z(G)$-module. The point of all this is that a group homomorphism of G into the group of automorphisms of an abelian group K determines a $Z(G)$-module structure on K.

9-10 COHOMOLOGY OF A GALOIS GROUP

Integral group rings are important in the cohomology of a Galois Group G of an extension field K of a field F. If we take the multiplicative group K^* of K (omit the zero element) then K^* is an abelian group and there exists a ready-made homomorphism ϕ of the Galois group G into the group of automorphisms of K^*. In fact G is a subgroup of this group of automorphisms. Thus K^* can be made into a $Z(G)$-module. With this preliminary step it is possible to show that the first cohomology group of G relative to this $Z(G)$-module is the unit group 1.

24. Let G be a multiplicative group and let M be a $Z(G)$-module. Define $C^o = M$, C^1 to be the set of mappings of $G \to M$, C^2 the set of mappings of $G \times G \to M$ and C^r the mappings of $G \times G \times \cdots \times G$($r$-fold) $\to M$. For each $r = 0, 1, 2, \ldots$ the mappings in C^r are called **r-cochains.**

(a) For $f, g \in C^r$, define the sum $f + g$ of the two r-cochains by

$$(x_1, x_2, \ldots, x_r)(f + g) = (x_1, \ldots, x_r)f + (x_1, \ldots, x_r)g, \quad x_1, \ldots, x_r \in G.$$

Prove that C^r becomes an abelian additive group.

(b) Prove that the mappings d_r of $C^r \to C^{r+1}$ defined, for $f \in C^r$, by

$$(x_1, x_2, \ldots, x_{r+1})(fd_r) = (x_2, x_3, \ldots, x_{r+1})f + \Sigma(-1)^i(x_1 x_2, \ldots, x_{i-1} x_{i+1},$$
$$\ldots, x_{r+1})f + (-1)^{r+1}(x_1, \ldots, x_r)f \cdot x_{r+1}$$

is a group homomorphism; that is, show $(fg)d_r = (fd_r)(gd_r)$. Note that the right-hand side of the definition is a sum of elements of M.

Prove that for $m \in C^o = M$,

$$x(md_o) = m - mx, \quad x \in G.$$

(c) Put $Z^r = \ker d_r$, $r = 0, 1, 2, 3, \ldots$ and $B^r = \operatorname{im} d_{r-1}$, $r = 1, 2, 3, \ldots$, $B^o = 0$. Now B^r and Z^r are subgroups of C^r. Show $B^r \subset Z^r$ by proving $d_{r-1} d_r = 0$. The elements of Z^r are called **r-cocycles** and the elements of B^r are called **r-coboundaries.**

The factor groups $H^r(G, M) = Z^r/B^r$, $r = 0, 1, 2, \ldots$ are called the **cohomology groups** of G relative to M, $H^r(G, M)$ being the **rth cohomology group.** $H^1(G, M)$ is called the **first cohomology group.**

In homology theory the sequence

$$C^0 \xrightarrow{\ d_0\ } C^1 \xrightarrow{\ d_1\ } C^2 \xrightarrow{\ d_2\ } \cdots \to C^r \xrightarrow{\ d_r\ } \cdots$$

of abelian groups and homomorphisms d_r such that $\operatorname{im} d_{r-1} \subset \ker d_r$ is called a **semiexact complex** of abelian groups.

chapter 10

Homological Algebra

Algebraic topology has transmitted to algebra many of its concepts and methods, and homological algebra has been described as an outgrowth of an "invasion of algebra by the methods of algebraic topology." An enterprise of algebraic topology is to characterize certain topological spaces by means of some associated algebraic systems, which are usually groups. One seeks to assign to each topological space a family of abelian groups, called its **homology groups**. The homology groups (attached to each space X) and their (induced) homomorphisms (attached to each continuous map $X \to Y$ of two spaces) furnish an algebraic description of topological spaces, and properties of the spaces or of their maps can often be derived from the properties of these groups and their homomorphisms. This, of course, is nothing but a translation of topological problems into algebraic problems, the hope being that the latter may be more easily solvable. The homology groups are topological invariants—that is, if two topological spaces are **homeomorphic** (there is a bijective mapping of one space into the other that is continuous and has a continuous inverse), then their homology groups are isomorphic. (However the converse is not in general true). Thus if the homological groups of two topological groups are not isomorphic, then the spaces are not homeomorphic. It has been said that this is usually the easiest way to prove that two topological spaces are not homeomorphic. A homeomorphism in topology is analogous to an isomorphism in algebra.

Similarly, the assignment of homology groups to other mathematical systems (the group, the associative algebra) can often yield information about these systems and their homomorphisms. In each instance the homology groups of the mathematical system can be calculated by means of a complex (a certain semiexact sequence of abelian groups fashioned out of objects that are all algebraic systems of the same type.) These objects and their appropriate homomorphisms form a "category" and the homology groups are the values of functors defined on this category whose range (codomain) is the category of abelian groups and their homomorphisms. More explicitly, the homology groups are these abelian groups. The concept of homology now assumes a purely algebraic flavor, virtually detaching itself from topology.

The methods and ideas of homological algebra have applications to other branches of algebra, notably in ring theory and in the theory of Lie algebras. For example, it has been proved that an associative ring R with identity is a semisimple ring with minimum condition on left ideals, if and only if every left R-module is projective. The reader is referred to Chapter 11 for a discussion of such rings.

The module proves to be a fundamental algebraic system in homological algebra. As we have seen, it includes the abelian group and the vector space. In particular the projective module is of special importance. The notion of an exact sequence of modules and their homomorphisms is basic to homological algebra and leads to the construction of diagrams of such sequences. A great deal of the work is facilitated by use of diagrams and many results are derived in terms of properties of sequences. In this branch of algebra, diagrams play such an important role that the technique of diagram chasing (following arrows around a diagram) is an activity of daily occurrence, attaining a fascinating incandescence.

Exact sequences, the splitting of exact sequences, projective and injective modules, categories and functors, are all rudimentary concepts in homological algebra, and we shall take them up in turn. Our purpose, in this chapter, is to provide a brief introduction, superficial as it is, to this important and recently developed branch of algebra.

10-1 THE SPLITTING OF EXACT SEQUENCES

Let capital letters $A, B, C, \ldots$ denote right unital modules over an associative ring R with unit element. Let Greek letters $\alpha, \beta, \gamma, \ldots$ denote module homomorphisms which we shall picture as arrows. ($A, B, C, \ldots$ can also be interpreted as groups and $\alpha, \beta, \gamma, \ldots$ as their group homomorphisms or as left unital modules and their homomorphisms.)

Let

$$(1) \qquad \cdots \to A \xrightarrow{\alpha} B \xrightarrow{\beta} C \to \cdots$$

denote a finite or infinite sequence of module homomorphisms.

Definitions. If im α = ker β then the sequence (1) is said to be **exact** at B. If im $\alpha \subseteq$ ker β the sequence is called **semiexact** at B.

If the sequence is exact at every $\ldots, A, B, C, \ldots$, it is called an **exact sequence**.

The homomorphism α is called the **input** homomorphism and β is called the **output** homomorphism.

An exact sequence of the form

$$(2) \qquad 0 \xrightarrow{\gamma} A \xrightarrow{\alpha} B \xrightarrow{\beta} C \xrightarrow{\delta} 0$$

where 0 denotes the zero module, is called a **short exact sequence.** Here γ maps 0 into 0 and hence im $\gamma = 0$; δ maps C into 0 and hence ker $\delta = C$. Thus α is injective and β is surjective.

THEOREM 1. The sequence (2) is a short exact sequence if and only if (i) α is injective, (ii) im $\alpha = $ ker β, and (iii) β is surjective.

Proof: The theorem follows at once from the last definition, since α is injective if and only if ker $\alpha = 0$ and β is surjective if and only if im $\beta = C$.

As a reminder to the reader, if a module $B = B_1 \oplus B_2$, then every $b \in B$ has a unique representation in the form $b = b_1 + b_2$, $b_1 \in B_1$, $b_2 \in B_2$. Moreover, for $b = b_1 + b_2$, $b + B_1 \rightarrow b_2$ is an isomorphism $B/B_1 \approx B_2$. Similarly, $B/B_2 \approx B_1$.

The homomorphisms p_i of $B \rightarrow B_i$, $i = 1, 2$ defined by $bp_i = b_i$ are called the **projections** of B on B_i, and the homomorphisms q_i of $B \rightarrow B_i$, $i = 1, 2$, defined by $b_i q_i = b_i$ are called the **injections** of B_i into B. Clearly, $q_i p_i = 1_{B_i}$, $i = 1, 2$ and $q_i p_j = 0$, $i \neq j$. All of this generalizes in an obvious way to

$$B = \sum_{i=1}^{n} \oplus B_i.$$

Definition. A sequence

(3) $$\cdots \rightarrow A \xrightarrow{\alpha} B \xrightarrow{\beta} C \rightarrow \cdots$$

which is exact at B is said to **split at** B if im $\alpha = $ ker β is a direct summand of B, that is if $B = $ im $\alpha \oplus D$ for some submodule D of B.

Definition. An exact sequence is called **split exact** if it splits at each of its nonterminal modules.

THEOREM 2. If the sequence (3) splits at B, then

$$B \approx \text{im } \alpha \oplus \text{im } \beta.$$

Proof: $B = $ im $\alpha \oplus D$. The mapping β of $B \rightarrow$ im β is an epimorphism. Hence

$$\text{im } \beta \approx B/\text{ker } \beta = B/\text{im } \alpha \approx D.$$

Hence $B \approx \text{im } \alpha \oplus \text{im } \beta$.

THEOREM 3. The sequence (3) splits at B if and only if there exists a homomorphism ϕ of $B \rightarrow A$ such that $\alpha\phi = 1_A$, the identity map on A.

Proof: Assume (3) splits at B. Then $B = \text{im } \alpha \oplus D$. Define the homomorphism ϕ of $B \to A$ by $b\phi = a, b \in B$, where $b = a\alpha + d$, $a\alpha \in \text{im } \alpha, d \in D$. Then for any $x \in A$, $x(\alpha\phi) = (x\alpha)\phi = x$, and therefore $\alpha\phi = 1_A$.

Conversely assume the existence of the homomorphism ϕ of $B \to A$ such that $\alpha\phi = 1_A$. Let b be any element of B and let $a = b\phi$. Write $b = a\alpha + (b - a\alpha)$. Then $(b - a\alpha)\phi = b\phi - a\alpha\phi = a - a = 0$. Hence $b - a\alpha \in \text{ker } \phi$. Therefore $B = \text{im } \alpha + \text{ker } \phi$.

For $x \in \text{im } \alpha \cap \text{ker } \phi$, $x = a\alpha$ and $x\phi = 0$. Hence $a\alpha\phi = a = 0$ and hence $x = 0$. Thus $\text{im } \alpha \cap \text{ker } \phi = 0$. Therefore

$$B = \text{im } \alpha \oplus \text{ker } \phi$$

and (3) splits at B.

The short exact sequence (2) is seen to split, trivially, at A and C. Hence if it splits at B it is a short split exact sequence, and by Theorem 2, we have $B \approx \text{im } \alpha \oplus \text{im } \beta$. Since α is injective, $\text{im } \alpha \approx A$, and since β is surjective $\text{im } \beta = C$. Hence for a short split exact sequence (2),

$$B \approx A \oplus C.$$

Example. Let A be a submodule of B and let ι be the inclusion mapping of $A \to B$ and let j be the natural epimorphism of $B \to B/A$. Then

(4) $$0 \to A \xrightarrow{\iota} B \xrightarrow{j} B/A \to 0$$

is a short exact sequence. It splits at B if and only if A is a direct summand of B.

In particular, if B is a finite-dimensional vector space then every subspace is a direct summand of B and so (4) is always a split exact sequence.

Theorem 3 of course holds for short exact sequences, but for these we also have the next theorem.

THEOREM 4. The short exact sequence (2) splits at B if and only if there exists a homomorphism ψ of $C \to B$ such that $\psi\beta = 1_C$.

Proof: Assume (2) splits at B. Then $B = \text{ker } \beta \oplus D$. Let $c \in C$. Since β is surjective, there exists at least one $b \in B$ such that $c = b\beta$. Also, we have $b = x + d, x \in \text{ker } \beta, d \in D$. Define ψ by $c\psi = d$, where $c = b\beta$ and $b = x + d$. ψ is well-defined, for if $c = b'\beta$ then $b\beta = b'\beta$ and $b - b' \in \text{ker } \beta$, and hence if $b' = x' + d', x' \in \text{ker } \beta, d' \in D$, then $d' = d$. Thus $b\beta = b'\beta$ implies $d = d'$. ψ is clearly a homomorphism. For each $c \in C$, $c\psi\beta = d\beta = (b - x)\beta = b\beta = c$. Thus $\psi\beta = 1_C$.

Conversely, suppose ψ exists such that $\psi\beta = 1_C$. Then ψ is injective. (For $x, y \in C, x\psi = y\psi$ implies $x\psi\beta = y\psi\beta$ implies $x = y$.) Hence $C \approx \text{im } \psi$.

Let $b \in B$. Write $b = b\beta\psi + (b - b\beta\psi)$. Now $(b - b\beta\psi)\beta = b\beta - b\beta(\psi\beta) = b\beta - b\beta = 0$. Therefore $b - b\beta\psi \in \ker \beta$, while $b\beta\psi \in \text{im } \psi$. Thus $B = \text{im } \psi + \ker \beta$.

Let $x \in \text{im } \psi \cap \ker \beta$. Then $x = c\psi, c \in C$ and $x\beta = 0$. Hence $c\psi\beta = c = 0$ and hence $x = 0$. Thus im $\psi \cap \ker \beta = 0$, and therefore

$$B = \ker \beta \oplus \text{im } \psi = \text{im } \alpha \oplus \text{im } \psi$$

which means that (2) splits at B.

Similarly, one can prove each of the following:

If $0 \to A \xrightarrow{\alpha} B$ is exact then im α is a direct summand of B if and only if there exists a homomorphism ϕ of $B \to A$ such that $\alpha\phi = 1_A$.

If $B \xrightarrow{\beta} A \to 0$ is exact then ker β is a direct summand of B if and only if there exists a homomorphism ψ of $A \to B$ such that $\psi\beta = 1_A$.

These exact sequences are often said to split if these conditions hold.

A short exact sequence $0 \to A \xrightarrow{\alpha} B \xrightarrow{\beta} C \to 0$ implies that in the exact sequence $A \xrightarrow{\alpha} B \xrightarrow{\beta} C$, α is injective and β is surjective, and that if it splits at B then $B \approx A \oplus C$. Because of its importance we re-state our results for this special case as a theorem.

THEOREM 5. If $A \xrightarrow{\alpha} B \xrightarrow{\beta} C$ is an exact sequence for which α is injective and β is surjective then the following are equivalent conditions:

  (i) The sequence splits at B.
  (ii) $B \approx A + C$.
  (iii) There exists a homomorphism ϕ of $B \to A$ such that $\alpha\phi = 1_A$.
  (iv) There exists a homomorphism ψ of $C \to B$ such that $\psi\beta = 1_C$.

No loss of generality ensues from following the usual practice of identifying isomorphic modules. It has the great advantage of simplifying the statements and proofs of theorems. For instance, in Theorem 5 if we identify im α with A and im β with C then we write $B = A \oplus C$ rather than B is isomorphic to $A \oplus C$. We shall follow this practice from now on.

10-2 PROJECTIVE MODULES

A projective module was defined in Chapter 7, where it was proved that a free module is projective. It was also proved there that any module

is isomorphic to the quotient of a free module. We use the properties
of exact sequences to prove the following theorem.

THEOREM 6. An R-module P is projective if and only if P is a direct
summand of a free R-module.

> **Proof:** Assume P to be projective. Now $P = F/Q$ where F is a free
> module. Hence the mapping ρ of $F \to P$, where ker $\rho = Q$, is an
> epimorphism.

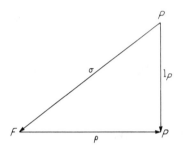

Since P is projective the homomorphism 1_P lifts to the homomor-
phism σ, and $\sigma\rho = 1_P$. If ι is the inclusion mapping of ker $\rho \to F$, then
the sequence ker $\rho \xrightarrow{\ \iota\ } F \xrightarrow{\ \rho\ } P$ is exact at F and moreover it splits at F,
by reason of the existence of σ. Hence P is a direct summand of F.

Conversely, let P be a direct summand of a free module F. Let ι be
the inclusion mapping of $P \to F$ and p the projection of $F \to P$. We
want to show P is projective.

Let γ be a homomorphism of $P \to C$ where β is an epimorphism of
$B \to C$. Then $\rho\gamma$ is a homomorphism of $F \to C$. Since F is projective,

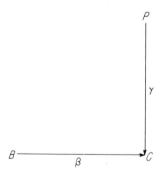

this lifts to a homomorphism σ of $F \to B$, and $\sigma\beta = \rho\gamma$. Hence $\iota\sigma\beta =
\iota\rho\gamma = \gamma$, since $\iota\rho = 1_P$. Thus γ lifts to the homomorphism $\iota\sigma$ of $P \to B$
and hence P is projective.

THEOREM 7. A module P is projective if and only if every short exact sequence $0 \to A \xrightarrow{f} B \xrightarrow{g} P \to 0$ splits.

Proof: Assume P is projective and that the sequence is exact. Then g is an epimorphism and hence 1_P lifts to h and $hg = 1_P$. The existence

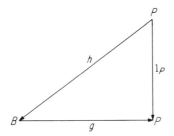

of such a homomorphism h implies (by Theorem 4) that the sequence splits at B.

Conversely assume every short exact sequence of the form $0 \to A \xrightarrow{f} B \xrightarrow{g} P \to 0$ splits. Now $P = F/K$ where F is a free module. Hence the mapping ρ of $F \to P$, where ker $\rho = K$, is an epimorphism. Thus $0 \to$ ker $\rho \to F \xrightarrow{\rho} P \to 0$, where ι is the inclusion mapping, is exact, and by hypothesis splits at F. Hence P is a direct summand of the free module F and by Theorem 6 it is therefore projective.

Let A and B be R-modules. Denote by Hom (A, B) the set of all module homomorphisms from $A \to B$. If $f, g \in$ Hom (A, B), $f + g$ is defined by $a(f + g) = af + ag$, $a \in A$. The zero homomorphism maps every $a \in A$ into 0, and $-f$ is defined by $a(-f) = -(af)$. With these definitions Hom (A, B) becomes an abelian group.

If R is a commutative ring with identity then scalar multiplication, rf, $r \in R$, $f \in$ Hom (A, b), can be defined in Hom (A, B) by $a(rf) = r(af)$, and Hom (A, B) becomes a unital R-module. This will not work however for a noncommutative ring, since for $r_1 r_2 \in R$, $r_1(r_2 f) \neq (r_1 r_2)f$.

THEOREM 8. A module P is projective if and only if for every exact sequence of modules $A \xrightarrow{\alpha} B \xrightarrow{\beta} C$, the sequence of abelian groups

$$(5) \qquad \text{Hom } (P, A) \xrightarrow{\gamma} \text{Hom } (P, B) \xrightarrow{\sigma} \text{Hom } (P, C)$$

is exact, where γ and σ are defined by

$$f\gamma = f\alpha, \quad f \in \text{Hom}\,(P, A) \text{ and } g\sigma = g\beta, \quad g \in \text{Hom}\,(P, B)$$

Proof: Note $P \xrightarrow{\ f\ } A \xrightarrow{\ \alpha\ } B$ and so $f\alpha \in \text{Hom}\,(P, B)$. Similarly $g\beta \in \text{Hom}\,(P, C)$. First we assume the condition and prove P is projective. Let α be an epimorphism of $X \to Y$, then $X \xrightarrow{\ \alpha\ } Y \to 0$ is exact. Then $\text{Hom}\,(P, X) \xrightarrow{\ \gamma\ } \text{Hom}\,(P, Y) \xrightarrow{\ \sigma\ } \text{Hom}\,(P, 0) = 0$ is exact and hence γ is surjective. Therefore for every

$$g \in \text{Hom}\,(P, Y),$$

there exists $f \in \text{Hom}\,(P, X)$ such that $g = f\alpha$. Hence P is projective.

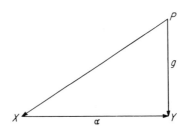

Conversely, assume P is projective and let $A \xrightarrow{\ \alpha\ } B \xrightarrow{\ \beta\ } C$ be exact. Form the sequence (5) with γ and σ defined as above. We want to show it is exact, that is im $\gamma = \ker \sigma$.

Let $f \in \text{Hom}\,(P, A)$, then $f\alpha = f\gamma \in \text{im}\,\gamma$. Since im $\alpha = \ker \beta$, $f\alpha\beta = 0$ and hence $f\alpha \in \ker \sigma$. Thus im $\gamma \subset \ker \sigma$.

Moreover, if $g \in \ker \sigma$, then $g\beta = 0$. Hence $g \in \ker \sigma$ if and only if $Pg \subset \ker \beta = \text{im}\,\alpha$.

Since P is projective, g lifts to $f \in \text{Hom}\,(P, A)$ and $f\alpha = g$. Hence $g \in \text{im}\,\gamma$. Hence $\ker \sigma \subset \text{im}\,\gamma$. Therefore im $\gamma = \ker \sigma$ and (5) is exact.

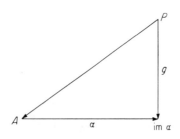

10-3 INJECTIVE MODULES

Definition. An R-module Q is called **injective** if it has the following property: For every injective homomorphism $A \xrightarrow{\alpha} B$ of two R-modules A and B and every homomorphism $A \xrightarrow{\beta} Q$, there exists a homomorphism $B \xrightarrow{\gamma} Q$ such that $\alpha\gamma = \beta$. The diagram illustrates the definition.

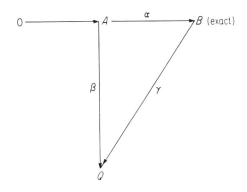

The exactness of the sequence in the top row implies α is injective and the diagram is required to be commutative, hence $\alpha\gamma = \beta$.

Thus if A is a submodule of a module B and if ι is the inclusion mapping $A \xrightarrow{\iota} B$, then a homomorphism of A into an injective module Q extends to a homomorphism $B \xrightarrow{\gamma} Q$, that is $\beta = \gamma$ on A.

Lemma 1. If Q is a direct summand of an injective module Q' then Q is an injective module.

Proof: Let p be the projection of $Q' \to Q$. Let the sequence $0 \to A \xrightarrow{\alpha} B$ be exact. If β is a homomorphism of $A \to Q$, then β is a homomorphism of $A \to Q'$. Since Q' is injective, there exists a homomorphism γ' of $B \to Q$ such that $\alpha\gamma' = \beta$.

Define a homomorphism γ of $B \to Q$ by $\gamma = \gamma'p$. Then for $a \in A$, $a\alpha\gamma = a\alpha\gamma'p = a\beta p = (a\beta)p = a\beta$, since $a\beta \in Q$. Therefore Q is injective.

THEOREM 9. If Q is an injective module then every short exact sequence of the form $0 \to Q \xrightarrow{\beta} A \xrightarrow{\alpha} B \to 0$ splits.

Proof: We need only show it splits at A.

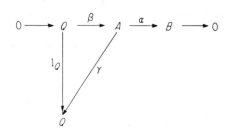

Since Q is injective, it is seen from the diagram that there exists γ such that $\alpha\gamma = 1_Q$. Hence by Theorem 3 the sequence splits at A.

The converse of this theorem is also true, although there appears to be no short elementary proof of it. It is a fact that every module can be proved to be a submodule of an injective module. If we assume this, then the converse can be proved easily as follows.

Let Q be a module such that every short exact sequence of the form $0 \to Q \to A \to B \to 0$ splits. Let Q be a submodule of the injective module Q'. Then $0 \to Q \xrightarrow{\iota} Q' \xrightarrow{j} Q'/Q \to 0$, where ι is the inclusion mapping and j is the natural epimorphism, is exact and hence, by hypothesis, it splits. Thus Q is a direct summand of the injective module Q' and hence by Lemma 1, Q is injective.

A theorem similar to Theorem 8 holds for injective modules. The proof is straightforward and left to the reader with this hint. Draw the necessary diagrams and chase them.

THEOREM 10. A module Q is injective if and only if for every exact sequence $A \xrightarrow{\alpha} B \xrightarrow{\beta} C$, the sequence

$$\text{Hom}\,(C, Q) \xrightarrow{f} \text{Hom}\,(B, Q) \xrightarrow{g} \text{Hom}\,(A, Q)$$

is exact, where for $\phi \in \text{Hom}\,(C, Q)$ and $\psi \in (B, Q)$, f and g are defined by $\phi f = \beta\phi$ and $\psi g = \alpha\psi$.

The reader should observe that injective and projective modules are "duals," in the sense that if the arrows in the diagram defining a projective module are reversed we have the diagram defining an injective module. The **dual of a diagram** is the diagram resulting from the reversal of the directions of all arrows, and if some concept is defined by a diagram the dual of the diagram defines the dual concept.

A coalgebra is a further example of dualization. In Chapter 8 we saw that an associative algebra A over a commutative ring R with 1 can be described as an R-module with a multiplication defined by a linear mapping ϕ of $A \otimes A \to A$ and satisfying the commutative diagram.

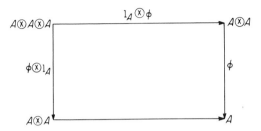

If we reverse all the arrows we get a *coalgebra*. A **coalgebra** is then an R-module and a linear mapping ψ of $B \to B \otimes B$ such that the diagram is commutative.

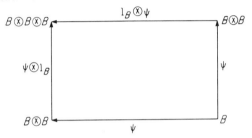

10-4 A TENSOR PRODUCT

We mention here some interesting ideas on tensor products and their relations with exact sequences. Many proofs are omitted but they are well within the scope of the book and the ability of the reader to carry them out. This section is offered as an encouragement of the reader to participate in some independent study.

For commutative rings we defined the tensor product of two modules in Sec. 7-11.

Now let R be an arbitrary associative ring with identity $1 \neq 0$. Let A be a right R-module and B a left R-module.

The *tensor product* $A \otimes B$ is defined as the additive abelian group generated by the symbols $a \otimes b$, $a \in A$, $b \in B$, subject to the conditions

$$(a + a') \otimes b = a \otimes b + a' \otimes b, \quad a, a' \in A, \quad b \in B,$$
$$a \otimes (b + b') = a \otimes b + a \otimes b', \quad a \in A, \quad b' \in B,$$
$$ar \otimes b = a \otimes rb, \quad a \in A, \quad b \in B, \quad r \in R.$$

If the mapping f of $A \times B \to A \otimes B$ is defined by $(a, b)f = a \otimes b, a \in A, b \in B$, then it is easy to show that $(A \otimes B, f)$ has the U.F.P. on $A \times B$ in the following sense. Let X be any additive abelian group and let g be any mapping of $A \times B \to X$ such that $(a + a', b)g = (a, b)g + (a', b)g$, $(a, b + b')g = (a, b)g + (a, b')g$, $(ar, b)g = (a, rb)g$. Then there exists a unique group homomorphism h of $A \otimes B \to X$ such that $g = fh$. In fact this U.F.P. can be used to define the abelian group $A \otimes B$ as the tensor product of the right R-module A and the left R-module B.

The tensor product does not in general preserve exactness. However if $A \to B \to C \to 0$ is an exact sequence of left R-modules and if D is a right R-module, then the tensored sequence

$$D \otimes A \to D \otimes B \to D \otimes C \to 0$$

of abelian groups can be proved to be exact by use of the U.F.P.

10-5 CATEGORIES AND FUNCTORS

The class of all homomorphisms for each type of algebraic system (group, module, etc.) have properties that are studied per se. We are going to consider a collection made up of "objects" (these are the algebraic systems themselves) and of "morphisms" (these are their homomorphisms). We shall subject this unusual collection to certain axioms and in this way obtain what is known as a category. In order to preserve full generality, we define a category abstractly, rather than specifying in concrete terms the particular natures of the objects and morphisms involved. However examples of categories follow right on the heels of the definition.

Functors are mappings of categories into categories; again, they are subject to certain axioms. Crudely speaking, functors on categories (called the domains of the functors) to categories (their codomains) are a sort of generalization of the notion of functions on sets to sets.

A **category** $\mathcal{C}$ is a collection of objects $A, B, C, \ldots$ which contains, for every pair of objects A and B, a set Mor (A, B) of mappings of $A \to B$, called **morphisms,** such that

(i) If $\alpha \in$ Mor (A, B) and $\beta \in$ Mor (B, C), then under map composition $\alpha\beta \in$ Mor (A, C).

(ii) If $\alpha \in$ Mor (A, B), $\beta \in$ Mor (B, C), $\gamma \in$ Mor (C, D), then $(\alpha\beta)\gamma = \alpha(\beta\gamma) = \alpha\beta\gamma$. (Thus $\alpha\beta\gamma$ is defined whenever $\alpha\beta$ and $\beta\gamma$ are defined.)

(iii) For every object $A \in \mathcal{C}$, $\mathcal{C}$ contains a unique morphism $1_A \in$ Mor (A, A) such that for every $\alpha \in$ Mor (A, B) then $1_A\alpha = \alpha$ and for every $\beta \in$ Mor (B, A), $\beta 1_A = \beta$. The morphism 1_A is called the identity on the object A.

Example. A collection of sets, as objects, and the mappings between any pair of them as morphisms form a category under map composition.

Example. A collection of modules as the objects and their homomorphisms as the morphisms is a category.

Functors of a Single Variable

Let $\mathcal{C}$ and $\mathcal{D}$ be categories. A function f of $\mathcal{C} \rightarrow \mathcal{D}$ mapping each object $X \in \mathcal{C}$ into an object $Xf \in \mathcal{D}$ is called a **covariant functor** (of a single variable) with domain $\mathcal{C}$ and codomain $\mathcal{D}$ if

(1) for $\alpha \in \text{Mor}\,(X, Y)$, $\alpha f \in \text{Mor}\,(Xf, Yf)$.

(2) $1_X f = 1_{Xf}$, for all $X \in \mathcal{C}$.

(3) whenever the product $\alpha\beta$ of morphisms in $\mathcal{C}$ is defined, then
$$(\alpha\beta)f = (\alpha f)(\beta f).$$

The function f of $\mathcal{C} \rightarrow \mathcal{D}$ mapping each object $X \in \mathcal{C}$ into an object $Xf \in \mathcal{D}$ is called a **contravariant functor** if

(1′) for $\alpha \in \text{Mor}\,(X, Y)$, $\alpha f \in \text{Mor}\,(Yf, Xf)$

(2′) $1X_f = 1_{Xf}$, for all $X \in \mathcal{C}$.

(3′) when $\alpha\beta$ is defined in $\mathcal{C}$ then $(\alpha\beta)f = (\beta f)(\alpha f)$.

Note that for a contravariant functor f, $X \xrightarrow{\alpha} Y$ is mapped into $Yf \rightarrow Xf$, the second arrow being reversed in direction. Also note the difference in (3) and (3′). Thus a contravariant functor is the dual of a covariant functor.

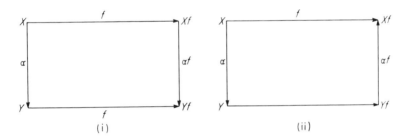

(i)   (ii)

Diagram (i) illustrates a covariant functor f and diagram (ii) a contravariant functor f.

Example. Let $\mathcal{C}$ be the category of sets and their mappings and let $\mathcal{D}$ be the category of abelian groups and their homomorphisms. Define a functor f of $\mathcal{C} \rightarrow \mathcal{D}$ as follows. For a set $S \in \mathcal{C}$ let Sf be the free abelian group generated by S. For each morphism α of $S \rightarrow T$ define αf as the unique homomorphism of $Sf \rightarrow Tf$ for which $\alpha f = \alpha$ on S. The

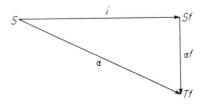

U.F.P. for (Sf, ι) where ι is the inclusion mapping and α is the mapping $S \xrightarrow{\alpha} T \xrightarrow{\iota} Tf$, $\alpha\iota = \alpha$, yields this unique homomorphism αf.

It can be easily verified that f is a covariant functor.   If

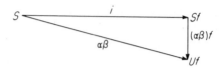

$S \xrightarrow{\alpha} T \xrightarrow{\beta} U$ in $\mathcal{C}$ then $(\alpha\beta)f = (\alpha f)(\beta f)$.   This follows from the U.F.P. We have $(\alpha\beta)f = \alpha\beta$ on S.

Example. Let $\mathcal{C}$ be the category of modules and their homomorphisms and let $\mathcal{D}$ be the category of regularly graded algebras A and their homomorphisms.

Define a functor f of $\mathcal{C} \to \mathcal{D}$ by $Mf = \wedge M$, where M is a module of $\mathcal{C}$ and $\wedge M$ is the exterior algebra of this module. If $T \in$ Hom (M, M'), define $Tf = \wedge T$, where $\wedge T$ is the unique algebra homomorphism of

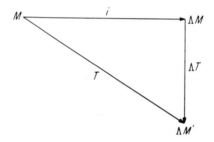

$\wedge M \to \wedge M'$ determined by the U.F.P. for $\wedge M$—that is, $\wedge T = T$ on M. This determines $\wedge T$ uniquely since M and $1 \in R$ generate $\wedge M$.

It is again readily verified that f is a covariant functor, that is that if $M_1 \xrightarrow{T_1} M_2 \xrightarrow{T_2} M_3$ then $\wedge(T_1 T_2) = \wedge T_1 \cdot \wedge T_2$.

Example. Let $\mathcal{C}$ be the category of modules and their homomorphisms.   Let $\mathcal{C}'$ be the category of abelian groups and their homomorphisms.

Define a functor f of $\mathcal{C} \to \mathcal{C}'$ as follows.   Let A be a fixed module. For a module X of $\mathcal{C}$ let $Xf =$ Hom (A, X), where Hom (A, X) is the additive abelian group of all homomorphisms of $A \to X$.   For the homomorphism α of $X \to Y$ in $\mathcal{C}$ define the homomorphism αf of

$$\text{Hom } (A, X) \to \text{Hom } (A, Y)$$

by
$$\gamma(\alpha f) = \gamma\alpha,$$
for all
$$\gamma \in \text{Hom}\,(A, X).$$

It is not hard to verify that f is a covariant functor, often written as Hom $(A, -)$.

Example. For the same two categories as in the preceding example define a functor f as follows. Let B be a fixed module. For a module X of $\mathcal{C}$ let $Xf = \text{Hom}\,(X, B)$. For a homomorphism α of $X \to Y$ in $\mathcal{C}$ define the homomorphism αf of Hom $(Y, B) \to \text{Hom}\,(X, B)$ by $\gamma(\alpha f) = \alpha\gamma$ for all $\gamma \in \text{Hom}\,(Y, B)$.

It can be readily checked that f is a contravariant functor. Again f is often written as Hom $(-, B)$.

Example. Let $\mathcal{C}$ be the category of modules M over a commutative ring R with identity 1 and their homomorphisms. Let $\mathcal{D}$ be the category of the dual modules M^* of M and their homomorphisms. Define a functor f of $\mathcal{C} \to \mathcal{D}$ as follows. For $M \in \mathcal{C}$, let $Mf = M^*$. For $\alpha \in$ Hom (M_1, M_2) let $\alpha f \in$ Hom (M_2^*, M_1^*) be defined by $\gamma(\alpha f) = \alpha\gamma$, $\gamma \in M_2^*$. This can be visualized by $M_1 \xrightarrow{\alpha} M_2 \xrightarrow{\gamma} R$.

For $M_1 \xrightarrow{\alpha} M_2 \xrightarrow{\beta} M_3$ and $\gamma \in M_3^*$—that is, $M_3 \xrightarrow{\gamma} R$, we have $\gamma(\beta f) = \beta\gamma$ and $\gamma(\beta f)(\alpha f) = \alpha\beta\gamma$. For $M_1 \xrightarrow{\alpha\beta} M_3$ and $M_3 \xrightarrow{\gamma} R$ we have $\gamma((\alpha\beta)f) = \alpha\beta\gamma$. Hence $(\alpha\beta)f = (\beta f)(\alpha f)$. This proves f is a contravariant functor.

Functors are not in general "exact," that is they do not preserve exact sequences. By this we mean that if f is a covariant functor of $\mathcal{C} \to \mathcal{D}$ and if $A \to B \to C$ is an exact sequence in $\mathcal{C}$, then it does not follow that $Af \to Bf \to Cf$ is an exact sequence in $\mathcal{D}$, or if f is a contravariant functor that $Cf \to Bf \to Af$ is an exact sequence in $\mathcal{D}$.

However, in the last example it is readily proved that if

$$0 \to A \xrightarrow{\alpha} B \xrightarrow{\beta} C \to 0$$

is a short exact sequence in $\mathcal{C}$, then

$$0 \to C^* \xrightarrow{\beta^*} B^* \xrightarrow{\alpha^*} A^* \to 0$$

is a short exact sequence in $\mathcal{D}$, where

$$\phi\beta^* = \beta\phi, \phi \in C^* \quad \text{and} \quad \rho\alpha^* = \alpha\rho, \rho \in B^*.$$

Thus this contravariant f is exact for short exact sequences.

10-6 BIFUNCTORS

Functors in several variables may be covariant in some of the variables and contravariant in the others. A **bifunctor** is a functor in two variables.

For example, if $\mathcal{C}_1, \mathcal{C}_1, \mathfrak{D}$ are categories, let f be the function on $\mathcal{C}_1 \times \mathcal{C}_2 \to \mathfrak{D}$ such that

(i) for objects $X_1 \in \mathcal{C}_1$, $X_2 \in \mathcal{C}_2$, $(X_1, X_2)f$ is an object in $\mathfrak{D}$.

(ii) if α_1 is a morphism of $\mathcal{C}_1$—that is, $\alpha_1 \in \text{Mor } (X_1, Y_1)$ and if α_2 is a morphism of $\mathcal{C}_2$, $\alpha_2 \in \text{Mor } (X_2, Y_2)$ then $(\alpha_1, \alpha_2)f$ is defined as the morphism $(Y_1, X_2)f \to (X_1, Y_2)f$ of $\mathfrak{D}$. (Note the direction of α_1 is reversed and that of α_2 is the same!)

(iii) $(1_{X_1}, 1_{X_2})f = 1_{(X_1, X_2)f}$

(iv) $(\alpha_1 \beta_1, \alpha_2 \beta_2)f = (\beta_1, \alpha_2)f \cdot (\alpha_1, \beta_2)f$ whenever $\alpha_1 \beta_1$ is defined in $\mathcal{C}_1$ and $\alpha_2 \beta_2$ is defined in $\mathcal{C}_2$.

Note that (iv) can be represented in diagrams by

$$X_1 \xrightarrow{\alpha_1} Y_1 \xrightarrow{\beta_1} Z_1, X_2 \xrightarrow{\alpha_2} Y_2 \xrightarrow{\beta_2} Z_2 \quad \text{together with}$$

(6) $$(Z_1, X_2)f \to (Y_1, Y_2)f \to (X_1, Z_2)f$$

which expresses the composite morphism on the right side of (iv). By diagram chasing, the reader can easily verify this, as well as the fact that (6) is the same as the left side $(\alpha_1 \beta_1, \alpha_2 \beta_2)f$ of (iv). Moreover, diagram chasing will also quickly verify that if $X_1 \in \mathcal{C}_1$ is held fixed, then f is a covariant functor of $X_2 \in \mathcal{C}_2$, while if X_2 is held fixed then f is a contravariant functor of X_1. Thus (i)–(iv) define f as a bifunctor, covariant in the second variable and contravariant in the first.

While not imperative for the study of this chapter, the following provides an amusing and useful exercise in the sport of diagram chasing.

If $\mathcal{C}_1, \mathcal{C}_2, \mathfrak{D}$ are categories then to each ordered pair (X_1, X_2) of objects $X_1 \in \mathcal{C}_1$, $X_2 \in \mathcal{C}_2$, let f be a function assigning the object $(X_1, X_2)f$ of $\mathfrak{D}$.

For a fixed $X_1 \in \mathcal{C}_1$ and $\alpha_2 \in \text{Mor } (X_2, Y_2)$ of $\mathcal{C}_2$ let $(X_1, \alpha_2)f$ denote the morphism $(X_1, X_2)f \to (X_1, Y_2)f$ of $\mathfrak{D}$ and for a fixed $X_2 \in \mathcal{C}_2$ and $\alpha_1 \in \text{Mor } (X_1, Y_1)$ of $\mathcal{C}_1$ let $(\alpha_1, X_2)f$ denote the morphism $(Y_1, X_2)f \to (X_1, X_2)f$ of $\mathfrak{D}$.

Then it is easy to see that for X_1 fixed, $X_2 \to (X_1, X_2)f$ and $\alpha_2 \to (X_1, \alpha_2)f$ define f as a covariant functor of $\mathcal{C}_2 \to \mathfrak{D}$.

Likewise for X_2 fixed,

$$X_1 \to (X_1, X_2)f \text{ and } \alpha_1 \to (\alpha_1, X_2)f$$

define f as a contravariant functor of $\mathcal{C}_1 \to \mathfrak{D}$.

If the accompanying diagram is commutative then the diagonal map

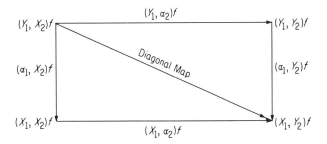

defines f as a bifunctor on $\mathcal{C}_1 \times \mathcal{C}_2 \to \mathcal{D}$ which is contravariant on $\mathcal{C}_1$ and covariant on $\mathcal{C}_2$. Moreover every mixed bifunctor of this kind can be obtained in this fashion.

The mapping functions $(X_1, \alpha_2)f$ and $(\alpha_1, X_2)f$ determine the mapping function $(\alpha_1, \alpha_2)f$ of the bifunctor f. For

$$(\alpha_1, \alpha_2)f = (1_{X_1}\alpha_1, 1_{X_2}\alpha_2)f$$
$$= (\alpha_1, X_2)f \cdot (X_1, \alpha_2)f \quad \text{by condition (iii).}$$

Example. Let $\mathcal{C}$ be the category of modules over a commutative ring R with identity, together with their homomorphisms, as its morphisms. Define a bifunctor on $\mathcal{C} \times \mathcal{C} \to \mathcal{C}$ by (1) $(A, B)f = A \otimes B$ for modules A and B (2) for homomorphisms $\alpha \in \text{Hom } (A, B')$, $\beta \in \text{Hom } (B, B')$, $(\alpha, \beta)f = \alpha \otimes \beta$. Thus $(\alpha, \beta)f$ is a homomorphism of $A \otimes B \to A' \otimes B'$ defined for $a \in A$, $b \in B$, by $(a \otimes b)(\alpha \otimes \beta) = a\alpha \otimes b\beta$. It is easy to see that f is a bifunctor covariant in both variables.

Example. If $\mathcal{C}$ is the category of modules and $\mathcal{C}'$ the category of abelian groups we saw earlier that $\text{Hom } (A, -)$ for A fixed is a covariant functor and $\text{Hom } (-, B)$ for B fixed is a contravariant functor from $\mathcal{C} \to \mathcal{C}'$. If we allow both A and B to vary we can obtain a bifunctor f of $\mathcal{C} \times \mathcal{C} \to \mathcal{C}$ which is contravariant in the first variable and covariant in the second variable.

To achieve this, let $\alpha \in \text{Hom } (A, A')$ and $\beta \in \text{Hom } (B, B')$ be given. Then each $T \in \text{Hom } (A', B)$ determines a unique $\alpha T \beta \in \text{Hom } (A, B')$. Hence the mapping H, defined by $TH = \alpha T\beta$, is seen to be a homomorphism $\text{Hom } (A', B) \to \text{Hom } (A, B')$ of these two abelian groups of $\mathcal{C}'$. If for the functor f we define $(\alpha, \beta)f = H$, then we find

$$(1_A, 1_B)f = 1_{(A, B)f}$$
$$(\alpha\alpha', \beta\beta')f = (\alpha', \beta)f \cdot (\alpha, \beta')f$$

whenever $\alpha\alpha'$ and $\beta\beta'$ are defined in $\mathcal{C}$.

Thus f is a mixed bifunctor, contravariant in the first variable and covariant in the second one.

10-7 UNIVERSAL ELEMENTS

Definition. Let $\mathbb{C}$ and $\mathbb{C}'$ be categories with objects $A, A', \ldots$ and morphisms $\alpha, \alpha', \ldots$, and let f be a covariant functor of $\mathbb{C} \to \mathbb{C}'$. An ordered pair (a', A), where A is an object of $\mathbb{C}$ and $a' \in A' = Af$, is called a **universal element** of the covariant functor f if for every object $X \in C$, and for any $x' \in Xf$, there exists a unique morphism $\gamma \in \text{Mor}(A, X)$ such that $a'(\gamma f) = x'$.

Note that $hf \in \text{Mor}(A', X')$ and is required to map the universal element $a' \in A'$ into $x' \in X'$.

Suppose (b', B), $b' \in B' = Bf$, is a second universal element of the same covariant functor f of $\mathbb{C} \to \mathbb{C}'$. Then there exists a unique morphism $\gamma \in \text{Mor}(A, B)$ such that $a'(\gamma f) = b'$ and a unique morphism $\rho \in \text{Mor}(B, A)$ such that $b'(\rho f) = a'$. Hence $a' = a'(\gamma f)(\rho f) = a'(\gamma \rho)f$, since f is covariant. But $a' = a'(1_A f)$. Since a' is universal, $\gamma \rho$ is unique and hence $\gamma \rho = 1_A$. In a similar way it follows that $\rho \gamma = 1_B$. Therefore the morphism γ is bijective (and so of course is ρ) and $a'(\gamma f) = b'$.

Hence if (a', A) and (b', B) are two universal elements of the same covariant functor f, there exists a bijection γ of $A \to B$ such that $a'(\gamma f) = b'$. This is called the **uniqueness property of universal elements.**

If S and X are sets, denote by X^S the set of all functions from $S \to X$—that is, S is the domain and X the codomain of these functions. We now give an example of a universal element which will be familiar to the reader. Let $\mathbb{C}'$ be the category whose objects are A^S, where S is a fixed set and A is an abelian group. The morphisms of $\mathbb{C}'$ are the mappings of $A^S \to B^S$ where A and B are abelian groups. Let $\mathbb{C}$ be the category of abelian groups and their homomorphisms.

Let f be the covariant functor of $\mathbb{C} \to \mathbb{C}'$ defined by $Af = A^S$, $A \in \mathbb{C}$, and for $\alpha \in \text{Hom}(A, B)$, αf is the mapping of $A^S \to B^S$, given by $\phi(\alpha f) = \phi \alpha$, $\phi \in A^S$. From the diagram $S \xrightarrow{\phi} A \xrightarrow{\alpha} B$ we see that $\phi \alpha \in B^S$.

A universal element of f is the pair (ρ, A), $\rho \in A^S$, $A \in \mathbb{C}$, where for every abelian group $X \in \mathbb{C}$ and every $\psi \in X^S$, there exists a unique homomorphism h of $A \to X$ such that $\rho h = \psi$. As a picture we have the commutative diagram.

Hopefully the reader will recognize that the universal element (ρ, A) is the free abelian group on the given set S. Moreover the uniqueness property of universal elements proves that free abelian groups on the same set are isomorphic. Note that this does not prove the existence of such a universal element.

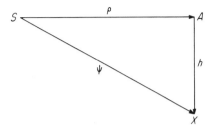

10-8 HOMOLOGY GROUPS

Homological algebra is properly the study of certain groups called **homology groups** and others called **cohomology groups.** The prefix "co" signifies a certain dualization of concepts, for example, coalgebras, important in applications to algebraic topology.

A **complex** C of unital R-modules (all left or all right modules) is a semiexact sequence

$$(7) \qquad \cdots \to C_n \xrightarrow{\ d_n\ } C_{n-1} \to \cdots - \infty < n < + \infty$$

of R-modules C and homomorphisms d_n. Semiexactness means im $d_{n-1} \subset$ ker d_n, for all n—that is, $d_{n-1} d_n = 0$. The groups

$$H_n(C) = \text{ker } d_n / \text{im } d_{n+1}$$

[they themselves are also modules] are called the **homology groups** of the complex. $H_n(C)$ is called the n**th homology group** of C. It is obvious that if the complex C is exact at C then $H_n(C) = 0$ and conversely. The **homology** $H(C)$ of C is the name given to the graded module

$$H(C) = \Sigma \oplus H_n(C).$$

An element of ker d_n is called an n-**cycle** of C and an element of im d_{n+1} is called an n-**boundary** of C. Two n-cycles are called **homologous** if they belong to the same coset of $H_n(C)$—that is, if and only if their difference is an n-boundary.

Let M be a fixed R-module and let C be the complex (7). The set Hom (C_n, M) is an abelian group of module homomorphisms, called **cochains,** of $C_n \to M$. Let Hom (C, M) denote the complex

$$\cdots \to \text{Hom } (C_n, M) \xrightarrow{\ d^{(n)}\ } \text{Hom } (C_{n+1}, M) \to \cdots$$

where $d^{(n)}$ is defined by the diagram $C_{n+1} \xrightarrow{\ d_{n+1}\ } C_n \xrightarrow{\ f\ } M$, that is for $f \in$ Hom $(C_n, M), f d^{(n)} = (-1)^{n+1} d_{n+1} f$. There is a purpose to this choice of sign, but it is beyond the scope of these introductory remarks. The reader can readily verify that Hom (C, H) deserves the name complex. For instance,

$$(d_{n+1} f) d^{(n+1)} = d_{n+1}(f d^{(n+1)}) = d_{n+1} d_{n+2} f = 0,$$

since C is a complex, which proves im $d^{(n)} \subset \ker d^{(n+1)}$.

Thus Hom $(-, M)$ applied to a complex C reverses the arrows. The homology groups of the complex $H(C, M)$ are called the **cohomology groups** of the complex C relative to M. The homology of Hom (C, M) is known as the **cohomology of the complex** C relative to M.

We offer as illustrations an informal introduction to the two funda-mental functors Ext and Tor. For simplicity let us assume that R is a commutative ring with identity.

A **projective resolution** of an R-module A is an exact sequence of the form

$$(8) \qquad \cdots \to P_n \xrightarrow{d_n} P_{n-1} \xrightarrow{d_{n-1}} \cdots \to P_1 \xrightarrow{d_1} P_0 \xrightarrow{\epsilon} A \to 0$$

in which each P_i is a projective module.

Since every R-module is the image of a free module and since a free module is projective, it follows that every R-module has a projective resolution.

We can form two important complexes out of a projective resolution of the module A. To get the first one apply Hom $(-, C)$ to (8), where C is a fixed module. We obtain the following sequence of abelian groups

$$(9)\, 0 \to \mathrm{Hom}\,(A, C) \xrightarrow{\epsilon'} \mathrm{Hom}\,(P_0, C) \xrightarrow{d_1'} \mathrm{Hom}\,(P_1, C) \xrightarrow{d_2'} \cdots$$

where $f\epsilon' = \epsilon f$, for $f \in \mathrm{Hom}\,(A, C)$, and $gd_1' = d_1 g$ for

$$g \in \mathrm{Hom}(P_0, C).$$

Moreover for $f \in \mathrm{Hom}(P_{n-1}, C)$, $f(d_n' d_{n+1}') = (d_n f) d_{n+1}' = d_{n+1} d_n f = 0$, so that (9) is a complex.

The homology of this complex is denoted by $\mathrm{Ext}^n(A, C)$, $n \geq 0$, that is $\mathrm{Ext}^n(A, C) = \ker d_{n+1}' / \mathrm{im}\ d_n'$ is the *nth homology group of the complex*. This is the definition of $\mathrm{Ext}^n(A, C)$. For each n, Ext^n is a bifunctor on the category of R-modules, covariant in C and contravariant in A, to the category of abelian groups.

The second complex is obtained from (8), by forming with a fixed module C the tensor products [i.e., tensoring C into (8)],

$$(10) \qquad \cdots \to P_n \otimes C \xrightarrow{d_n \otimes 1_C} P_{n-1} \otimes C \to$$

$$\cdots \to P_1 \otimes C \to C \to P_0 \otimes C \to A \otimes C \to 0$$

It is a complex of abelian groups whose homology groups are denoted by

$$\text{Tor}_n^R(A, C) = \ker d_n \otimes 1_C / \text{im } d_{n+1} \otimes 1_C$$

(For an arbitrary ring R the tensor product of two modules is at least an abelian group. Actually in this case where R is commutative, the product forms an R-module. See Chapter 7.) For each n, Tor is a covariant bifunctor on the category of R-modules to the category of abelian groups.

It can be shown that Ext and Tor are independent of the particular projective resolution used to define them.

Homological algebra is the story of Hom, $\otimes$, Ext, and Tor.

Some Algebraic Topology

In the hope that the reader may develop an interest in acquiring some comprehension of the link between homological algebra and algebraic topology, we introduce some basic topological ideas. We can hardly aspire to be thorough.

A *topological space* is a set of points for which certain subsets are designated as "open" sets, and these open sets constitute what is called the **topology** of the space. More precisely, if X is a nonempty set, then a topology T on X is a family F of subsets of X such that (i) the union of members of F is a member of F, (ii) the intersection of a finite number of members of F is a member of F, and (iii) F contains X itself and the empty set. It is the members of F that are called the **open sets.** For instance if R is the real line, the **usual** (as it is called) **topology** on R is that for which the open sets are those that are unions of open intervals. If X and Y are two topological spaces, then a map f of $X \rightarrow Y$ is said to be **continuous** if and only if the inverse image $f^{-1}V$ of an open set V of Y is an open set of X. Thus a map f of $R \rightarrow R$ is continuous if the inverse image of an open interval of R is an open set of R. The reader should not have any serious difficulty in convincing himself that this is equivalent to the traditional definition of continuity given in the calculus. The notion of a continuous map in turn leads to the concept of a bijective map of two topological spaces, which is continuous both ways, that is, continuous from each space to the other. Such a map is called a **homeomorphism,** and two topological spaces that are homeomorphic share many topological properties in common. Such properties are known as **topological invariants.** For example, if two spaces are homeomorphic then an open set of one maps under the homeomorphism into an open set of the other, that is the topology is "preserved."

Homology theory itself has a geometrical motivation. It originated in the attempt to attach a numerical criterion or measurement to the bounding effects on a surface caused by closed curves drawn on the surface (i.e., whether a surface is or is not divided by closed curves), and the extension of such a measurement to more general topological spaces. Homology theory evolves into an algebraic study of the connectivity properties of a topological space, leading to topologically invariant properties that provide a test for distinguishing topological spaces. It has now been developed into a highly abstract part of algebra. Homology theory has been successfully axiomatized and constitutes the foundation of algebraic topology.

If X is a topological space, then a topology on a subset A of X can be defined by requiring the open sets of A to be the intersection of A with the open sets of X. A then becomes, by definition, a *subspace* of X.

If X is a topological space and A is a subspace of X, then (X, A) forms what is called a **topological pair.** Let $\mathcal{C}$ be the category of all topological pairs (these pairs are its objects) whose morphisms f of $(X, A) \rightarrow (Y, B)$ are continuous maps of $X \rightarrow Y$, satisfying $Af \subset B$. A **homology theory** on $\mathcal{C}$ is by definition a set of three functions which satisfy seven axioms (the Eilenberg-Steenrod axioms). The first function assigns to each topological pair and to each integer n an abelian group (called the nth **homology group of** X **modulo** A), the second function assigns to each morphism of two topological pairs and to each integer n an (induced) homomorphism of their nth homology groups, the third function assigns to each pair (X, A) and each integer n a homomorphism ∂ (called the **boundary operator**) of the nth homology group of (X, A) into the $(n - 1)$st homology group of the subspace A (here A stands for the pair (A, ϕ), where ϕ is the empty subspace of A). The first two of the seven axioms prove that the first two functions constitute a covariant functor on the category $\mathcal{C}$ to the category of all abelian groups and their homomorphisms. It is called the nth **homology functor.** Thus a homology theory on $\mathcal{C}$ can be described as a collection of pairs, one for each integer n, each pair consisting of a homology functor and a boundary operator, and all satisfying the necessary axioms.

Another triple of functions satisfying an analogous set of axioms defines what is called a **cohomology theory** on the category $\mathcal{C}$. There is a duality in the two theorems. Their difference is essentially that the cohomology functor (the analogue of the homology functor) is contravariant.

The existence of a homology theory on what is known as an admissible category can be proved by means of the singular homology theory on the category of all topological pairs. We shall not define an admissible category but merely make the pronouncement that the category of all topological pairs is the largest admissible category. It contains all admissible categories as subcategories.

We are now going to indicate in a very brief way how the (singular) homology groups of a topological space X are calculated by means of a complex. We shall identify X with the topological pair (X, ϕ), where ϕ denotes the empty subspace of X.

Let E_{n+1} be $(n + 1)$-dimensional Euclidean space. By definition, this means E_{n+1} is the set of all ordered n-tuples $(x_0, x_1 \ldots x_n)$ of real numbers, with the distance $d(x, y)$ between two points $x = (x_0, x_1, \ldots, x_n)$ and $y = (y_0, y_1, \ldots, y_n)$ of E_{n+1} defined by

$$d(x, y) = \left[\sum_{i=0}^{n} (x_i - y_i)^2 \right]^{1/2}$$

This makes E_{n+1} what is called a **metric space** with the metric d. A topology is defined on E_{n+1} by means of this metric. The open sets of E_{n+1} are defined as those subsets that are the unions of open balls, an open ball with center y and radius $r > 0$ being the set $\{x \in E_{n+1} \mid d(x, y) < r\}$. This makes E_{n+1} a topological space. For instance, an open ball in E_1 is an open interval.

A **convex** subset Δ of E_{n+1} is a set such that if $x, y \in \Delta$, then Δ contains all points on the line segment joining the points x and y. For all $n \geq 0$, we are going to define certain convex subspaces of E_{n+1}, called simplexes. These Euclidean simplexes are often picturesquely described as the basic building blocks for the construction of spaces.

For all $n \geq 0$, the **unit n-simplex** Δ_n in E_{n+1} is defined as the **subspace** of E_{n+1} consisting of all points $(x_0, x_1, \ldots, x_n)$ for which

$$\sum_{i=0}^{n} x_i = 1, x_i \geq 0, i = 0, 1, \ldots, n.$$

The points $(1, 0, 0, \ldots, 0), (0, 1, 0, \ldots 0), \ldots, (0, 0, \ldots, 0, 1)$ are called the **vertices** of Δ_n. In fact, Δ_n is the smallest convex set containing these vertices (the **convex hull** of the vertices).

The subspace $\Delta_n^{(i)} = \{(x_0, x_1, \ldots, x_n) \in \Delta_n \mid x_i = 0\}$ of Δ_n is called the ith face of Δ_n, $i = 0, 1, \ldots, n$.

For example, Δ_1 is the line segment joining the points $(1, 0)$ and $(0, 1)$ of the plane; Δ_2 is the triangle (with its interior) whose vertices are $(1, 0, 0), (0, 1, 0), (0, 0, 1)$ in 3-dimensional space. The faces of Δ_2 are the three sides of the triangle. For $n = 3$, Δ_3 is a tetrahedron.

Denote by τ_i the map of $\Delta_{n-1} \to \Delta_n$ defined by

$$(x_0, x_1, \ldots, x_{n-1})\tau_i = (x_0, \ldots, x_{i-1}, 0, x_i, \ldots, x_n).$$

Thus $\Delta_{n-1}\tau_i = \Delta_n^{(i)}$, the ith face of Δ_n.

Let X be an arbitrary topological space and assume $n \geq 0$. A continuous map ϕ of $\Delta_n \to X$ is called a *singular n-simplex* of X.

For $n > 0$, the composite map $\tau_i \phi$ of $\Delta_{n-1} \to X$ is a singular $(n - 1)$-simplex called the ith face $\phi^{(i)}$ of ϕ.

Now form the free abelian (additive) group $C_n(X)$ generated by the set $S_n(X)$ of all singular n-simplexes on a topological space X. The elements of the group $C_n(X)$ are finite sums of the form $m_1\phi_1 + \cdots + m_r\phi_r$, where the ϕ_i are continuous maps of $\Delta_n \to X$ and the m_i are integers. They are called *singular n-chains.* A singular n-chain is therefore *not* a sum of points of X (in general, meaningless) but a sum of continuous maps of Δ_n into X.

By the U.F.P. for this free abelian group,

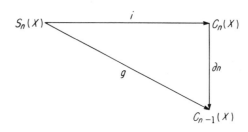

where i is the inclusion map and the map g is defined by

$$\phi g = \sum_{i=0}^{n} (-1)^i \phi^{(i)}, \qquad \phi \in S_n(X),$$

we obtain a unique homomorphism ∂_n such that $i\partial_n = g$. Thus $\phi\partial_n = \Sigma(-1)^i\phi^{(i)}$ and ∂_n maps an n-chain into an $(n-1)$-chain. The homomorphism ∂_n is called the *nth boundary operator.*

For $n < 0$, define $C_n(X) = 0$ and define ∂_n, for $n \leq 0$, as the zero homomorphism. With this understanding, we can thus form a singular chain complex of the space X,

$$(11) \quad \cdots \to C_{n+1}(x) \xrightarrow{\partial_{n+1}} C_n(X) \xrightarrow{\partial_n} C_{n-1}(X) \to \cdots,$$

is a semiexact (im $\partial_{n+1} \subset$ ker ∂_n) sequence of free abelian groups. The semiexactness follows from the fact that it can be proved that $\partial_{n+1}\partial_n = 0$.

For every n-chain $\alpha \in C_n(X)$, the $(n-1)$-chain $\alpha\partial_n$ of $C_{n-1}(X)$ is called the *boundary* of α.

The quotient group

$$(12) \qquad\qquad H_n(X) = \text{ker } \partial_n / \text{im } \partial_{n+1}$$

.is called the *nth singular homology group of the topological space X.*

The kernel of ∂_n is called the group $Z_n(X)$ of *singular n-cycles of X* and the image of ∂_{n+1} is called the group $B_n(X)$ of *singular n-boundaries of X.*

We can describe two n-cycles of X as being *homologous* if and only if their difference is an n-boundary—that is, if and only if they belong to the same coset of B_n in Z_n. This homology relation is an equivalence relation on the set of n-cycles and the equivalence classes are called n-dimensional **homology classes** of X. Of course the nth homology group $H_n(X)$ is the group of the n-dimensional homology classes. These are the cosets of $H_n(X) = Z_n(X)/B_n(X)$.

If A is a subspace of X, we can form the subgroup $C_n(A)$ of $C_n(X)$. Let us define

$$C_n(X, A) = C_n(X)/C_n(A).$$

Moreover we see that $C_{n-1}(A) = C_n(A)\partial_n$, and therefore ∂_n induces a homomorphism

$$\overline{\partial}_n : C_n(X, A) \to C_{n-1}(X, A).$$

Again it can be shown that $\overline{\partial}_{n+1}\overline{\partial} = 0$, and we arrive at another semiexact sequence

(13) $$\cdots \to C_{n+1}(X, A) \xrightarrow{\overline{\partial}_{n+1}} C_n(X, A) \xrightarrow{\overline{\partial}_n} C_{n-1}(X, A) \to \cdots$$

called the *singular chain complex of the topological pair* (X, A).

The group

(14) $$H_n(X, A) = \ker \partial n / \operatorname{im} \partial_{n+1}$$

is called the *nth singular homology group of the topological pair* (X, A).

If A is the empty subspace ϕ of X and if we identify X with the pair (X, ϕ), then (13) and (14) are seen to reduce to (11) and (12) respectively.

It can be proved that definition (14) of the homology groups, together with their induced homomorphisms and boundary operators, constitute a homology theory on the category of all topological pairs.

We mentioned earlier that the homology groups of a topological space can yield information about the space itself. We cite the following illustration.

Let $[0, 1]$ be the closed unit interval, regarded as a subspace of the real line R (Euclidean one-dimensional space). A *path* in a topological space X joining two points x and y of X is defined to be a continuous map f of $[0, 1] \to X$, such that $f(0) = x$ and $f(1) = y$. The space X is said to be *pathwise connected* if there is a path in X joining any pair of its points. Pathwise connectedness is a topological invariant.

For example, Euclidean space E_2 is pathwise connected. If (x_1, y_1) and (x_2, y_2) are any two points of E_2, then the map f of $[0, 1] \to E_2$ defined by

$$f(t) = ((1 - t)x_1 + tx_2, (1 - t)y_1 + y_2), \qquad t \in [0, 1],$$

is continuous and $f(0) = (x_1, y_1), f(1) = (x_2, y_2)$.

Indeed, Euclidean space of any finite dimension is pathwise connected.

Our illustration (given without proof) is this: A topological space X is pathwise connected if and only if the singular homology group $H_0(X)$ is an infinite cyclic group.

EXERCISES

1. R is an associative ring with identity 1. Prove that every R-module is projective if and only if every short exact sequence $0 \to A \to B \to C \to 0$ of R-modules splits.

2. If R is an associative ring with identity 1, prove that every R-module is projective if and only if every R-module is injective.

3. Let A be a submodule of the module B. Prove that the sequence $0 \to A \xrightarrow{i} B \xrightarrow{j} A/B \to 0$, where i is the inclusion and j the natural epimorphism, is exact. Show that if it splits then the sequence

$$0 \to B/A \xrightarrow{f} B \xrightarrow{g} A \to 0,$$

where

$$jf = 1_B, \ ig = 1_A,$$

is exact and conversely.

4. If the exact sequence $0 \to A \xrightarrow{f} B \xrightarrow{g} C \to 0$ splits, show there exist homomorphisms g' and f' such that the sequence $0 \to C \xrightarrow{g'} B \xrightarrow{f'} A \to 0$ is exact.

5. Let $0 \to A \xrightarrow{T} B$ be an exact sequence of finite dimensional vector spaces. If C is a finite dimensional vector space, choose bases for these spaces to prove that the sequence $0 \to A \otimes C \xrightarrow{T \otimes 1_C} B \otimes C$ is exact. Note

$$(a \otimes c)T \otimes 1_C = aT \otimes c.$$

6. Prove that the (external) direct sum $\sum_{i \in I} \oplus P_i$ of a family of R-modules is projective if and only if each direct summand is projective.

7. Do the same thing for injective modules.

8. If f is a functor of $\mathcal{C} \to \mathcal{C}'$ and f' a functor of $\mathcal{C}' \to \mathcal{C}''$, prove that $f \circ f'$ is a functor of $\mathcal{C} \to \mathcal{C}''$ and prove that if f and f' are both covariant or both contravariant then $f \circ f'$ is covariant, whereas if one is covariant and the other contravariant then $f \circ f'$ is a contravariant functor.

9. Write down the conditions to be satisfied in order that $\mathcal{D}$ be a subcategory of a category $\mathcal{C}$. Prove your result.

10. In the accompanying diagram of modules and homomorphisms if the

$$0 \longrightarrow A \xrightarrow{\alpha} B \xrightarrow{\beta} C \longrightarrow 0$$
$$\downarrow{\gamma} \qquad \downarrow{\delta} \qquad \downarrow{\sigma}$$
$$0 \longrightarrow A' \xrightarrow{\alpha} B' \xrightarrow{\beta} C' \longrightarrow 0$$

two rows are exact and the diagram is commutative prove
    (1) If γ and σ are isomorphisms, then so is δ.
    (2) If γ and σ are injective, then so is δ.
    (3) If γ and σ are epimorphisms, then so is δ.
(This is called the Short Five Lemma.)

11. Define the (external) direct sum of two R-modules A_1 and A_2. Let i_1, i_2 be the injections of A_1 and A_2 into $A_1 \oplus A_2$ and let p_1, p_2 be the projections of $A_1 \oplus A_2$ on A_1 and A_2. Prove

$$i_j p_k = 0, \quad j \neq k$$
$$i_j p_j = 1_{A_j}, \quad j = 1, 2$$
$$p_1 i_1 + p_2 i_2 = 1_{A_1 \oplus A_2}.$$

This leads to the direct-sum diagram

$$A_1 \xrightarrow{i_1} A_1 \oplus A_2 \xleftarrow{i_2} A_2.$$
$$\xleftarrow{\quad} \qquad \xrightarrow{\quad}$$
$$p_1 \qquad\qquad p_2$$

12. If in the diagram $A_1 \underset{f'}{\overset{f}{\rightleftarrows}} B \underset{g''}{\overset{g}{\rightleftarrows}} A_2$ the homomorphisms satisfy the three identities in Exercise 11, prove there is exactly one isomorphism of $B \to A_1 \oplus A_2$.

13. If $A \xrightarrow{\alpha} B \xrightarrow{\beta} C \to 0$ is an exact sequence of R-modules and D is any R-module, show that homomorphisms f and g can be defined so that the induced sequence $0 \to \text{Hom}\,(C, D) \xrightarrow{f} \text{Hom}\,(B, D) \xrightarrow{g} \text{Hom}\,(A, D)$ is exact. [$\text{Hom}\,(C, D)$, for instance, is the abelian group of all module homomorphisms of $C \to D$.]

14. If $0 \to A \xrightarrow{\alpha} B \xrightarrow{\beta} C$ is exact, find f and g so that the induced sequence

$$0 \to \text{Hom}\,(D, A) \xrightarrow{f} \text{Hom}\,(D, B) \xrightarrow{g} \text{Hom}\,(D, C) \text{ is exact.}$$

15. Let R be a commutative ring with identity 1. Let $\mathcal{C}$ be the category of all R-algebras (with identities) for objects and their algebra homomorphisms for morphisms. Let M be a fixed R-module. Let f be a mapping of $\mathcal{C}$ into a category $\mathcal{C}'$ and let f map an algebra $A \in \mathcal{C}$ into the set $\text{Hom}\,(M, A) \in \mathcal{C}'$. $\text{Hom}\,(M, A)$ is the set of all module homomorphisms of $M \to A$ (A is also an

R-module). Let f map the algebra homomorphism α of $A \to A'$ into the morphism $\bar{\alpha}$ of Hom $(M, A) \to$ Hom (M, A') defined by $\phi\bar{\alpha} = \phi\alpha$, $\phi \in$ Hom (M, A).

Prove that f is a covariant functor and show that a universal element of this functor f and category $\mathcal{C}$ is $(\rho, T(M))$, the tensor algebra on M. Here ρ is the module homomorphism of $M \to T(M)$.

16. Find a functor and category for which the free module on a set S is the universal element. (Model your construction after the method used for a free abelian group.)

The importance of universal elements is by now quite evident.

chapter 11

Elementary Structure Theory of Rings

In this chapter our objective is twofold. We wish to give an intro-duction into some deeper results in ring theory as well as to acquaint the reader with some of its terminology and methods. Secondly we prove an important theorem in the structure of rings. This is done by means of the density theorem. An alternative proof for this structure theorem, using homological methods, can be found in James P. Jans, *Rings and Homology*, Holt, New York, 1964.

11-1 THE RADICAL

We first define an ideal **J(R)**, called the (Jacobson) radical of a ring, and derive its important properties. The vanishing of this radical char-acterizes a very important type of ring, called a semisimple ring, and a fundamental result is that the quotient ring $R/J(R)$ is a semisimple ring.

Let R be an arbitrary associative ring. We emphasize that R is not assumed commutative nor to have an identity.

Definition. A *modular right ideal B* of R is a right ideal for which there exists an element $a \in R$ such that $ar - r \in B$, for all $r \in R$.

Clearly any right ideal containing a modular right ideal is modular. Note that if R has an identity e, then every right ideal B of R is modular. For $er - r = 0 \in B$ for all $r \in R$.

Lemma 1. Every modular right ideal B is contained in a modular maximal right ideal.

Proof: A maximal right ideal is one that is not strictly contained in any proper right ideal of R. A modular maximal right ideal is simply a maximal right ideal that is modular.

Let $ar - r \in B$, for all $r \in R$. Now $a \notin B$, for if it did, then $ar \in B$ and $r \in B$, and hence $B = R$. If we apply Zorn's axiom to the family F for all right ideals that contain B but not the element a (F is a poset under inclusion), it is easy to prove that there exists a maximal right ideal C containing B but not a. Since $C \supset B$, C is modular. Moreover, C is a maximal right ideal. For if $C \subset D$, where D is a right ideal, then $a \in D$. Otherwise $D \in F$ which would contradict C being a maximal element of F. But $a \in D$ implies $D = R$. Hence C is a modular maximal right ideal that contains B.

If B is a right ideal of R then R/B is not a ring, but R/B is a right R-module.

Lemma 2. If B is a modular maximal right ideal then R/B is an irreducible R-module.

Proof: The submodules of R/B would have to have the form C/B, where $B \subset C$ and C is a right ideal. Since B is maximal, the only submodules of R/B are 0 and R/B. Next $(R/B)R \neq 0$. For if $(R/B)R = 0$, then $(R/B)R \subsetneqq B$ and $R^2 \subset B$. Since B is modular, this implies $B = R$.

Lemma 3. If M is an irreducible R-module, then $M \approx R/B$, for some modular maximal right ideal B.

Proof: If M is irreducible, then M is cyclic and hence $M = mR$, $m \neq 0$. Define a right ideal B by $B = \{r \in R \mid mr = 0\}$. Then B is maximal. Since if $\mathrm{B} \subset C$ then $mC \neq 0$ and, since mC is a submodule of M, $mC = M = mR$. This implies $C = R$. (For if x is any element of R, then there is an element $t \in C$ such that $mt = mx$. Hence $m(x - t) = 0$ and so $x - t \in B \subset C$. Thus $x \in C$, and hence $C = R$.)

Next we show B is modular. Since $M = mR$, there exists $a \in R$ such that $m = ma$ and therefore $m(ar - r) = 0$, for all $r \in R$. Hence $ar - r \in B$ for all $r \in R$.

Finally, the mapping $r + B \to mr$ is readily proved to be an isomorphism of R/B to $mR = M$.

The results of the last two lemmas prove that an R-module M is irreducible if and only if it is isomorphic to R/B for some modular maximal right ideal B. We have in R/B, where B ranges over all modular maximal right ideals of R, isomorphic copies of all irreducible modules over the ring R.

We are going to define a certain ideal of R, called its radical, that is of fundamental importance in the structure theory of rings.

If M is an R-module, then the **annihilator** $A(M) = \{r \in R \mid Mr = 0\}$ is a two-sided ideal of R.

Definition. The (Jacobson) **radical** $J(R)$ of the ring R is the ideal

$$J(R) = \cap A(M)$$

where the intersection is taken over the *annihilators* of all irreducible R-modules.

If R has no irreducible R-modules then we define $J(R) = R$.

First note that, as the intersection of ideals, $J(R)$ is an ideal. Moreover $x \in J(R)$ if and only if $Mx \equiv (0)$ for every irreducible R-module M.

Definition. If B is a right ideal of a ring R, the *quotient ideal* $(B:R)$ is defined by

$$(B:R) = \{x \in R \mid Rx \subset B\}.$$

Note that $(B:R)$ is a two-sided ideal. In the notation of Sec. 7-7, this ideal would be written $(R:B)$.

If B is a modular maximal right ideal then $(B:R)$ is nothing but the annihilator of the irreducible R-module R/B. For $(R/B)x = 0$, $x \in R$, if and only if $Rx \subset B$.

Since all irreducible R-modules are isomorphic to the irreducible R-modules R/B, for B a modular maximal right ideal of R, it follows that

THEOREM 1. $J(R) = \cap (B:R)$.

If B is a modular right ideal of R then $(B:R)$ is the largest ideal contained in B. For let $ax - x \in B$ for all $x \in R$. If $x \in (B:R)$ then $ax \in B$ and hence $x \in B$. Thus $(B:R) \subset B$. If C is an ideal and $C \subset B$, then $Rc \subset C \subset B$ for $c \in C$. Hence $c \in (B:R)$ and $C \subset (B:R)$.

THEOREM 2. $J(R) = \cap B$ where the intersection is taken over all the modular maximal right ideals B of R.

Proof: Since each $(B:R)$ is the largest ideal in the modular maximal right ideal B, it follows that

$$J(R) = \cap (B:R) \subset \cap B.$$

Now let $x \in \cap B$. The $(R/B)x = 0$ for all irreducible R-modules R/B. Hence $x \in \cap(B:R) = J(R)$. Thus $\cap B \subset \cap(B:R)$ and therefore $J(R) = \cap B$.

Corollary. If R is a commutative ring with identity, then $J(R) = \cap B$ over all maximal ideals B of R.

We now derive an important internal or annular property of the radical. It can serve to characterize the radical.

THEOREM 3. For every $a \in J(R)$ there exists $b \in J(R)$ such that

$$a + b = ab.$$

Proof: Let $a \in J(R)$. If no such b exists, then the right ideal B defined by $B = \{ar - r \mid r \in R\}$ does not contain a. B is modular, and hence there exists a modular maximal right ideal of R that does not contain a. This contradicts Theorem 2.

Moreover, since $b = ab - a$ and $a \in J(R)$, it follows that $b \in J(R)$.

Definition. An element $x \in R$ for which there exists $y \in R$ such that $x + y = xy$ is called **right-quasi-regular** and y is called the **right-quasi inverse** of x.

Definition. A right ideal B in R is called **right-quasi-regular** if each of its elements is right-quasi-regular.

Hence if B is a right-quasi-regular right ideal then if $x \in B$, there is a $y \in R$ such that $x + y = xy$. Since $xy \in B$, it follows that $y \in B$. Hence B contains the right-quasi inverses of its elements.

Thus $J(R)$ is a right-quasi-regular ideal. This of course does not mean that $J(R)$ contains all right-quasi-regular elements of R. For example the maximal ideals of the ring Z of integers are (p), where p is a prime. Hence $J(Z) = (0)$. However 2 is a right-quasi-regular element of Z.

THEOREM 4. $J(R)$ contains all right-quasi-regular right ideals of R.

Proof: Let B be a right-quasi-regular right ideal and let M be any irreducible R-module. If $MB \neq (0)$, then $MB = M$. Let $m \neq 0$ generate M. Then $m = mb$ for some $b \in B$. Now $b + c = bc$, $c \in B$. Hence $m(b + c - bc) = 0$, that is $m + mc - mc = m = 0$. This contradiction implies $MB = (0)$. Hence $B \subset J(R)$.

Definitions. An element $x \in R$ is called **left-quasi-regular** if there exists $z \in R$ such that $x + z = zx$ and z is called the **left-quasi-inverse** of x.

Lemma 4. If an element of R has a right-quasi-inverse and a left-quasi-inverse then the two are equal.

The proof of this lemma is left as an exercise.

THEOREM 5. Every element of $J(R)$ is left-quasi-regular and hence $J(R)$ is a left-quasi-regular ideal.

Proof: Let $a \in J(R)$. Then there exists $b \in J(R)$ such that $a + b = ab$. Also there exists $c \in J(R)$ such that $b + c = bc$.

Thus a is the left-quasi-inverse and c is the right-quasi-inverse of the same element b. Hence by Lemma 4, $c = a$ and therefore $a + b = ab = ba$ and hence a is left-quasi-regular.

THEOREM 6. $J(R)$ contains all left-quasi-regular left ideals.

The proof of this theorem is left as an exercise.

Conceivably we might have arrived at a different radical had we used left ideals and left R-modules. Analogous results to those above could have been proved inclusive of one analogous to Theorem 6. But this analogue and Theorem 6 prove the two radicals are the same

THEOREM 7. $J(R/J(R)) = (0)$.

Proof: Write $J(R) = J$. The radical of the ring R/J would be an ideal $\bar{J} = J'/J$, since J' is an ideal in R and $J \subset J'$. For any $r \in J'$ there exists $s \in J'$ such that $\bar{r} + \bar{s} = \bar{r}\bar{s}$, where $\bar{r} = r + J$, $\bar{s} = s + J$. But this implies $r + s - rs \in J$. Write $r \circ s = r + s - rs$ and observe that this "circle" operation is associative. Since $r \circ s \in J$, therefore there exists $t \in R$ such that $(r \circ s) \circ t = r \circ (s \circ t) = 0$. Hence J' is a right-quasi-regular ideal and therefore $J' \subset J$. Hence $J' = J$ and $J(R(R/J)) = (0)$.

With the convention that $J(R) = R$ when R has no irreducible R-modules, we see that $J(R)$ is an ideal in R that can range from being the zero ideal to being the ring itself.

Definition. A ring R for which $J(R) = (0)$ is called a **semi-simple ring.**

Definition. A ring R for which $J(R) = R$ is called a **radical ring.**

If R is a division ring or a field, then $J(R) = (0)$. For every nonzero element of a division ring or a field is a unit and hence the only ideals are (0) and R. This means (0) is the only maximal ideal in a division ring or field.

If R is a ring with identity that contains nonunits, then R has maximal right ideals. For if $a \in R$ and $a \neq 0$, then, if a is not a unit, aR is a right ideal. Moreover $aR \neq R$. (For $aR = R$ implies a is a unit.) Now apply Zorn's axiom to the family F of all proper right ideals of R containing aR (F is a poset under inclusion and the union of a totally ordered subset of F is an upper bound of the subset) and we obtain a maximal right ideal C for the family. But C is a maximal right ideal in R, since if $C \subset D$ then $D \notin F$ and therefore $D = R$. We use this fact in the proof of the next theorem.

THEOREM 8. If R is a ring with identity 1 then $J(R) \neq R$.

Proof: For $J(R) = (0)$ if R is a division ring or a field, and if R is neither then R contains a maximal right ideal. If $J(R) = R$ then $1 \in J(R)$ and hence 1 would have to be in the maximal right ideal, which is a contradiction. Therefore $J(R) \neq R$.

Definition. A right ideal B is called a **nil right ideal** if for each $b \in B$ there exists a positive integer $n = n(b)$, that is depending on b, such that $b^n = 0$.

Definition. A right ideal B is called a **nilpotent right ideal** if $B^n = (0)$ for some positive integer n.

Thus a nilpotent right ideal is nil.

Lemma 5. Every nil right ideal B is in the radical.

Proof: Let $b \in B$, and let $b^n = 0$. Put

$$a = -b - b^2 - b^3 - \cdots - b^{n-1}.$$

Then $b + a = ba$ (and $b + a = ab$), and hence b is quasi-regular. Hence B is quasi-regular and therefore $B \subset J(R)$.

Corollary. Every nilpotent right ideal is in the radical.

Many different types of radicals have been introduced in ring theory, of which the prime radical appears to be one of the most important. If R is a ring with identity, the intersection of all prime ideals of R is called the **prime radical.** We recall that an ideal P in R is prime if, for any ideals A and B of R, $AB \subset P$ implies $A \subset P$ or $B \subset P$.

EXERCISES

1. B is a right ideal in a ring R and $a \in R$, but $a \notin B$. F is the set of all right ideals of R that contain B but not the element a. Then F is a poset under inclusion. Prove that F contains a maximal element. (Use Zorn's axiom and note that the union of a totally ordered subset of F is an element of F and is an upper bound of the subset.)

2. Define a left-quasi-regular left ideal and prove that it contains all left-quasi-inverses of its elements.

3. Prove Lemma 4.

4. Prove that $J(R)$ contains all left-quasi-regular left ideals.

5. If f is an epimorphism $R \rightarrow R'$ of two rings prove that $J(R)f \subseteqq J(R')$.

6. If B is a two-sided ideal of a ring R prove that $J(B) = B \cap J(R)$.

7. An element e of a ring R is called an *idempotent* if $e^2 = e$. Prove $J(R)$ has no nonzero idempotent.

8. B is a maximal right ideal in a ring R.

(a) Prove R/B is an irreducible R-module.

(b) If $R^2 \not\subset B$ prove B is modular.

(c) If $R^2 \not\subset B$, find an element $a \in R$ such that $ax - x \in B$, for all $x \in R$. (If $r \notin B$, $r + B$ generates R/B, hence there exists $a \in R$ such that $(r + B)a = r + B$. Show $r(ax - x) \in B$ for all $x \in R$. Next use the automorphism $rx + B \to x + B$, $x \in R$, of R/B to prove that $ax - x \in B$ for all $x \in R$.)

9. Prove that the radical $J(R)$ can be defined by

$$J(R) = \{a \in R \mid aR \text{ is right-quasi-regular}\}.$$

10. Let $a \in R$. Prove the modular right ideal $B = \{ax - x \mid x \in R\} = R$ if and only if a is right-quasi-regular.

11. If $J(R) \neq R$ and $a \notin J(R)$, then there exists a modular maximal right ideal of R which does not contain the element a.

12. Let R be a commutative ring with identity 1. An element $r \in R$ is called a **unit** if $rs = 1 = sr$, for some $s \in R$. Prove that R has a unique maximal ideal M if and only if the nonunits form an ideal M. (A ring R with exactly one maximal ideal is called a **local ring.** Local rings are used as components in some of the structure theory of rings. They are important in algebraic geometry, in the study of local properties of an algebraic variety. The most important property of such a ring is that its nonunits form an ideal of the ring. In geometry however a local ring is also required to be a ring that is Noetherian.)

13. Let R be an arbitrary ring with $0 \neq 1$. R is called a **local ring** if R contains exactly one maximal right ideal. (This generalizes the definition of local ring.) Prove that a ring R is a local ring if and only if the quotient ring $R/J(R)$ is a division ring.

11-2 PRIMITIVE RINGS

The primitive ring is fundamental in the study of the structure theory of rings. It includes the division ring as a special case. In fact if commutative, a primitive ring is a field. The definition of a primitive ring illustrates the close relationship in ring theory that exists between a ring and its modules.

Definition. An R-module M is called **faithful** if $Mr = (0)$ (that is, $mr = 0$ for every $m \in M$), $r \in R$, implies $r = 0$. Thus an R-module M is faithful if its annihilator $A(M) = (0)$.

Definition. A ring R that has a faithful irreducible R-module is called a **primitive ring.**

A division ring R is a primitive ring, for R is its own faithful, irreducible R-module. However a primitive ring is not in general a division ring.

The reader should verify the following claims.

(1) If M is an R-module and if B is a subring of R then M can be regarded as a B-module.

(2) If B is an ideal in R and M is an R/B-module, then $mr = m(r + B)$, $m \in m$, $r \in R$, defines M as an R-module.

(3) Conversely, if M is an R-module and B is an ideal in R, then $m(r + B) = mr$, $m \in M$, $r \in R$, defines M as an R/B-module.

(4) If f is an epimorphism $R \to R'$ of two rings then $R' \approx R/\ker f$. Hence an R-module can be defined as an R'-module and conversely.

(5) If B is an ideal in R then the submodules of M as an R-module are the same as those of M as an R/B-module.

If M is an irreducible R-module and if $A(M)$ is its annihilator then the ring $R/A(M)$ is a primitive ring. For M is a faithful irreducible module over $R/A(M)$—that is, M is an $R/A(M)$-module.

If B is a modular maximal right ideal of R then $R/(B:R)$ is a primitive ring. For $(B:R)$ is the annihilator of the irreducible R-module R/B.

THEOREM 9. R is a primitive ring if and only if $(B:R) = (0)$ for some modular maximal right ideal B of R.

Proof: If there is a modular maximal right ideal B such that $(B:R) = (0)$, then, since $R/(B:R)$ is a primitive ring, R is primitive.

Conversely, if R is a primitive ring then it has a faithful irreducible R-module R/B, where B is a modular maximal right ideal. Clearly $(R/B)(B:R) = (0)$ and, since R/B is faithful, we have $(B:R) = (0)$.

THEOREM 10. A primitive ring is semisimple.

Proof: Now $J(R) = \cap (B:R)$ over all modular maximal right ideals B. Hence by Theorem 9, we have $J(R) = (0)$.

Since $(B:R)$ is the largest ideal contained in the modular maximal right ideal B it follows that a ring R is primitive if and only if it has a modular maximal right ideal that contains no nonzero ideal of R.

We have the hierarchy

Field $\subset$ division ring $\subset$ primitive ring $\subset$ semi-simple ring.

Definition. A ring of endomorphisms E of an abelian group A is called **irreducible** if and only if the only subgroups B of A that are invariant under E are 0 and A. (A subgroup B is *invariant* under E if $BE \subset B$—that is, if $B\gamma \subset B$ for all $\gamma \in E$.)

We remind the reader that a representation of a ring R is a homomorphism of R into a ring of endomorphisms of an abelian group M.

THEOREM 11. The ring E of endomorphisms of the abelian group A is irreducible if and only if for any $a \neq 0$ in A, $A = aE$.

Proof: Let $A = aE$ for any $a \neq 0$. Let B be a subgroup of A. If $B = 0$ then $BE \subset B$. Let $B \neq 0$ then there exists $a \neq 0$ in B and $aE = A$. Now $BE \subset B$. Hence $aE = B$. Hence $B = A$. Conversely let $a \neq 0$. Then aE is a subgroup of A. But aE is invariant under E. Since $aE \neq 0$, $aE = A$ (since the only subgroups that are invariant are 0 and A).

THEOREM 12. A ring R is primitive if and only if it is isomorphic to an irreducible ring of endomorphisms of some abelian group.

Proof: Let M be a faithful irreducible R-module. Then $M = mR$, $m \neq 0$, (since M is irreducible). M defines a representation ρ of R into a ring E of group endomorphisms of M. This is $\rho : r \to \phi$ where $m\phi = mr$. Now $\phi = 0$ if and only if $r = 0$ (since M is faithful). Hence $M = mE$ and hence E is irreducible. Since $\ker \rho = 0$, ρ is injective and R is isomorphic to E. Conversely, let R be isomorphic to an irreducible ring E of endomorphisms of some abelian group M. This representation ρ determines a module structure on M by $m\phi = mr$, $\phi \in E$. Since ρ is injective, $Mr = 0$ if and only if $r = 0$— that is, M is faithful. It is irreducible, since E is irreducible, so that $M = mE$, $m \neq 0$, that is $M = mR$. (Any submodule of M would have the form mS, where S is a right ideal in R, and mS would be a subgroup of M. Since $mSE \subset mS$, this means $mS = 0$ or $ms = M$.)

Corollary 1. Any ring that is isomorphic to a primitive ring is primitive.

Corollary 2. A ring R is isomorphic to an irreducible ring of group endomorphisms of an abelian group M if and only if the module M determined by this representation is a faithful irreducible R-module.

The two corollaries follow at once from the theorem.

Theorem 12 provides an alternative definition of a primitive ring. It is a ring that is isomorphic to an irreducible ring of endomorphisms of some abelian group.

THEOREM 13. A commutative primitive ring is a field and conversely.

First Proof: The converse follows at once, since a division ring is a primitive ring.

Now let R be a commutative primitive ring. Let $a \neq 0 \in R$, and let M be a faithful irreducible R-module. Hence there exists $m \in M$ such that $ma \neq 0$ and therefore $maR = M$. Hence there exists $b \in R$ such that $mab = m$. Now we also have $M = mR$.

Since R is commutative, then for every mr it follows that

$$mr(ab) = m(ab)r = mr.$$

Since M is faithful, this implies ab is a unit element of R and b is the inverse of the nonzero element a. Hence R is a field.

Second Proof: Let R be a commutative primitive ring. Then R contains a modular maximal ideal B that contains no nonzero ideal. Let $b \in B$. Then $Rb \subset B$. But $(B:R) = (0)$. Hence $Rb \subset B$ implies $b = 0$. Hence $B = (0)$ and therefore (0) is a maximal ideal. Thus R contains only the ideals (0) and R. Moreover if $a \neq 0 \in R$, then $(a) = R$. Hence there exists $e \in R$ such that $ae = ea = a$. Let x be any element of R. Then $x = ay$ for some $y \in R$. Hence $xe = ex = eay = ay = x$, and e is a unit element. A commutative ring R with an identity whose only ideals are (0) and R is a field. For if $b \neq 0 \in R$ then $Rb \neq (0)$ and hence $Rb = R$. Therefore there exists $c \in R$ such that $cb = bc = e$.

Definition. An ideal P of a ring R is called a **primitive ideal** if R/P is a primitive ring.

THEOREM 14. For any ring R, $J(R) = \bigcap P$ over all primitive ideals P of R.

Proof: Let P be a primitive ideal and let M be a faithful irreducible module over R/P. Then $M(r + P) = (0)$ implies $r \in P$. Hence $P = A(M)$, the annihilator of the irreducible module M.

Conversely, if $P = A(M)$ where M is an irreducible R-module, then M is a faithful irreducible module over the ring $R/A(M)$. Hence $R/A(M)$ is a primitive ring and therefore $P = A(M)$ is a primitive ideal.

Thus $J(R) = \bigcap A(M) = \bigcap P$, taken over all primitive ideals P in R.

EXERCISES

1. B is an ideal in a ring R. If f is the natural epimorphism of $R \to R/B$, show that an R-module M can be defined as an R/B-module and conversely.

2. Show that the ring of all endomorphisms of a cyclic group of order n is isomorphic to the ring of residue classes modulo n.

3. If $A(M)$ is the annihilator of an R-module M, prove that the ring $R/A(M)$ is isomorphic to a ring of endomorphisms of the abelian group M. (Consider the ring of endomorphisms f_a, $a \in R$, defined by $mf_a = ma$.)

4. Show that the ring Z of integers is not primitive.

5. A ring R is called a **prime ring** if $aRb = (0)$, $a, b \in R$, implies that $a = 0$ or $b = 0$. Prove that a ring R is prime if and only if the zero ideal is a prime ideal of R.

6. Show that a primitive ring is always prime.

7. Give an example to show a prime ring need not be primitive.

8. If R is a simple ring show $R^2 = R$.

9. Prove that a simple ring with an identity is primitive.

10. If the only ideals of a ring R are (0) and R and if $R^2 = (0)$, show that R is a finite ring with a prime number of elements.

11. B is an ideal in a ring R. Show how an R/B-module M can be defined as an R-module, and show that the submodules of M are the same for M regarded as an R/B-module or as an R-module.

12. Any nonzero ideal of an irreducible ring of endomorphisms of an abelian group G is also an irreducible ring of endomorphisms of G.

11-3 THE d.c.c. AND SIMPLE RINGS

The descending chain condition (d.c.c.) on right ideals is an example of a finiteness condition and it is frequently imposed on a ring. As we have seen earlier it is equivalent to the minimum condition on right ideals. Much of the classical theory of rings is concerned with rings which satisfy the d.c.c. Theorem 15 is an illustration of the effect on the radical of this assumption. Moreover with this condition, primitive rings become semi-simple rings, and vice versa. The ring of integers does not satisfy the d.c.c., for the descending chain of ideals

$$(2) \supset (4) \supset (8) \supset (16) \supset \cdots$$

does not terminate. Another finiteness condition that we have used earlier is the ascending chain condition which was used to define Noetherian rings.

Definition. A **simple ring** R is one whose only ideals are (0) and R and for which $R^2 \neq (0)$.

Definition. A **minimal right ideal** of a ring R is a nonzero right ideal that does not properly contain any nonzero right ideal. A minimal right ideal of a family of right ideals is one that does not properly contain any other right ideal of the family.

Definition. A ring R is said to satisfy the **descending chain condition** *on right ideals* (d.c.c.) if every descending chain of right ideals

$$B_1 \supset B_2 \supset \cdots \supset B_n \supset \cdots.$$

terminates, that is if for some N, $B_N = B_{N+1} = \cdots.$

We know from our previous study (Chapter 7) that the d.c.c. is equivalent to the minimum condition on right ideals; that is, that every nonempty family of right ideals of R contains at least one minimal right ideal.

THEOREM 15. If R has the d.c.c. on right ideals, then $J(R)$ is nilpotent.

Proof: $J \supset J^2 \supset J^3 \supset \cdots \supset J^n = J^{n+1} = \cdots$. Let W be the ideal for which $WJ^n = (0)$. If $W \supset J^n$, then $J^{2n} \subset WJ^n$. Hence $J^{2n} = (0)$, and hence $J^n = (0)$ and hence J is nilpotent. We show $W \supset J^n$. Let ϕ be the natural endomorphism of $R \to \bar{R} = R/W$, that is $r\phi = r + W$. Then $\bar{J} = J\phi \subset J(\bar{R})$ and ϕ maps rJ^n into $\bar{r}\bar{J}^n$. Now $\bar{r}\bar{J}^n = (\bar{0})$ implies $rJ^n \subset W$ and hence that $rJ^{2n} = (0)$. This, in turn, implies $rJ^n = (0)$, and hence if $r \in W$ then $\bar{r} = \bar{0}$. Therefore $\bar{r}\bar{J}^n = (\bar{0})$ if and only if $\bar{r} = \bar{0}$.

Now suppose $J^n \neq (\bar{0})$—that is, that $J^n \not\subset W$. Then $\bar{J}^n$ contains a minimal right ideal ρ (by the d.c.c. applied to the homomorphic image $\bar{R}$ of R). But $\bar{\rho}$ is an irreducible R-module and $\bar{J}^n \subset J(\bar{R})$. Hence $\rho\bar{J}^n = (\bar{0})$ $[J(\bar{R})$ annihilates irreducible modules]. Hence $\bar{\rho} = (\bar{0})$ $[\bar{r}\bar{J}^n = (\bar{0})$ if and only if $\bar{\rho} = \bar{0}]$. This is a contradiction. Hence $\bar{J}^n = (\bar{0})$ and hence $J^n \subset W$ and hence $J^n = (0)$ and therefore J is nilpotent.

THEOREM 16. A simple ring R with d.c.c. on right ideals is primitive.

Proof: Now $J = (0)$ or $J = R$. Also J is nilpotent. If $J = R$ then R is nilpotent. However, the ideal $R^2 \neq (0)$ and so $R^2 = R$. This implies $R^n = R$ and, therefore, R is not nilpotent. Hence $J = (0)$ and hence R is semisimple. Hence R must contain proper primitive ideals [for if it had none, then by definition $J(R) = R$]. Since R is simple, any such ideal is (0) and hence R is primitive.

EXERCISES

1. R is a ring with d.c.c. If f is an epimorphism of $R \to R'$, where R' is a ring, prove that R' has the d.c.c.

2. Prove that the d.c.c. and the minimum condition on right ideals are equivalent.

3. If the ring R has the d.c.c. on right ideals prove that any nil right ideal is nilpotent.

4. If B is a nilpotent right ideal in a ring R, then RB is a nilpotent two-sided ideal.

5. The sum of two nilpotent ideals is a nilpotent right ideal.

11-4 THE DENSITY THEOREM

Let n be a positive integer. A ring R of linear transformations of a vector space V over a division ring D is said to be *n-fold transitive* if for any given ordered set of n linearly independent vectors $x_1, x_2, \ldots, x_n$ and every ordered set of n arbitrary vectors $y_1, y_2, \ldots, y_n$ there exists $r \in R$ such that $x_i r = y_i$, $i = 1, \ldots, n$. R is called *dense* if it is n-fold transitive for every positive integer n. If R is k-fold transitive, $k > 1$, then it is t-fold transitive for all $t < k$. For example the set R of all linear transformations of an arbitrary vector space V over D is dense. [For there exists a basis of V containing an independent set $x_1, \ldots, x_k$. Then $x_i a = y_i$, $i \leq k$, $x_i a = 0$, $i > k$ defines a linear transformation.]

If V is an n-dimensional vector space over a division ring, the set R of all linear transformations of $V \to V$ is clearly a dense ring of linear transformations on V, that is it is k-fold transitive up to the dimension of V. (This is true of any vector space.) If U is a proper subspace of V the subring S consisting of all linear transformations mapping $U \to U$ is a dense ring on U. However, the subring T of all linear transformations of $V \to U$ is not a dense ring on V. If β_1, β_2 is a basis of two-dimensional vector space V, the set of all linear transformations T such that $\beta_1 T = \beta_2 T$ is a ring which is not even 1-fold transitive, since $(\beta_1 - \beta_2) T = 0$ for every T in the ring.

Lemma 6. The set Δ of module endomorphisms of an irreducible R-module M is a division ring and M is a right vector space over Δ.

Proof: Let $f \in \Delta$. Then ker f is a submodule of M. Since M is irreducible, ker $f = 0$ or ker $f = M$. If ker $f = M$, f is the zero endomorphism––that is, the endomorphism which maps all of M into the zero element of M. If ker $f = 0$ then f is injective and, since im f is a submodule, im $f = M$. Hence f is an automorphism of M. Thus all endomorphisms of M, with the exception of the zero endomorphism, are automorphisms of M and hence have inverses which are automorphisms of M. Thus under ordinary map composition (multiplication) and map addition, Δ is a division ring.

Now M is a right vector space over Δ. The endomorphisms are the scalars. For if $m \in M$, $f \in \Delta$, then $mf \in M$. Also $(mr)f = (mf)r$, $r \in R$, and $(m_1 + m_2)f = m_1 f + m_2 f$, $m_1, m_2 \in M$. It is easily verified that M satisfies all the axioms for a right vector space over Δ.

THEOREM 17 (Density Theorem). A primitive ring R is a dense ring of linear transformations on a faithful irreducible R-module M, regarded as a vector space over the division ring of its module endomorphisms.

Proof: The theorem will be proved if we can show that for a finite dimensional subspace V of M, if $m \in M$ and $m \notin V$ there exists $r \in R$ such that $Vr = (0)$ but $mr \neq 0$. For then $mrR = M$ (since M is irreducible). Hence there exists $s \in R$ such that mrs is arbitrary, yet $Vrs = (0)$. Hence if $x_1, x_2, \ldots, x_n$ are independent vectors of M and $y_1, y_2, \ldots, y_n$ are arbitrary vectors of M, then there exists $r_i \in R$ such that $x_j r_i = 0$, $j \neq i$, and $x_i r_i = y_i$. [Here V is the vector space with the basis $x_1, \ldots, x_{i-1}, x_{i+1}, \ldots, x_n$.] Then for

$$r = r_1 + r_2 + \cdots + r_n, x_i r = y_i, i = 1, 2, \ldots, n.$$

We use induction on the dimension of the finite dimensional vector space V, that is we assume that for each $w \notin V$ there exists $r \in R$ such that $Vr = (0)$ and $wr \neq 0$. Hence $m\mathcal{A}(V) = (0)$, where $m \in M$ and $\mathcal{A}(V)$ is the right annihilator of V (it is a right ideal), implies $m \in V$. Thus $w\mathcal{A}(V) \neq (0)$. (Note the induction hypothesis is obviously true for the case dim $V = 0$, and so we take dim $V \geq 1$.) Hence $M = w\mathcal{A}(V)$, since M is irreducible. Thus for each $x \in M$, $x = wa$ for some $a \in \mathcal{A}(V)$.

Let $W = V + w\Delta$, where Δ is the division ring. (Then dim $W = 1 + $ dim V.) Let $m \notin W$ and suppose that $Wr = (0)$ implies $mr = 0$. The mapping ϕ of $M \to M$ defined by $x\phi = ma$, where $x = wa \in M$ and $a \in \mathcal{A}(V)$ is well-defined (for if $wa = 0$ then $Wa = Va + wa\Delta = 0$ and hence $ma = 0$). Moreover, clearly $\phi \in \Delta$. Hence $(m - w\phi)a = 0$. This is true of each x of M and therefore $(m - w\phi)\mathcal{A}(V) = 0$.

Hence $m - w\phi \in V$. But this implies $m \in W$, a contradiction. Hence if $m \notin W$ there exists r such that $Wr = (0)$ and $mr \neq 0$. This completes the induction. [Note the contradiction proves there exists $r \neq 0$ such that $Wr = (0)$ while $mr \neq 0$. For if $Wr = (0)$ if and only if $r = 0$, then $Wr = (0)$ would imply $mr = 0$. Note also that R cannot be a ring of linear transformations on the irreducible module M unless M is faithful. For this ensures that $r = 0$ is the only linear transformation mapping M into 0.]

Observe that the division ring Δ is the set of all group endomorphisms on M that commute with every linear transformation $r \in R$. Δ is often called the *commuting ring of R on M*.

If R is a primitive ring, then there exists a modular maximal right ideal ρ of R such that $(R:\rho) = (0)$. Then R/ρ is a faithful irreducible R-module. Let Δ be the division ring of its module endomorphisms. Hence R/ρ is a vector space over Δ. Put $\bar{x} = x + \rho$, $x \in R$. Let $\delta \in \Delta$. Then $(\bar{x} + \bar{y})r = \bar{x}r + \bar{y}r$, $(\bar{x}\delta)r = (\bar{x}r)\delta$, $r \in R$, and so r is a linear transformation on the vector space R/ρ. The density theorem asserts that R is a dense ring of linear transformations on the vector space R/ρ (that is it is k-fold transitive for all k up to the dimension of R/ρ). To be linearly independent means: $\overline{x}_1\delta_1 + \cdots + \overline{x}_k\delta_k = \rho, \delta_i \in \Delta$, implies $\overline{x}_i\delta_i = \rho$ for each i—that is, $\delta_i = 0$ for each i, for $\overline{x}_i \neq \rho$ (an independent vector cannot be 0). If $\delta_i \neq 0$ then $\overline{x}_i\delta_i \neq \rho$, since δ_i would then be a module automorphism of R/ρ.

The density theorem is due to Jacobson and is of great importance in the study of the structure of rings. By means of it we shall prove the important structure theorem for semisimple rings with the d.c.c.

If R is a 1-fold transitive ring of linear transformations on a right vector space V over a division ring D, then for every $x \neq 0 \in V$, $xR = V$. Hence V is an irreducible R-module. If $r \neq 0 \in R$, im $r = V$ and hence V is also a faithful R-module. Thus R is a primitive ring and, by the density theorem, it is a dense ring of linear transformations on the vector space V over the division ring Δ of all module endomorphisms of V, considered as an R-module. This division ring Δ, however, may not be the same as D, although of course $D \subset \Delta$. Thus R is not necessarily a dense ring of linear transformations on the vector space V over D.

Illustrative Example. Let V be the two-dimensional vector space of ordered pairs (x, y) over the rational field F, with the usual vector addition and scalar multiplication. Take the basis to be $(1, 0)$ and $(0, 1)$.

Let R be the set of all linear transformations T on V of the form $(1, 0)T = (x, y)$ and $(0, 1)T = (y, x + y)$. It is easy to prove that R is a commutative ring and that R is 1-fold transitive. (Clearly R is not 2-fold transitive.) Hence V is a faithful irreducible R-module. The non-zero linear transformations of R are seen to be automorphisms of V, hence, since R is commutative, they are elements of Δ, the division ring of all module endomorphisms of the R-module V. Those linear transformations of R for which $y = 0$ correspond to elements of F regarded as module endomorphisms of the R-module V. Hence $F \subset R$ and hence $F \subset \Delta$, but $F \neq \Delta$.

However, if R is 2-fold transitive then $D = \Delta$ and we have the following converse of the density theorem.

THEOREM 18. If R is a 2-fold transitive ring of linear transformations on the vector space V over the division ring D, then D is the division ring of all module endomorphisms on the R-module V and R is a dense ring of linear transformations on the vector space V over D.

Proof: Since R is a fortiori 1-fold transitive, R is primitive. Let Δ be the division ring of module endomorphisms on the faithful irreducible R-module V. Clearly $D \subset \Delta$. If $\delta \in \Delta$ and $\delta \notin D$ and if $x \neq 0 \in V$, then x and $x\delta$ are linearly independent vectors over D. (For if $x = x\delta a, a \in D$, then $x(1 - \delta a) = 0$. Now

$$1 - \delta a \ne 0,$$

since $\delta = a^{-1}$ implies $\delta \in D$. Since $1 - \delta a \in \Delta$ it is an automorphism of the R-module V and hence $x = 0$, a contradiction.) Since R is 2-fold transitive, there exists $r \in R$ such that $xr = 0$ and $(x\delta)r = x \ne 0$. But $(x\delta)r = (xr)\delta = 0$. This contradiction is caused by x and $x\delta$ being independent vectors over D, which they would not be if $\delta \in D$. Hence $D = \Delta$ and R is a dense ring on the vector space V over D. Thus 2-fold transitivity implies R is dense on V over D.

Ideals in a Matrix Ring. Let R be a ring with identity 1 and let R_n be the ring of all $n \times n$ matrices over R. R_n is called the **complete matrix ring.** What are the ideals of the ring R_n?

Let H denote an ideal in the ring R_n and let S be the set of all elements of R that occur as entries in the matrices belonging to H.

For $a \in R$ let a_{ij} denote the matrix of R_n with the entry a in the ith row and jth column and with zeros everywhere else. Let T be a matrix in H, and let $x \in S$ occur in the pth row and qth column of T. Then

$$1_{ip}T1_{qj} = x_{ij} \in H \text{ (since } H \text{ is an ideal)}.$$

Hence $S_n \subset H$, where S_n is the set of all $n \times n$ matrices over the set S. But $H \subset S_n$, by the definition of H. Hence $H = S_n$. Furthermore S is an ideal in R. For if $x \in S$, $x_{pq} \in H$ and hence

$$r_{ip} x_{pq} 1_{qj} = (rx)_{ij} \in H.$$

Hence $rx \in S$ for all $r \in R$. Similarly $xr \in S$ for all $r \in R$. Obviously if $x, y \in S$ then $(x - y) \in S$ and hence S is an ideal.

Thus all ideals of R_n have the form S_n where S is an ideal in R.

If R is a division ring, then R has only the ideals (0) and R, that is R is simple ($R \ne 0$). Hence R_n is a simple ring. The only right ideals of a division ring are (0) and the whole ring. Thus R_n has the d.c.c.

The ring of all linear transformations on an n-dimensional vector space over a division ring Δ (called the **complete ring** of such transformations) is isomorphic to Δ_n (see Chapter 3) and hence is a simple ring with the d.c.c. Thus we have proved

THEOREM 19. If any ring R with the d.c.c. is isomorphic to the complete ring of linear transformations on a finite dimensional vector space over a division ring, then R is a simple ring (hence is also primitive).

As an application of the density theorem we can prove the converse of Theorem 19. This converse is the basic structure theorem for simple rings with minimum condition.

THEOREM 20. A primitive ring R with the d.c.c. is isomorphic to the complete ring of linear transformations on a finite dimensional vector space over a division ring.

Proof: Since R is primitive, there exists a faithful irreducible R-module M, and M is a vector space over the division ring Δ of its module endomorphisms.

If M is not finite-dimensional, there exists an infinite set $x_i, i = 1, 2, 3, \ldots$ of linearly independent vectors of M.
Define

$$L_k = \{r \in R \mid x_1 r = x_2 r = \cdots = x_k r = 0\}, \qquad k = 1, 2, 3, \cdots.$$

Then the L_k are right ideals. Now $L_1 \supseteq L_2 \supseteq L_3 \supseteq \cdots$. By the density theorem there exists $b_k \in R$ such that $b_k \in L_k$ and $b_k \notin L_{k+1}$. Hence $L_1 \supset L_2 \supset L_3 \supset \cdots$ contrary to the d.c.c. Thus M is finite-dimensional.

By the density theorem R is isomorphic to a dense ring R' of linear transformations on M. Since M is finite dimensional, say of dimension n, and since R' is n-fold transitive, it follows that R' is the complete ring (ring of all) linear transformations on M. This proves the theorem.

Thus $R \approx \Delta_n$, the complete matrix algebra of $n \times n$ matrices over Δ. As an algebra (or as a vector space) R is of dimension n^2.

Corollary 1. The dimension n is unique.

Proof: For if D and D' are two division rings, then $D_n \approx D'_m$ implies $n = m$ and $D \approx D'$. (The details are left as an exercise.)

Corollary 2. A primitive ring R with the d.c.c. has an identity.

Proof: $R \approx R'$ and R' has an identity. Hence R has one.

Corollary 3. A primitive ring R with the d.c.c. is simple.

Proof: This follows at once from Theorem 19 and Theorem 20.
Since a simple ring with the d.c.c. is primitive, Theorem 20 is essentially a converse of Theorem 19.

THEOREM 21. A primitive ideal P in a ring R with the d.c.c. on right ideals is maximal.

Proof: R/P is a primitive ring and, being a homomorphic image of R, R/P has the d.c.c. on right ideals. Hence by Corollary 3, Theorem 20, R/P is a simple ring. Let Q be an ideal in R such that $P \subset Q, P \neq Q$. Then Q/P is an ideal in R/P. Since R/P is simple, $Q = R$, and therefore P is maximal.

EXERCISES

1. D_n is the matrix algebra of $n \times n$ matrices over the division ring D and D'_m is the matrix algebra of $m \times m$ matrices over the division ring D'. Prove that if $D_n \approx D_m$, then $n = m$ and $D \approx D'$.

2. Substantiate all the claims made in the illustrative example in Sec. 11-4.

3. V is a two-dimensional vector space over a division ring. If R is the ring of linear transformations of $V \to V$, prove that V is an irreducible faithful R-module. What does this prove about R? Prove that R is an irreducible ring of endomorphisms of V, regarded as an abelian group.

4. The *rank* of a linear transformation f of $V \to V$, where V is an arbitrary vector space over a division ring, is defined to be the dimension of im f. Prove that the linear transformations of V of finite rank form an ideal A in the complete ring R of linear transformations of V. Prove A is a dense ring of linear transformations of V. How do we know that A is a proper ideal in R?

5. An (algebra) right ideal B of an algebra A over a field F is a subspace of A such that $bx \in B$ for all $b \in B$ and all $x \in A$. An (algebra) ideal of A is a subspace of A which is also a (ring) ideal of A. Since a ring ideal of A is not in general a subspace of A, it is conceivable that the algebra A could have two distinct radicals, a ring radical and an algebra radical. Prove the following:

(a) A modular maximal right (ring) ideal of A is also an (algebra) right ideal of A.

(b) The radical of A as an algebra is the same as the radical of A as a ring.

11-5 THE PRINCIPAL STRUCTURE THEOREM FOR SEMISIMPLE RINGS WITH THE d.c.c.

In this section we are going to prove that a semisimple ring with the d.c.c. (on right ideals) is the direct sum of a finite number of simple rings. The proof given is that of Nathan Jacobson in *Structure of Rings*, American Mathematics Society, 1964.

If a ring R has a primitive ideal $P = (0)$ then clearly R is a primitive ring, and if, in addition, R has the d.c.c. then R is already a simple ring. It should be kept in mind that in any ring with the d.c.c. a primitive ideal is a maximal ideal.

Assume henceforth that R is a semisimple ring with the d.c.c. for which no primitive ideal is (0). The radical of R is (0) and hence $\cap P = (0)$, where the intersection is taken over all primitive ideals P of R.

Lemma 7. If R is a semisimple ring with the d.c.c. then there exists a finite set P_i, $i = 1, 2, \ldots, n$, of primitive ideals in R such that $\bigcap_{i=1}^{n} P_i = (0)$, whereas any intersection of a subset of these P_i is *not* zero.

Proof: If the intersection of some pair of primitive ideals is (0), then this pair is the finite set required. If no pair has zero intersection then consider intersections of three primitive ideals. If one of these is (0), then again we have the finite set required. If all of them are not (0), then proceed to intersections of four primitive ideals. Continuing in this way we must arrive at a finite set $P_1, P_2, \ldots, P_n$ of primitive ideals whose intersection is (0) and for which any intersection of a subset is not zero. For otherwise we would obtain a non-terminating descending chain of nonzero ideals. This would violate either the d.c.c. or the hypothesis that R has zero radical.

Definition. The **direct sum** $B = \sum_{i=1}^{n} \oplus B_i$ of a finite number of ideals $\beta_i, i = 1, 2, \ldots, n$, is an ideal B such that each $b \in B$ has a unique representation in the form $b = b_1 + b_2 + \cdots + b_n$, where $b_i \in B_i$ for $i = 1, \ldots, n$.

As before this condition can be seen to be equivalent to the condition that for each i,

$$B_i \cap \left(\sum_{j \neq i} B_j \right) = (0).$$

THEOREM 22 (Structure Theorem). A semisimple ring R with the d.c.c. is the direct sum

$$R = \sum_{i=1}^{n} \oplus S_i$$

of a finite number of simple rings S_i. Moreover the decomposition is unique.

Proof: Let $P_1, P_2, \ldots, P_n$ be the finite set of primitive ideals of R satisfying the conditions of Lemma 7.

Define $S_i = \bigcap_{j \neq i} P_j$. Then S_i is a nonzero ideal and $S_i \cap P_i = (0)$. Since P_i is maximal, $R = S_i + P_i$ and hence $R = S_i \oplus P_i$.

Put $B_r = P_1 \cap P_2 \cap \cdots \cap P_r$. Then B_r is an ideal. Observe that

$$S_{m+1} = \bigcap_{j=2}^{n-m} B_m \cap P_{m+j}.$$

Next we prove, by induction, that

$$R = S_1 \oplus \cdots \oplus S_k \oplus B_k.$$

Note that this is true for $k = 1$, since $R = S_1 \oplus P_1$. Assume it true for $k = m$. Consider the ideal B_m. By the second isomorphism theorem for rings (Sec. 5-4):

$$\frac{B_m}{B_m \cap P_{m+j}} \approx \frac{B_m + P_{m+j}}{P_{m+j}}, \quad j = 1, 2, \ldots, n - m.$$

By the induction hypothesis, $R = S_1 \oplus \cdots \oplus S_m \oplus B_m$. Moreover, $P_{m+j} \supset S_i$ for each $i = 1, 2, \ldots, m$. Hence $S_1 \oplus \cdots \oplus S_m \subset P_{m+j}$. Hence $R = B_m + P_{m+j}$. Therefore

$$\frac{B_m}{B_m \cap P_{m+j}} \approx \frac{R}{P_{m+j}}.$$

Since the P_{m+j} are maximal in R, it follows that the $B_m \cap P_{m+j}$ are maximal in B_m, $j = 1, 2, \ldots, n - m$. Also

$$\bigcap_{j=1}^{n-m} B_m \cap P_{m+j} = \bigcap_1^n P_i = (0)$$

and no proper subset of the $B_m \cap P_{m+j}$ has zero intersection.

$$B_{m+1} = B_m \cap P_{m+1}$$

and hence is maximal in B_m. Hence

$$B_m = B_{m+1} \oplus \bigcap_{j=2}^{n-m} (B_m \cap P_{m+j}) = B_{m+1} \oplus S_{m+1},$$

and

$$\left(B_{m+1} \cap \bigcap_{j=2}^{n-m} B_m \cap P_{m+j}\right) = (B_m \cap P_{m+1}) \cap \left(\bigcap_{j=2}^{n-m} B_m \cap P_{m+j}\right)$$

$$= \bigcap_{j=1}^n (B_m \cap P_{m+j}) = \bigcap_1^n P_i = (0).$$

Hence

$$R = S_1 \oplus \cdots \oplus S_m \oplus B_m$$
$$= S_1 \oplus \cdots \oplus S_{m+1} \oplus B_{m+1}$$

and the induction is completed. Hence

$$R = S_1 \oplus \cdots \oplus S_{n-1} \oplus B_{n-1}$$
$$= S_1 \oplus \cdots \oplus S_{n-1} \oplus S_n \text{ (since } B_{n-1} = S_n).$$

From $R = S_i \oplus P_i$, it is seen that $S_i \approx R/P_i$, and hence S_i is a primitive ring. Moreover, R has the d.c.c. and hence the homomorphic image R/P_i of R has the d.c.c. Thus S_i has the d.c.c. and hence S_i is simple.

Another way of proving that S_i is simple is the following: $S_i = \bigcap_{j \neq i} P_j$ and hence $S_i^2 = S_i$. Moreover, P_i is a maximal ideal and hence

R/P_i has no proper ideals. Since S_i and R/P_i are isomorphic rings, it follows that S_i can have no proper ideals. Thus S_i is simple.

We have therefore proved that R is the direct sum of the finite number of simple rings S_i. The S_i are ideals of R. The uniqueness of this direct sum decomposition follows from Lemma 8 below.

Note that the S_i are primitive rings but are not primitive ideals of R. In fact if they were primitive ideals they would then be maximal ideals of R. This would contradict the next lemma.

Lemma 8. The ideals S_i of Theorem 22 form the set of all minimal ideals of R.

Proof: Let T be a minimal ideal of R. We show that $T = S_i$ for some i. Now $R = \sum_{i=1}^{n} \oplus S_i$ implies $RT = \sum_{i=1}^{n} \oplus S_i T$.

Now $Rt \neq (0)$. For if $RT = (0)$ then $T^2 = (0)$ and T would be a nilpotent ideal. Since R has the d.c.c. then by the corollary to Lemma 5, we have $T \subset J(R) = (0)$, hence $T = (0)$. This contradicts T being a minimal ideal. Hence $RT \neq (0)$.

Hence there exists at least one S_i such that $S_i T \neq (0)$. Now $S_i T \subset S_i$ and S_i is simple. Hence $S_i T = S_i$. Also $S_i T \subset T$. Since T is minimal, $S_i T = T$. Hence $T = S_i$.

Note in Theorem 22 that if $x_i \in S_i$ and $x_j \in S_j$, $i \neq j$, then $x_i x_j \in S_i \cap S_j$, and hence $x_i x_j = 0$. Thus a product of an element of one S_i and an element of a second distinct S_j is always zero.

THEOREM 23. Every semisimple ring with the d.c.c. has an identity.

Proof: The ideals S_i in the structure theorem are primitive rings with the d.c.c. and hence have identities 1_i, $i = 1, 2, \ldots, n$ (Theorem 20, Corollary 2).
Set

$$1 = \sum_{i=1}^{n} 1_i.$$

Since $1_i a_i = a_i = a_i 1_i$ and $1_i a_j = 0 = a_j 1_i$, $a_i \in S_i$ and $i \neq j$, it follows that 1 is the identity for the ring R.

We remark that the following converse of the structure theorem can be proved: if R is a ring and $R = S_1 \oplus \cdots \oplus S_n$, where each S_i is a simple ring, and if each S_i is isomorphic to a complete ring of linear transformations of a finite-dimensional vector space over a division ring, then R is a semisimple ring with the d.c.c.

Bibliography

The following books are recommended for their general usefulness, and in some instances also for their excellence, ascertained by direct experience with them. They are in two groups, the first has been selected for supplementary reading and epexegesis, while the second is recommended for more advanced study.

The reader should explore the shelves of a good library for additional material. He will then be able to make his own choices and to acquaint himself with many types of exposition. Moreover, different authors emphasize different aspects of the subject, even of the same topic.

SUPPLEMENTARY READING

ARTIN, E. *Galois Theory.* University of Notre Dame Press, Notre Dame, 1944.

―――. *Geometric Algebra.* Interscience Publishers Inc., New York, 1957.

BIRKHOFF, G., and S. MACLANE. *A Survey of Modern Algebra.* Third Edition. Macmillan, New York, 1965.

DUBREIL, P., and M. L. DUBREIL-JACOTIN. *Lecons d'Algèbre Moderne.* Dunod, Paris, France, 1961.

HERSTEIN, I. N. *Theory of Rings.* University of Chicago Press, Mathematics Lecture Notes. Chicago, 1961.

HU, S. T. *Elements of General Topology.* Holden-Day, San Francisco, 1964.

―――. *Elements of Modern Algebra.* Holden-Day, San Francisco, 1965.

―――. *Homology Theory.* Holden-Day, San Francisco, 1966.

JACOBSON, N. *Lectures in Abstract Algebra.* Volume I (*Basic Concepts*), 1951; Volume II (*Linear Algebra*), Van Nostrand, Princeton, 1953.

JANS, J. P. *Rings and Homology.* Holt, Rhinehart and Winston, New York, 1964.

KUROSH, A. G. *The Theory of Groups.* Vol. I and Vol. II. Translated by K. A. Hirsch. 2d English edition. Chelsea, New York, 1960.

LANDAU, E. *Foundations of Analysis.* Chelsea, New York, 1951.

MACLANE, S., and G. BIRKHOFF. *Algebra.* Macmillan, New York, 1967.

MASSEY, W. S. *Algebraic Topology, An Introduction.* Harcourt, Brace and World, New York, 1967.

McCoy, N. H. *Rings and Ideals.* ("Carus Mathematical Monograph," No. 8). The Mathematical Association of America, Buffalo, New York, 1948.

———. *The Theory of Rings.* Macmillan, New York, 1964.

Mostow, G. D., J. H. Sampson, and J. Meyer. *Fundamental Structures of Algebra.* McGraw-Hill, New York, 1963.

Nickerson, H. K., D. C. Spencer, and N. E. Steenrod. *Advanced Calculus.* Van Nostrand, Princeton, 1959.

Van der Waerden, B. L. *Modern Algebra.* Vol. I and Vol. II (English translation). Ungar, New York, 1949 and 1950.

Zariski, O., and P. Samuel. *Commutative Algebra.* Vol. I, Van Nostrand, Princeton, 1958.

ADVANCED READING

Bourbaki, N. *Éléments de Mathématique.* Herman, Paris, France. (This is a treatise on various branches of mathematics, published in several volumes, over the last quarter of a century. The attention of the reader is particularly directed to the sections on algebra and topology.)

Cartan, H., and S. Eilenberg. *Homological Algebra.* Princeton University Press, Princeton, 1956.

Chevalley, C. *Fundamental Concepts of Algebra.* Academic Press, New York, 1956.

Curtis, C. W., and I. Reiner. *Representation Theory of Finite Groups and Associated Algebras.* Interscience Publishers, Inc., New York, 1962.

Jacobson, N. *Lectures in Abstract Algebra.* Vol. III (*Theory of Fields and Galois Theory*). Van Nostrand, Princeton, 1964.

———. *Lie Algebras.* Interscience Press, New York, 1962.

———. "Structure of Rings," *American Mathematical Society Colloquium Publications,* Vol. XXXVII (revised edition). American Mathematical Society, Providence, 1964.

Lang, S. *Algebra.* Addison-Wesley, Reading, Mass., 1965.

MacLane, S. *Homology.* Springer Verlag, Berlin, 1963.

Index